DISCOVERING NUMBER THEORY

Jeffrey J. Holt
University of Virginia

John W. Jones
Arizona State University

W. H. FREEMAN AND COMPANY / NEW YORK

The authors acknowledge funding for this project from the National Science Foundation.

Sponsoring Editor: Craig Bleyer
Executive Marketing manager: Claire Pearson
Project Editor: Vivien Weiss, Christopher Miragliotta
Cover Design: Michael Minchillo
Production Coordinator: Julia DeRosa
Composition: John Jones
Manufacturing: Quebecor Printing

CATALOGING-IN PUBLICATION DATA AVAILABLE FROM THE LIBRARY OF CONGRESS

ISBN: 0-7167-4284-5

Printed in the United States of America

First printing 2000

To Kathy, Beth, Katie, Tom, Mike, Laura, and Mike

Contents

Preface   **xi**
    Chapter Interdependence xviii

Course Guide   **1**
    Welcome 1
    Course Organization 2
    How to Assemble This Book 4
    Using the CD-ROM 4
    How to Talk Like a Mathematician 6
    Tips for Writing Proofs 9
    Sample Maple Lab: Prime Numbers 18
    Sample Lab Report: Prime Numbers 20
    Maple Lab: Introduction to Using Maple 24
    Mathematica Lab: Introduction to Using Mathematica 29
    Web Lab: Introduction to Active Web Pages 34

1. Divisibility and Factorization   **41**
    Prelab 41
    Maple Lab 45
    Mathematica Lab 56
    Web Lab 65
    Summary 85
    Homework 91

2. The Euclidean Algorithm and Linear Diophantine Equations   **95**
    Prelab 95
    Maple Lab 97
    Mathematica Lab 106
    Web Lab 114
    Summary 131
    Homework 143

3. Congruences   **145**
    Prelab 145
    Maple Lab 148

Mathematica Lab 155
Web Lab 161
Summary 177
Homework 187

4. Applications of Congruences   189
Prelab 189
Maple Lab 195
Mathematica Lab 204
Web Lab 212
Summary 233
Homework 243

5. Solving Linear Congruences   249
Prelab 249
Maple Lab 251
Mathematica Lab 264
Web Lab 275
Summary 301
Homework 311

6. Primes of Special Forms   315
Prelab 315
Maple Lab 316
Mathematica Lab 320
Web Lab 324
Summary 333
Homework 343

7. The Chinese Remainder Theorem   345
Prelab 345
Maple Lab 347
Mathematica Lab 352
Web Lab 357
Summary 367
Homework 379

8. Multiplicative Orders   383
Prelab 383
Maple Lab 385
Mathematica Lab 392
Web Lab 398

Summary 411
Homework 421

9. The Euler ϕ-function   423

Prelab 423
Maple Lab 424
Mathematica Lab 431
Web Lab 438
Summary 453
Homework 463

10. Primitive Roots   467

Prelab 467
Maple Lab 468
Mathematica Lab 479
Web Lab 488
Summary 507
Homework 519

11. Quadratic Congruences   523

Prelab 523
Maple Lab 526
Mathematica Lab 533
Web Lab 539
Summary 551
Homework 561

12. Representation Problems   565

Prelab 565
Maple Lab 566
Mathematica Lab 572
Web Lab 578
Summary 587
Homework 597

13. Continued Fractions   599

Prelab 599
Maple Lab 604
Mathematica Lab 611
Web Lab 617

Summary 631

Homework 643

Index   **647**

Preface

This text is designed to be used for a one-semester upper-division undergraduate course in number theory. There are no specific prerequisites beyond basic algebra and some ability in reading and writing mathematical proofs. The content of *Discovering Number Theory* is similar to that of many other number theory texts. However, the organization of the subject matter and the method of presentation are substantially different from other texts.

A major goal in developing this text was to produce course materials that would allow students to work in a manner similar to that of a mathematician conducting research. Rather than having the key results laid out for them, students work to discover many of the important concepts and theorems themselves. After reading introductory material on a particular subject, they work through electronic materials that contain additional background, exercises, and Research Questions. The Research Questions prompt them to use the computer to generate numerical data and to use this data as a guide for forming conjectures. Once they have a conjecture, students are expected (in most cases) to find a proof.

By developing the subject matter on their own, students gain valuable insight into some of the realities of mathematical research. Instead of simply being presented with a tidy package containing theorems stated in the most efficient form, students experience firsthand the difficulties that are frequently a part of conducting mathematical research. On the other hand, they also get to experience the personal satisfaction that comes from finding the exact theorem that brings order to seemingly chaotic numerical data.

This text consists of a printed volume and interactive computer software. The software is provided in three formats: *Mathematica* notebooks, *Maple* worksheets, and HTML with Java applets for computations.[1] The electronic notebooks are an integral part of the text and contain a substantial portion of the subject matter. Each chapter of the text consists of several parts: a prelab, an electronic notebook, a chapter summary, and a homework set. The student volume includes the prelabs, printed copies of each version of the notebooks together with electronic versions on an included CD, and the homework sets. The chapter summaries, which contain solutions to Research Questions posed in the lab notebooks, are included only in the instructor's guide.

---

[1]Throughout this section, we shall refer to these generically as *electronic notebooks*.

Anatomy of a Chapter As mentioned earlier, a key feature of this text is that it allows students to learn about new topics by working on Research Questions *before* being exposed to the corresponding polished theorems and proofs. Thus this text is organized differently than a traditional number theory text. Below are details of each chapter component, along with a description of the implementation that we have used in pilot courses. Naturally, most aspects of the implementation can be adjusted to suit a particular instructor's requirements.

Prelab This section contains introductory discussion and relevant definitions for the topic to be studied. The prelab also contains a collection of short exercises that reflect the types of computations carried out later by the computer. Completing the prelab exercises by hand reduces the degree with which students view the lab computations as "black boxes". The prelab exercises are designed to be completed prior to the commencement of work in the computer lab.

Electronic notebook After completing the prelab section, the next step is lab work. The lab notebooks contain all remaining discussion about the current topic(s), lab exercises, Research Questions, and software to generate examples. The lab exercises are usually specific in nature and often computational. The Research Questions are typically more open ended and require students to respond with a conjecture and proof. The lab work culminates in the preparation of a lab report. The lab report will typically consist of responses to lab exercises, statements of conjectures, and proofs.

Chapter summary Once lab reports have been submitted, the chapter summary can be distributed to students. (Recall that it isn't included with the student version.) The summary contains the target theorem associated with each Research Question, along with a proof of the theorem. Students can compare their own theorems and proofs with more-polished versions. The chapter summary may also include additional topics that aren't suitable for presentation in the computer lab.

Homework set The homework set contains both computational problems and questions requiring proofs. Problems typically have students apply the results of the Research Questions. Although the electronic materials may be useful for some homework problems, most do not require the use of a computer.

The chapter summaries are a key part of the text. While they may look similar to a traditional text, students read them with a much different background since they have worked on the material explained therein. Moreover, while there is value in even unsuccessful attempts by students in the discovery phase of a chapter, much learning takes place when they compare their own efforts with the chapter summary. Finally, providing chapter summaries allows results of Research Questions in one chapter to be used in later chapters.

Some electronic notebooks and chapter summaries contain sections labeled Going Farther.[2] These are optional sections which may be assigned at the instructor's discretion. Going Farther sections of the electronic notebooks contain applications of the material which lend themselves to interactive use, such as various cryptographic schemes. Going Farther sections of chapter summaries typically contain more-detailed analysis of the subject matter at hand. For example, in an early chapter, students find the general solution to a linear diophantine equation in two variables. In the Going Farther section of the chapter summary, a description of how to extend the solution that they have found to the general problem of such equations in three variables is given, followed by an indication of how to further extend to n variables. When Going Farther sections occur, there are often corresponding exercises in the homework sections as well.

Topical Table of Contents Provided below is a list of chapters, together with a brief indication of the topics covered in each chapter.

Course Guide Topics: General mathematical terminology; Help on writing proofs; Sample lab and lab report; Introductions to the different forms of electronic notebooks.

1. **Divisibility and Factorization** Topics: Unique factorization of integers; Introduction of greatest common divisors and the Division Algorithm; Number of divisors of an integer; Going Farther (summary): Sum of divisors of an integer.

2. **The Euclidean Algorithm and Linear Diophantine Equations** Topics: Euclidean Algorithm; Integer solutions to $ax + by = c$; Going Farther (summary): Linear diophantine equations in 3 or more variables; Going Farther (summary): Proof of the Fundamental Theorem of Arithmetic.

3. **Congruences** Topics: Introduction of congruences from several points of view (congruence classes, as remainders, and in terms of divisibility); Well-defined arithmetic mod m; Additive analog of Wilson's Theorem; Additive orders; Going Farther (summary): Further analysis of additive orders; Formula for $\sum_{d|n} \phi(d)$.

4. **Applications of Congruences** Topics: Check digits (e.g., ISBN numbers); Divisibility tests; Rock game; Going Farther (summary): Further analysis of divisibility tests; Going Farther (lab): Introduction to cryptography; Going Farther (lab): Shift ciphers.

---

[2]The term *Going Farther* comes from the vision of number theory as a vast territory filled with treasures to be discovered, as represented by the cover art on this text.

5. **Solving Linear Congruences** Topics: Multiplicative inverses modulo m; Solving $ax \equiv b \pmod{m}$ in general; Wilson's Theorem; Additive orders; Going Farther (lab): Affine ciphers; Going Farther (lab): Basic cryptanalysis; Going Farther (summary): Solving linear systems of congruences.

6. **Primes of Special Forms** Topics: Mersenne numbers; Perfect numbers; Fermat numbers; Primes in arithmetic progressions; Going Farther (lab): Prime Number Theorem, introduced and used in a heuristic manner to predict whether there are a finite number or infinite number of Mersenne/Fermat primes; Going Farther (summary): Constructing regular polygons.

7. **Chinese Remainder Theorem** Topics: Standard Chinese Remainder Theorem; Two CRT-type congruences when the moduli may not be relatively prime; Explicit formulas for solving CRT; Going Farther (summary): CRT as a bijective function; Going Farther (summary): Using the CRT for multiprecision computer arithmetic.

8. **Multiplicative Orders** Topics: Definition of (multiplicative) order of an integer modulo m; Limitations on orders; Fermat's Little Theorem and Euler's generalization; Going Farther (summary): Using Fermat's Little Theorem for compositeness testing; Going Farther (summary): Binary exponentiation algorithm.

9. **The Euler ϕ-function** Topics: Standard formula for computing the Euler ϕ-function; Going Farther (lab): RSA public key cryptosystem; Going Farther: Rabin-Miller compositeness test.

10. **Primitive Roots** Topics: Connecting multiplicative orders with additive orders; Definition and existence of primitive roots; All multiplicative orders in the presence of a primitive root; Going Farther (lab): Discrete logarithms in cryptography (e.g., the Diffie-Hellman protocol); Going Farther (summary): Explanation of part of the Pohlig-Hellman Algorithm for computing discrete logarithms; Going Farther (summary): Proofs for primitive roots.

11. **Quadratic Congruences** Topics: Basics of quadratic residues/nonresidues, Legendre symbol; Euler's criterion; Computing the Legendre symbol for 2 over p and -1 over p; Quadratic reciprocity; Going Farther (summary): Computing Legendre symbols quickly; Going Farther (summary): Solution of a general quadratic congruence modulo an odd prime.

12. **Representation Problems** Topics: Representing integers as the difference of two squares; Representing integers as the sum of two squares; Pythagorean triples; Representing all integers as a sum of squares; Going Farther (summary):

Fermat's Last Theorem, proof for exponent 4; Going Farther: Complete proof of Fermat's Last Theorem (just kidding).

13. Continued Fractions Topics: Introduction of continued fractions as good rational approximations; Continued fraction expansions for square roots; $x^2 - dy^2 = 1$.

Course Benefits This text integrates its novel features with a traditional curriculum for undergraduate number theory.

Students have a Research Experience In many fields of study, all students can be involved in research as undergraduates. This may involve routine tasks such as cleaning test tubes or monitoring bacteria cultures, but it gives them an idea of what research is like in their chosen area. Research experiences are important for all students. It is particularly important for those who are considering graduate study, and for preservice mathematics teachers who should know the nature of the subject they will someday teach.

Undergraduate research opportunities exist in mathematics, but few students get to participate. From its basic design, use of this text gives undergraduates an idea of what research is like in mathematics.

Students Improve Proof Writing Skills We include a section in the Course Guide on how to write proofs. Students who are already proficient at proof writing can skim or skip this section. Others may find it especially helpful. In particular, it makes it possible to use this text with students with different levels of proof writing skills.

The system of writing lab reports helps students improve their proof writing skills throughout the course. Discussing the lab report with other students provides a good opportunity for students to work out problems they are having in writing proofs. By having the lab reports corrected by their instructor, and by reading the chapter summaries, students get direct feedback on how they could have written their responses.

Students Improve their Problem Solving Skills All math classes (should) improve students problem solving skills. This text goes a little further than most. Research Questions in early chapters lead students slowly through the process of gathering and analyzing their data. Later chapters progressively let the students control more of the structure of their investigations. See, for example, the progression of Research Questions in Chapter 1 on the number of divisors of an integer, Research Questions in Chapter 9 on a formula for the Euler ϕ-function, and Research Question 3 in Chapter 12.

Students Learn Core Material As can be seen from the topical table of contents above, the basic subject matter is unchanged. Congruences play a dominant role in this text, but material on elementary diophantine equations (Chapters 2, 12, and 13), primes of special forms (Chapter 6), and the beginning of diophantine approximation (i.e., continued fractions in Chapter 13) are covered as well. The goal of this book is *not* to change the content of a number theory course; it is to teach the standard content in a more effective manner while giving students important auxiliary experiences.

Applications to cryptography are finding their way into many number theory texts, and this text is no exception. Several number theoretic cryptosystems and their cryptanalysis are studied. In some cases, this is part of the lab materials so that students can actively work with the cryptosystem.

The Fundamental Theorem of Arithmetic is an important theorem for elementary number theory. This text gives more emphasis on how to apply the theorem than most other number theory books. Students often recognize that this is supposed to be an important theorem, yet they find it too cumbersome to apply in a specific proof. We provide a complete proof at the end of Chapter 2, but this takes place after students have spent time learning how to apply the Fundamental Theorem of Arithmetic in Chapter 1.

The treatment of the Prime Number Theorem in a Going Farther section of Chapter 6 is somewhat novel. In addition to stating the theorem and giving numerical examples, we apply the theorem to make conjectures as to whether there are a finite number or an infinite number of primes of various special forms. This material is not typically covered in elementary number theory texts.

Students Improve General Job Skills For students headed to the general workforce, the skills employers currently rank as most important are the ability to work as part of a team, communication skills, and computer skills. With this text, students not only need to use a computer, they need to make decisions regarding what kinds of computations are best done with the computer. Students need to work in teams and come to a consensus regarding what will appear in the group's lab report. Finally, talking with other students as part of the group work and writing up the group report helps students develop their communication skills.

Acknowledgements The development of *Discovering Number Theory* course materials was supported by a grant from the National Science Foundation (DUE 9554970). We would also like to thank Sid Graham, Andrew Bremner, and David Roberts for ideas and suggestions along the way.

Next, we would like to thank Michelle Julet, Craig Bleyer, Maria Epes, Melanie Mays, Vivien Weiss, Peter Hamlin, and Vicki Tomaselli, and all of the other people at W. H. Freeman who helped produce this book.

We would like to thank Stephen Adams for writing a BigInt.java (©1996 MIT), and Daeron Meyer for writing PackerLayout.java (©1995). They are both used in the Java applets of the web notebooks.

We would like to thank the instructors who provided reviews of the manuscript: Wayne Aitken, California State University at San Marcos; Richard Blecksmith, Northern Illinois University; Antonie Boerkoel, Texas A&M University-Kingsville; G. Dresden, Washington and Lee University; Kevin Ford, University of South Carolina; Farshid Hajir, California Technical Institute; Louis W. Kolitsch, University of Tennessee-Martin; Stephanie Kolitsch, University of Tennessee-Martin; Robbin O'Leary, Seattle Pacific University; Paul Pasles, Villanova University; Rama Rao, University of North Florida; Jonathan W. Sands, University of Vermont; and Brett Tangedal, College of Charleston. We are indebted to the faculty who class-tested this text: Robbin O'Leary, Seattle Pacific University; Libby Krussell, University of Montana; and John Boardman, Hillsdale College.

Finally, we would like to thank the students who used the book.

Chapter Interdependence

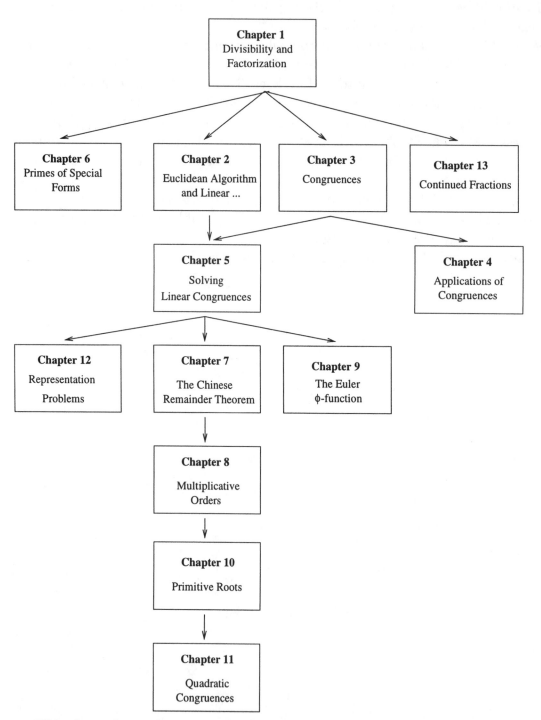

This chart shows the main relationships between chapters. Here we will point out some minor dependencies between chapters:

- In some cases, a homework exercise may depend on material from an earlier chapter which is not indicated as a prerequisite by this chart.

- Chapter 8 on Multiplicative Orders and Chapter 13 on Continued Fractions contain discussions of periodic sequences. ·Basic terminology is contained in Chapter 4 on Applications of Congruences. An instructor who feels that his or her students would benefit from reading an introduction to periodicity with basic terminology explained can assign just the Prelab section for Chapter 4.

- Cryptographic applications run through the book as a theme in Going Farther sections. Depending on the level of the class, an instructor may wish his or her students to read earlier sections which explain background information.

DISCOVERING NUMBER THEORY

Course Guide

Welcome

C ongratulations! As of this moment, you are a mathematician. For the duration of this course, you will be expected to conduct yourself as a mathematician. Lucky for you, your instructor is quite experienced at this and can help guide you along the way.

You may be aware that a principal activity of mathematicians is to prove theorems. What may not be as clear is how this process really looks from the perspective of a mathematics researcher. Here are the facts of life:

- Mathematicians are a special breed of scientist and accordingly follow an analogue of the scientific method.

- When an area is first studied, not only are the answers unclear, but often the right questions are also not apparent. In other words, frequently a mathematician must discover both the statement of a theorem as well as its proof.

- It is often possible to guess the correct statement of a theorem by doing experiments, just like any scientist. In science, a guess worthy of further testing is called a hypothesis; in mathematics it is called a *conjecture*.

- In science, carefully controlled experiments are often the best evidence one can provide for a conjecture. In mathematics, a complete mathematical proof is the **only** way to establish a conjecture as being true.

In addition to learning undergraduate number theory, an important goal of this course is to provide insight into how mathematical theory evolves. You will be working in a manner similar to that of many practicing mathematicians, performing numerical experiments to help find and refine conjectures, and then searching for proofs of your conjectures. Thus, just as a mathematician must often do, you will be discovering statements *and* proofs of number theoretic theorems. Along the way, you will be confronted with related questions, such as what is a *good* theorem, when is one statement of a theorem better than another, how do seemingly minor shadings in the wording of a conjecture affect whether or not it is true?

If you think that the above scenario sounds challenging, it is. However, the course is organized in a manner designed to increase your chances of success. First of all, you will be pointed in the direction of some known theorems so that there will always be something of interest to discover. Second, you will be working in small groups throughout the course. Discussing your ideas with others is an excellent way to develop them. Indeed, a significant amount of mathematical research is conducted by mathematicians working in groups. If working together helps the professionals, then it should also help you. Finally, when your group seems to be stuck and confused, the instructor is there to help point you back in the right direction. Have fun, and best of luck!

Course Organization

Each chapter has five components: the prelab worksheet, lab work, lab report preparation, postlab class discussion, and homework. These components are discussed in detail below:

Prelab worksheet Each chapter will begin with the prelab worksheet. It provides an introduction to the ideas and material to be studied in the chapter. The worksheet will typically include general discussion, examples, definitions, and some exercises. Exercises look like this:[3]

  1. **Determine all integers n such that $n^2 = 1$.**

  Reading over the worksheet and working through the exercises will help prepare you for the deeper study to be done during the remainder of the chapter.

Lab work The majority of the material in each chapter is presented in interactive computer software. Three formats are available: *Maple* worksheets, *Mathematica* notebooks, and HTML with Java applets.[4] Each electronic notebook will typically contain discussion and examples, as well as two types of problems:

  - Exercises, which ask specific questions and are often computational.
  - Research Questions, which are open ended and require numerical experimentation. These types of problems require you to formulate conjectures on the basis of your experimentation, and then to find proofs of your conjectures. (Or, to show that your conjecture is incorrect, you will need to provide a counterexample.)[5]

---

[3]You do not have to turn in a solution to this sample question.

[4]We use the generic term *electronic notebook* to refer to any format of computer software. The format used in your course will depend on what is available on your school's computers.

[5]The role of proofs and counterexamples is discussed in the "How to Talk Like a Mathematician" section of this chapter.

During lab meetings (and at other nonclass times as well), you will be working with a partner at the computer, completing lab exercises, performing numerical experiments, forming conjectures, and working on proofs. This work will form a preliminary basis for the material to be submitted in the lab report.

Lab report preparation Once your lab team has worked through the lab and has some preliminary results, you will meet with your research team to discuss the lab report. Typically, each research team will consist of two lab teams. The lab teams should discuss their respective findings, and iron out any differences that may exist. Once the research team feels that the final results are ready to be written up, they need to decide what to include in the lab report. Each team will submit one lab report for a shared grade, and all team members are responsible for understanding the contents of the report. One team member will be designated as having the final responsibility for assuring that the report is complete and submitted by the deadline. (This duty will rotate to another team member for the next chapter.) Lab reports should be carefully prepared, with explanations given in complete sentences that are free of spelling and punctuation errors. Each report should also include (at minimum) the following sections:

1. An introduction, which should contain a summary of the important ideas and concepts covered in the chapter.

2. The results of findings related to the Research Questions. You should include all relevant experimental work, statements of your conjectures, and proofs or counterexamples as appropriate.

3. The answers to all exercises.

You might want to write your lab report as a *Maple/Mathematica* notebook. To provide a guide, the *Maple/Mathematica* notebooks `sample.mws` / `sample.nb` and `samplereport.mws` / `samplereport.nb` provide a sample lab and lab report, respectively.[6]

Chapter summaries After lab reports have been submitted, your instructor will distribute the chapter summary. It contains model answers for the lab report including conjectures and proofs. The chapter summary usually includes additional material which may (optionally) be assigned for reading.

Homework The last phase of a chapter is the homework assignment.

---

[6]Both *Maple* and *Mathematica* forms of the labs and lab reports are included on the accompanying cd-rom. Only the *Maple* versions are printed in the book.

Answering open-ended questions in a math course may be a new experience for you. It is not clear cut what you need to do to get an A. However, research in any field involves open-ended investigation, and mathematics is no different. Your work will be judged the way any mathematical research is judged: by the quality of the statements, and by how much can be proven. When you are answering Research Questions,

- Try to state the best conjecture you can, even if you have no idea of how to prove it.

- Prove as much of your conjectures as you can.

- Provide numerical evidence for any part of a conjecture that you are unable to prove.

The relative emphasis which will be given to conjectures as opposed to proofs is up to your instructor. In any case, it is better to have a great conjecture with little proof than it is to only conjecture the part you can prove.

How to Assemble This Book

When you first receive this book, it may seem that there are big chunks missing; there are no chapter summaries! That is because your instructor will make each chapter summary available at the end of the chapter. A chapter would not be as much fun if you had the answers at the beginning.

The pages of this book are perforated so that you can tear them out. We recommend getting a three-ring binder to keep them in. As you accumulate chapter summaries, they can be inserted between each chapter's prelab and homework pages in your binder. If you like, you can add printed copies of the lab notebooks as well. By the end of the term, you will have the complete book.

Using the CD-ROM

This book comes with a cd-rom computer disk. Be sure to see the `ReadMe.txt` file on the cd-rom; it contains important information which is not repeated here.

The cd-rom contains copies of the electronic notebooks in different formats. Most people would only use one format (Maple, Mathematica, or Web), so they can ignore the other formats.

Maple To use this version of the electronic notebooks, you need a copy of the program *Maple* on your computer. This is a commercial piece of software from Waterloo Maple which must be purchased. The cd-rom contains files for use in different versions of Maple (Maple V5R4, Maple V5R5, and Maple 6). Maple version 5.1 is compatible with Maple V5R5.

Once you start Maple on your computer, choose "Open" from the "File" menu. Then navigate your way on your computer to the Maple worksheet you want to use. Depending on how your computer has been configured, it may be possible to have Maple open a particular file by double clicking on an icon for the file.

Mathematica To use this version of the electronic notebooks, you need a copy of the program *Mathematica* on your computer. This is a commercial piece of software from Wolfram Research which must be purchased. The notebooks on the cd-rom provided work with Mathematica versions 3 or 4.

Once you start Mathematica on your computer, choose "Open" from the "File" menu. Then navigate your way on your computer to the Mathematica notebook you want to use. Depending on how your computer has been configured, it may be possible to have Mathematica open a particular file by double clicking on an icon for the file.

Web To use this version of the electronic notebooks, you need a web browser capable of running Java applets. Versions of the two most common web browsers, Netscape and Internet Explorer, released since 1998 have been capable of running Java applets. Other web browsers can run Java as well. Note, the ability of a browser to execute Javascript is not sufficient.[7]

If your web browser cannot run Java applets, consider downloading a newer version of your browser. Most web browsers (including the two listed above) are free to download and use.

If you have difficulty running Java applets, check the *Options* or *Preferences* settings for your web browser. Your problem could be as simple as the fact that Java has been turned off for your browser.

It is *not* necessary to have an internet connection to use the web version. Your web browser can use the files directly off of a disk in your computer (it can be a hard disk, a floppy disk, or a cd-rom disk). Once the web browser has started, choose "Open" (or "Open Page") from the "File" menu. You should then be able to click on a button which allows you to browse for a file on your own computer to open. Click on that button, and navigate your way to wherever

---

[7] *Javascript* sounds similar to *Java*, and both are programming languages associated with web pages, but they are not interchangeable.

you have the web version of the electronic notebooks stored. Open the file `index.html`. It has links to the files for each chapter.

Regardless of whatever format you decide to use for the electronic notebooks, we suggest copying the corresponding folder (directory) to your computer's hard disk. In the Maple and Mathematica versions, you can make changes to the electronic notebook and save the changes for future reference. It is not possible to save the files to the cd-rom disk, so this can only be done on a copy. The files fit on a floppy disk, so they can be carried around from one computer to another.

It is not possible to make changes to the files in the web version. However, these files also fit on a floppy disk so they can be easily transported for use on different computers.

How to Talk Like a Mathematician

Upon entering this course, you officially became a mathematician, and so it's important that you talk like one. By talking like a mathematician, we don't mean mumbling and cracking silly puns, although a certain amount of this is allowed. Instead, we mean that you should know the meaning of the important mathematical terminology and how it is used, so that you can communicate clearly with the other mathematicians that you encounter. Below we describe a few mathematical terms that will be used constantly in this course.

Definition

In mathematics, a *definition* fulfills the same role as in everyday language: to provide the meaning of a word or phrase. Mathematical theory is built up in layers, with the bottom layer consisting of definitions.[8] The first step towards comprehending any mathematical statement is understanding the underlying definitions. Here's an example of a definition that we will be using later in this course:

Definition An integer n is *even* if and only if n can be expressed in the form $n = 2k$, where k is an integer.

---

[8]In fact, definitions are not *really* the bottom layer; axioms are. This is to get around the following difficulty. Imagine you are writing a dictionary from scratch, and need to define your first word. What words would you be able to use in the definition? The answer, of course, is that there are none. The same problem occurs in making mathematical definitions. To work around this problem, the foundation of mathematics is made up of *axioms*, that is, basic assumptions. They play the role of definitions for *undefined* terms. Rather than give a definition of a term, axioms describe its properties. The axioms underlying the material in this course come from set theory and will not be used directly here. Instead, we will assume some basic properties of integers and real numbers. For example, we will assume that the sum, difference, or product of two real numbers is another real number.

Note that this definition requires knowledge of the term *integer*. As described in footnote 8, the definition of integer is left to a course in set theory. We will simply think of integers as follows:

> The *integers* are the set of all positive and negative "counting numbers", together with 0:
>
> $$\ldots, -3, -2, -1, 0, 1, 2, 3, \ldots.$$

We will assume that integers satisfy some familiar properties, such as that the sum, difference, and product of two integers is an integer.

Conjecture

A mathematical statement that is not known to be true or false is called a *conjecture*. Here are two examples of simple conjectures:

1. The sum of two even integers is an even integer.

2. For all nonnegative integers k, the number n given by

$$n = 2^{2^k} + 1$$

is a prime number.[9]

A major part of this course will involve forming and investigating conjectures. Once a conjecture is formed, then your job will be to determine if it is true or false. To show that a conjecture is true, you must provide a rigorous mathematical proof. Once such a proof is found, the conjecture is then promoted to the status of theorem (see below). To show that a conjecture is false, you must find a counterexample (see below).

Theorem

A mathematical statement which has been rigorously proved to be true is called a *theorem*. Let's take another look at the first conjecture given above:

1. The sum of two even integers is an even integer.

This conjecture sounds reasonable enough. Here are a few examples:

$$2 + 4 = 6, \quad 28 + 12 = 40, \quad 123456 + 765432 = 888888.$$

In each of these cases, the conjecture holds, which suggests that it might be true. However, remember that **examples supporting a conjecture do not constitute a proof!** Since there are infinitely many even integers, you can't hope to verify the conjecture by checking all possible cases. Instead, we return to the definition of *even integer* to allow us to fashion a proof:

---

[9]As you probably already know, a number $p > 1$ is *prime* if the only positive divisors of p are 1 and p.

Proof. Let n_1 and n_2 be even integers. Then $n_1 = 2k_1$ and $n_2 = 2k_2$ for some integers k_1 and k_2. Therefore

$$\begin{aligned} n_1 + n_2 &= 2k_1 + 2k_2 \\ &= 2(k_1 + k_2), \end{aligned}$$

and so $n_1 + n_2$ is also an even integer. $\square$

Thus, we see that our first conjecture is now promoted to the status of theorem. Writers of mathematics use different names for theorems. Here is a short rundown:

Lemma *Lemmas* (or, if you're stuffy, *lemmata*) are mathematical statements that are used to provide an intermediate step towards proving a theorem.

Corollary *Corollaries* are mathematical statements that follow immediately from a theorem.

Proposition A *proposition* is essentially equivalent to a theorem. If an author uses both terms, then *theorem* is reserved for the most important results.

Lemmas, corollaries, and propositions require the same rigorous proofs as theorems. Although they all are important, in this course this distinction in terminology will play a relatively minor role.

Counterexample

To show that a conjecture is false, you need to come up with a *counterexample*. This is nothing more than one specific example that contradicts the conjecture, and thus shows that the conjecture is false. The second conjecture above states that $2^{2^k} + 1$ is prime for all $k = 0, 1, 2, \ldots$. Here are the first few cases:

$$2^{2^0} + 1 = 3, \quad 2^{2^1} + 1 = 5, \quad 2^{2^2} + 1 = 17.$$

Each of these are clearly prime, and it isn't too hard to show that

$$2^{2^3} + 1 = 257 \quad \text{and} \quad 2^{2^4} + 1 = 65537$$

are also prime. Looks good so far, doesn't it? Unfortunately for this conjecture

$$2^{2^5} + 1 = 4294967297 = 641 \cdot 6700417.$$

Thus $2^{2^5} + 1$ is not prime, which provides a counterexample to the second conjecture, thereby disproving it.

Tips for Writing Proofs

One of the most difficult things you will attempt in a mathematics course is to *invent* the proofs to various propositions. While there is no set algorithm to follow, there are certain guidelines you can use. Some are listed below. You may want to refer back to this section as the course progresses.

These guidelines are far from a cure-all, but at least they can orient you correctly. They are used by everyone from students to great mathematicians either consciously or subconsciously. These techniques are described in vague terms in an attempt to be generally applicable. To bring things back down to earth, they are followed by examples.

Use the definitions As simple as it may sound, this is the most basic technique of proof. It is indispensable, not to mention very *very* important. Needless to say, you should master this technique. Proficiency with definition use in proofs is frequently the difference between passing and failing.

The idea is that the objects in the statement of a proposition can usually be broken down into simpler concepts by use of definitions. Sometimes this can be repeated. Since all definitions are *if and only if*, this process can be safely applied to both the hypothesis and the conclusion of the proposition. By doing so, you will have transformed the statement into an equivalent one which is almost always easier to prove. When a concept has just been introduced, its definition may be all you have at your disposal and you are forced to apply this technique.

If and only if statements Some of the most useful statements in mathematics come in the form of an if and only if statement. For example, consider the Factor Theorem from algebra:

Theorem *If a is a number and $f(x)$ is a polynomial, then $f(a) = 0$ if and only if $f(x) = (x - a)q(x)$ for some polynomial $q(x)$.*

Proving the Factor Theorem involves writing two proofs. First, we have to prove: "If $f(a) = 0$, then $f(x) = (x - a)q(x)$ for some polynomial $q(x)$." Next, we have to prove: "If $f(x) = (x - a)q(x)$ for some polynomial $q(x)$, then $f(a) = 0$."

Sometimes it is possible to prove both directions of an if and only if statement at once (by having a chain of statements where each statement is equivalent to the next). Proofs of that sort are very efficient to write up, but you should not start the proof process by trying to produce such a proof. Always view if and only if statements as two statements, and approach the proofs separately. After you can prove each direction, then you may want to go back and see if the two proofs can be polished, or combined as described above. Trying for a clever proof from the start usually takes *much more* time.

Showing two sets are equal Since virtually all mathematical objects are built out of sets, many propositions are basically of the form $A = B$, where A and B are sets. Sometimes, one can apply the definitions of the sets A and B so that their equality is immediately clear. Otherwise, you should show each is a subset of the other.

Your proof will start with, "Let $a \in A, \ldots$" after which you deduce " $\ldots$ therefore $a \in B$" (and so $A \subseteq B$). Then you start again with "Let $b \in B \ldots$ " and deduce " $\ldots$ therefore $b \in A$" (and so $B \subseteq A$). You may then conclude that $A = B$. Note that this approach is a special case of the preceeding discussion on if and only if statements: we prove that $A = B$ by proving $x \in A$ if and only if $x \in B$.

Indirect proof To prove a statement indirectly, you assume that the statement is false and derive a contradiction. The classic example of this type of reasoning in everyday life is the alibi. For example, suppose Clarence Darrow is trying to defend Brutus for the murder of Julias Caesar. Autopsy reports show that Caesar died of a knife wound in his home at approximately 9 P.M. on the fateful night. So, Clarence might argue that if Brutus had committed the murder, he would have had to have been at the crime scene around 9 P.M. However, several people have sworn that Brutus was playing golf in his own backyard from 8 until 11 P.M. on the night in question, a contradiction. Therefore, Brutus did not murder Caesar. Clarence would have made a fine mathematician.

In mathematics, the statements we prove often come in the form of an if-then statement; they have a hypothesis and a conclusion. If we consider the statement *If P, then Q* and take its negation, we get the logical statement *It is not the case that (If P, then Q)*. This is logically equivalent to the simpler combination: *P and not Q*.[10] Thus, when applying the method of indirect proof to an if-then statement, you should assume the hypothesis (P) and the negation of the conclusion (*not Q*), and then try to derive a contradiction.

You should not be shy to attempt proofs indirectly, even if you think a direct proof is possible. It is **always** a better idea to have one complete proof, and then refine it, than it is to attempt to save time by devising the best proof from the start. This advice is worth repeating: trying for a clever proof from the start usually takes *much more* time.

There are a couple of situations which should immediately suggest an indirect proof. They are:

---

[10]This equivalence is a result from formal logic. It can be proved easily using truth tables.

- You are trying to prove that something does not exist. Arguing indirectly lets you assume that this mythical object does exist, and manipulate it using definitions and theorems until you have a contradiction.

- You are trying to prove that something is unique. There will typically be some list of properties the object must have. So, you assume that it is not unique. Therefore, there are at least two objects which both possess all of the prescribed properties. Use them to deduce that these "two objects" are really one and the same (thus, a contradiction).

Existence Another type of conclusion you are frequently asked to reach is the existence of an object which satisfies certain conditions. This can rarely be done indirectly (existence proofs by contradiction are usually quite sophisticated). Instead, the number one technique is to construct the desired object directly. This usually amounts to writing down the definition of a set or a function, and then verifying that it has all of the right properties. Note, the properties will usually contain clues for how to set up the definition.

Advanced use of definitions Using definitions is important enough to discuss it a second time. Some problems involve concepts which have several equivalent definitions. You may have to experiment to see which one works the best. An example of this phenomenon is the definition of a continuous function $f\colon \mathbf{R} \to \mathbf{R}$. It can be defined equivalently by any of the following statements:

1. For all $c \in \mathbf{R}$, $\lim_{x \to c} f(x) = f(c)$.

2. For any $c \in \mathbf{R}$ and any sequence c_n such that $\lim_{n \to \infty} c_n = c$, $\lim_{n \to \infty} f(c_n) = f(c)$.

3. The inverse image under f of any open interval is a union of open intervals.

When you have several equivalent ways of defining something, you may have to try each form of the definition before you find the one which works best.

Simplify the problem This is one of the toughest techniques to get students to apply. The idea is to give yourself extra assumptions, and see if you can prove the new, easier proposition. If you are successful, you then try to remove the extra assumptions. To your surprise, this may be quite easy. On the other hand, it may be hard. It all depends on how much extra you let yourself assume. In any case, it is frequently easier to have a proof which you are trying to generalize, than to solve the original problem from scratch.

This technique is also useful if carried to an extreme. By this we mean, **doing examples**. When stuck, it is an excellent idea to try the proposition out in a few examples. These verifications will not constitute a proof, but they may lead you to one. We know several working mathematicians who live by this technique.

In this course, it will be hard to overlook this approach. You will usually need to compute many examples just to discover the correct statement of a theorem.

Induction No discussion of techniques of proof would be complete without a mention of induction. We do not plan to go into the ins and outs of induction here. The idea is usually that the proof has some part which needs to be iterated some number of times. The general idea may be clear, but induction is what makes it rigorous. A sample proof below gives both the formal and "casual" versions of an induction proof. It is up to your instructor to decide the level of rigor required for your class.

There are some instances when induction is indispensable. For example, the parameter N in the formula

$$\sum_{i=1}^{N} i^2 = \frac{N(N+1)(2N+1)}{6}$$

makes it a natural choice for an inductive proof, and one is given below. Note that the induction is easy to execute, although the proof is not very enlightening.

The big theorem At times, the key to getting unstuck is to look for a related theorem in the preceding material. Depending on the situation, you may be able to apply the theorem, or mimic its proof in your problem.

Where to focus We all know that in a proof, you assume the hypothesis and work toward the conclusion. So where should you focus when starting a proof? The conclusion! Imagine walking from your home to the grocery. You start at home and think about where you are going. If you keep focused on where you are, you are likely to wander in circles, and it is unlikely that you will stumble into the grocery. The same is true in proofs. Being mindful of what sort of conclusion you must reach starts you in the right direction. Notice that several of the guidelines given above are keyed on particular types of conclusions.

It can also be useful to work backwards from the conclusion. This is something of a mathematician's secret. After all, assuming the conclusion in the presentation of a proof is a fatal flaw.[11]

---

[11]If you are not familiar with why assuming the conclusion in a final proof is so bad, suppose we want to prove that $1 = 2$. If we assume that $1 = 2$, then we know that $1 = 2$ and the statement is proved!

Another (erroneous approach) is to assume the conclusion and deduce a true statement. In our example, we start with $1 = 2$ and multiply both sides by 0 to get $0 = 0$, a true statement. Thus, $1 = 2$.

In reality, if we could assume the conclusion, we could prove *anything*.

Try to work from the conclusion back to your hypothesis and from the hypothesis forward to the conclusion. If you can get the two ends to meet, you might have a proof on your hands. You still need to check that all of the implications work properly starting from the hypothesis and ending with the conclusion. Finally, be sure to write your proof so that it flows in the proper direction.

The Obvious and the Trivial

In both the reading and writing of proofs, you must come to grips with the question "When is something obvious?" The claim that a statement is obvious or trivial has a different meaning in mathematics than it has anywhere else. Proofs would become very long, and excruciatingly boring if the reasoning were reduced to pure logic at every stage. So, the author leaves some things to the reader on the grounds that "anyone" should be able to fill in the gaps (if he or she wanted to). The big question is who constitutes "anyone"? A good rule of thumb is that "anyone" should be the expected audience, the author of the proof, and his or her (mathematical) peers. So, when you want to claim something as trivial, it should be straightforward for the reader, your classmates, and of course yourself.

Examples

We now give some sample proofs and comment on where the above techniques are used. You may want to try to prove the propositions first before looking at our proofs.

Proposition *Let A, B, and C be sets. Then*

$$(A \cup B) \cap C = (A \cap C) \cup (B \cap C).$$

Proof. Let $x \in (A \cup B) \cap C$. Then $x \in C$ and either $x \in A$ or $x \in B$. If $x \in A$, then $x \in A \cap C$. If $x \in B$, then $x \in B \cap C$. In either case, $x \in (A \cap C) \cup (B \cap C)$. Now suppose that

$$x \in (A \cap C) \cup (B \cap C).$$

Then either $x \in A \cap C$ or $x \in B \cap C$. So, either $x \in A$ and $x \in C$, or $x \in B$ and $x \in C$. Note that in either case $x \in C$. Furthermore, either $x \in A$ or $x \in B$. So we know that $x \in C$ and that $x \in A \cup B$. That is, $x \in (A \cup B) \cap C$.

This is a classic example of showing that two sets are equal to each other. Following the basic form suggested above, the rest amounts to applying the definitions of union and intersection repeatedly.

Proposition *Every finite, nonempty set of real numbers has a largest element.*

Proof. Let S denote our finite set of real numbers. We apply induction on n, the number of elements in the set. Since the set is nonempty, we start with $n = 1$.

If $n = 1$, then $S = \{a\}$ for some $a \in \mathbf{R}$. Then since $a \geq a$, a is the largest element.

Now assume that the statement is true for all sets of order n (i.e., with n elements), with $1 \leq n \leq N$. Let S be a set of order $N + 1$. Since $N + 1 > 0$, S has an element, call it a. If $a \geq x$ for all $x \in S$, we are done. Otherwise, there exists $b \in S$ with $b > a$. Furthermore, $S - \{a\}$ has N elements, and so by the induction hypothesis, it has a largest element c, such that $c \geq x$ for all $x \in S - \{a\}$. But $b \in S - \{a\}$, so $c \geq b > a$. Thus, c is the largest element of S.

The proposition may appear obvious. An attempt to spell out a proof may look something like this:

> *Since the set is nonempty, pick an element. If it is the biggest, you are done. If not, pick an element which is bigger. Repeat the process. It must end some time since the set is finite. That element is clearly larger than the discarded elements and larger than the remaining ones. So, it is the largest element.*

This would be fine as a casual argument. However, the statement to repeat the process until you have a largest element is not rigorous. To make it so, one uses induction as illustrated to the left.

Proposition *For any two nonzero integers n and m, there exists a unique integer $d > 0$ such that*

1. *$d \mid n$ and $d \mid m$.*

2. *If $c \mid n$ and $c \mid m$, then $c \leq d$.*

This number d is called the *greatest common divisor* of n and m.

Proof. First we prove existence. Since $n \neq 0$, $\{a \in \mathbf{Z} : a \mid n\}$ is finite. Therefore,

$$S = \{a \in \mathbf{Z} : a \mid n\} \cap \{a \in \mathbf{Z} : a \mid m\}$$

is finite and clearly contains 1. So, by the previous proposition, S has a greatest element, d. Since $1 \in S$ and d is the largest element of S, we have $1 \leq d$. Finally, if $c \mid n$ and $c \mid m$, then $c \in S$; this implies that $c \leq d$.

The finiteness of this set would require some proof, but we don't want to get too bogged down in the example. This proposition would be considered obvious by most texts. However, in light of the comments above on what is obvious, this merely means that filling in the details should be easy. Here we have given the details.

Proposition *Every interval (a, b) with $a < b$ has no maximum element.*

Proof. Suppose not. Then there exists $m \in (a, b) = \{x \in \mathbf{R} : a < x < b\}$ which is maximum. Therefore, for all $x \in (a, b)$, $x \leq m$. But since $m \in (a, b)$, $m < b$. So,

The conclusion is that something does not exist. So, we assume it does and argue indirectly.

$$a < m < \frac{m + b}{2} < b.$$

But this means that $(m + b)/2 \in (a, b)$ and $m < (m+b)/2$ which contradicts the maximality of m. Therefore (a, b) does not have a maximum element.

Next we prove the formula for the sum of consecutive squares, as promised above.

Proposition *For all integers $N \geq 1$,*

$$\sum_{i=1}^{N} i^2 = \frac{N(N + 1)(2N + 1)}{6}.$$

Proof. This is the induction proof promised earlier. First we check to see that it is true for $N = 1$. We get

$$1 = \frac{1 \cdot (1 + 1)(2 + 1)}{6}$$

which is true.

Now we assume that it is true for a sum with N terms for some positive integer N and prove that it is true for $N + 1$ terms. Since the formula holds true for a sum of N terms, we have

$$\sum_{i=1}^{N} i^2 = \frac{N(N + 1)(2N + 1)}{6},$$

and we want to deduce that

$$\sum_{i=1}^{N+1} i^2 = \frac{(N + 1)(N + 2)(2(N + 1) + 1)}{6}.$$

Now, the left-hand side is equal to

$$
\begin{aligned}
\sum_{i=1}^{N} i^2 + (N+1)^2 &= \frac{N(N+1)(2N+1)}{6} + (N+1)^2 \\
&= \frac{2N^3 + 3N^2 + N}{6} + \frac{6N^2 + 12N + 6}{6} \\
&= \frac{2N^3 + 9N^2 + 13N + 6}{6} \\
&= \frac{(N+1)(N+2)(2(N+1)+1)}{6}.
\end{aligned}
$$

So, the equation is true for all N.   □

The algebra could be made simpler here. However, we want to emphasize that clever algebra skills are not the key to this proof. Once you realize that you have to prove that

$$
\frac{N(N+1)(2N+1)}{6} + (N+1)^2 = \frac{(N+1)(N+2)(2(N+1)+1)}{6},
$$

both sides can be expanded and simplified until they match. This is an example of discovering a proof by working with the conclusion, but then writing the final proof in the correct order.

The previous examples have been relatively easy to prove. So, we finish with a more difficult proposition. Since there are no comments, you should see how the above techniques are being used.

Proposition (The Division Algorithm) *Let a be an integer and b be a positive integer. Then there exist unique integers q and r such that*

$$
a = bq + r
$$

with $0 \leq r < b$.

Proof. We will prove uniqueness first. Assume that there are two pairs of integers q, r and q', r' such that

$$
a = bq + r \quad \text{and} \quad a = bq' + r'
$$

with $0 \leq r < b$ and $0 \leq r' < b$. Then, eliminating a from the equations, we get $bq + r = bq' + r'$, which implies that

$$
b(q - q') = r' - r. \tag{1}
$$

Now, $0 \leq |r' - r| < b$ by the inequalities on r and r'.[12] Therefore, $b \cdot |q - q'| < b$. Since $b > 0$, we can divide by b to get $|q - q'| < 1$. Since q and q' are integers, this implies that $q = q'$. Equation (1) then implies that $r = r'$.

Now for existence. The idea is to construct either q or r so that $a = bq + r$ with the given condition on r. If we had our q and r, then $a/b = q + r/b$ with $0 \leq r/b < 1$.

So, we define q to be the greatest integer less than or equal to a/b, which implies $0 \leq a/b - q < 1$. Multiplying these inequalities by b, we get that $0 \leq a - bq < b$. Letting $r = a - bq$, we are done. $\square$

The file names on the cd-rom for the sample labs are:

Maple electronic notebook `00-sample.mws`

Mathematica electronic notebook `00-sample.nb`

The sample report for the sample lab is:

Maple electronic notebook `00-samplereport.mws`

Mathematica electronic notebook `00-samplereport.nb`

The introductions to Maple, Mathematica, and active web pages are:

Maple electronic notebook `00-intro.mws`

Mathematica electronic notebook `00-intro.nb`

Web electronic notebook Start with the web page `index.html`

---

[12]Both r and r' lie between 0 and $b - 1$, so the largest that the distance between them can be is $b - 1$. The distance between r and r' is given by $|r' - r|$, so the inequality as shown follows.

Sample Maple Lab: Prime Numbers

■ How Many Primes Are There?

A prime number is an integer $p > 1$ whose only positive divisors are 1 and p. Prime numbers are important for factoring other integers, and they arise quite often in number theory.

One of the first basic questions we can ask is: How many prime numbers are there? Will we run out of primes at some point, or are there infinitely many primes? You may already know the answer to this question, namely that there are infinitely many prime numbers. Here is a quick sketch of a proof which was formulated by Euclid.

■ Proof that there are infinitely many prime numbers

We will describe a method that takes a list of prime numbers, and always produces some new primes. As we shall see, we can repeat this as often as we like, and so we can generate as many prime numbers as we like. Therefore, there must be infinitely many primes.

Suppose we have a list of prime numbers, $p_1, p_2, \ldots, p_n$. To construct new primes, we simply factor the number $m = p_1 p_2 \cdots p_n + 1$. The result will be a product of primes. Here is a Maple function which does this:

```
> newprimes:= proc(plist)
    local m, j;
    m := mul(plist[j], j=1..nops(plist))+1;
    printf('m = %d', m);
    printf('\nFactorization of m: ');
    ifactor(m);
  end:
```

Let's try it with some lists of primes:

```
[ > newprimes([2,3,5,7,11,13]);
```

We started with the first six primes as input. Multiplying them together and adding 1 produces 30031. Note, 30031 is not prime. However, when we factor it, we get the primes 59 and 509. Let's see what happens when we add 59 and 509 to our list of primes:

```
[ > newprimes([2,3,5,7,11,13,59,509]);
[ >
```

This time the method produced just one new prime, 901830931. What is important is that the method always produces a new prime. In fact, *each* of the primes which divide m are not on the original list. We give a formal proof of this statement now.

Proof: Let $m = p_1 p_2 \cdots p_n + 1$, and suppose that q is a prime number in the factorization of m, that is, m is a multiple of q. We want to show that q cannot be one of the original p_i. However, since m is a multiple of q, the remainder when m is divided by q is 0. On the other hand, for each p_i, the number m is 1 more than a multiple of p_i. Therefore, the remainder when m is divided by p_i is 1, and so q cannot be any of the original p_i.

18

> **Exercise 1**
>
> Use the function `newprimes` to generate at least 5 new primes starting from the set { 2, 3 }. You may have to use the function more than once.

Primes of the form $4k + a$

Now that we have seen that there are infinitely many prime numbers, we look at the number of primes of the forms $4k$, $4k + 1$, $4k + 2$, and $4k + 3$. To look at primes of these different forms, we use the following function:

```
> primeseeker:= proc(a, howmany)
    local j, p;
    for j from 1 to howmany do
     p:= 4*j+a;
     if isprime(p) then printf('%d is prime\n',p) fi;
    od;
  end:
```

To check the first 30 integers of the form $4k + 1$ for primes, we use

```
> primeseeker(1,30);
>
```

Use `primeseeker` to answer the following research questions. In each case, the phrase "how many primes are there" should be interpreted so that "infinitely many" is a possible answer.

Research Question 1

How many primes are there of the form $4k$?

Research Question 2

How many primes are there of the form $4k + 1$?

Hint: The proof of the right conjecture is quite difficult. We are mainly looking for a good conjecture in this case.

Research Question 3

How many primes are there of the form $4k + 2$?

Research Question 4

How many primes are there of the form $4k + 3$?

Hint: A proof of the right conjecture can be constructed along the lines of Euclid's proof described above.

Sample Lab Report: Prime Numbers

Submitted by:

Julie Barnes
Pete Cochran
Lincoln Hayes

September 23, 1968

▣ Introduction

Instructor comments are given in this report in italics.

This section of the report should contain a brief but precise summary of what is contained in the report. It should include an indication of the question (or questions) that were studied, as well as a summary of the results. In addition, it should contain a summary of the important ideas and concepts included in the chapter. Here's an example that works for this report.

In this report, we considered the question of the number of primes of the forms $4k + a$ where a is 0, 1, 2, and 3. For two values of a, namely $a = 0$ and $a = 2$, there is at most one prime. In the other cases, we conjecture that there are infinitely many prime numbers. We were unable to prove this, but we provide some numerical evidence to support the conjecture.

The chapter also addressed a proof that there are infinitely many prime numbers altogether, and presented an algorithm for taking a finite list of prime numbers and generating new primes (not on the original list).

▣ Research Questions

Below is a sample of the type of discussion and work which will typically accompany your conjectures and proofs. Conjectures should be stated precisely in this part.

▣ Research Question 1

Theorem: There are no primes of the form $4k$.

Proof: For any number of the form $4k$, we note that $4k = 2(2k)$. Neither of these factors is 1, and so numbers of the form $4k$ always have a nontrivial factoring. Therefore, they are never prime.

▣ Research Question 2

Conjecture: There are infinitely many primes of the form $4k + 1$.

We were not able to find a proof to this conjecture. Using the function `primeseeker`, we were always able to find more primes of the form $4k + 1$ by just increasing the search range. For example, from the first 100 numbers of the form $4k + 1$ we found:

```
> primeseeker(1,100);
5 is prime
13 is prime
17 is prime
29 is prime
37 is prime
41 is prime
```

```
 53 is prime
 61 is prime
 73 is prime
 89 is prime
 97 is prime
101 is prime
109 is prime
113 is prime
137 is prime
149 is prime
157 is prime
173 is prime
181 is prime
193 is prime
197 is prime
229 is prime
233 is prime
241 is prime
257 is prime
269 is prime
277 is prime
281 is prime
293 is prime
313 is prime
317 is prime
337 is prime
349 is prime
353 is prime
373 is prime
389 is prime
397 is prime
401 is prime
```

We did not find a proof that there are infinitely many primes of this form.

This conjecture is in fact true. To see a proof, we will have to wait until the end of the course before we will have built up the right tools for proving it!

◼ Research Question 3

Theorem: There is exactly one prime of the form $4k + 2$.

Proof: Any number of the form $4k + 2$ can be factored $4k + 2 = 2(2k + 1)$. Since we have a factor of 2, the only way $4k + 2$ can be prime is if this factoring is trivial, that is, $4k + 2 = 2$. Note, 2 is of the form $4k + 2$: $2 = 4(0) + 2$. Thus, there is exactly one prime of the form $4k + 2$.

Note, the statement of the conjecture is better than saying that there are finitely many primes of the form $4k + 2$, because it asserts more information (and is still correct). One of the things which is important in phrasing conjectures is to make the best possible statement. If you find a really good conjecture but can only prove a weaker statement, you should give both statements, and the proof of the weaker statement.

◼ Research Question 4

Conjecture: There are infinitely many primes of the form $4k + 3$.

The proof given here is really for reference, to see how Euclid's proof can be modified to prove this conjecture. In practice, it would be too difficult for a student to think up on the first day of class. In particular, it uses some of the tools developed in the first few chapters.

Proof: We will loosely follow the approach for proving that there are infinitely many primes.

Suppose we have n primes of the form $4k + 3$: $p_1, p_2, \ldots, p_n$. If the prime number 3 is in this list, we remove it (we will see why in a minute). We then let $m = 4\, p_1 p_2 p_3 \bullet \bullet \bullet p_n + 3$. First, we see that m is of the form $4k + 3$. We imagine it factored as a product of primes, and consider the form of the prime divisors of m. By the division algorithm (applied with division by 4), every integer is of one of the forms $4k$, $4k + 1$, $4k + 2$, or $4k + 3$. Since m is odd and $4k$ and $4k + 2$ are even, the prime divisors of m must be either of the form $4k + 1$ or $4k + 3$. We claim m has at least one prime divisor of the form $4k + 3$.

First, we note that the product of two numbers of the form $4k + 1$ is again of the form $4k + 1$. To see this, we multiply $(4k + 1)(4l + 1) = 16kl + 4(k+l) + 1 = 4(4kl + k + l) + 1$. So, we could not produce m (which has form $4k + 3$) by multiplying only numbers of the form $4k + 1$. So, m is divisible by at least one prime of the form $4k + 3$.

Now, suppose that m is a multiple of some p_i. Then $m - 4\, p_1 p_2 p_3 \bullet \bullet \bullet p_n$ is also a multiple of p_i. Since $m - 4\, p_1 p_2 p_3 \bullet \bullet \bullet p_n = 3$, we know that 3 is a multiple of p_i. This forces $p_i = 3$. However, we explicitly removed 3 from the list of p_i above, so this is a contradiction. Therefore, none of the p_i can be a divisor of m. Moreover, m cannot be a multiple of 3 (or we can show, by similar reasoning, that 3 is one of the p_i). So, m has a prime divisor of the form $4k + 3$ which is not equal any of the p_i nor is it equal to 3. In this way, we can construct as many primes of the form $4k + 3$ as we like.

Exercises

Any exercises from the lab are answered here.

Exercise 1

We started with $\{2, 3\}$, and used the function newprimes to find some new primes:

```
> newprimes([2,3]);
m = 7
Factorization of m:
```
$$(7)$$

It found the prime 7, so we repeated with $\{2, 3, 7\}$:

```
> newprimes([2,3,7]);
m = 43
Factorization of m:
```
$$(43)$$

This time we found the prime 43, so we repeat:

```
> newprimes([2,3,7,43]);
m = 1807
Factorization of m:
```
$$(13)(139)$$

This time we found two new prime numbers, 13 and 139. Finally, we ran newprimes one more time (so that we would have at least five new primes):

```
> newprimes([2,3,7,13,43,139]);
m = 3263443
Factorization of m:
```
$$(3263443)$$

We found one more prime, 3263443. So the five primes we found by this method are 7, 43, 13, 139, and 3263443.

Maple Lab: Introduction to Using Maple

▣ Maple Basics Click on the plus sign in the box in the left margin to open the group.

If you're reading this text now, then you've mastered the second hardest thing about Maple -- figuring out how to open a closed collection of groups.

The hardest thing about Maple is remembering to type semicolons at the end of each input line to be evaluated. For example, to have Maple evaluate 17 + 34, we type

```
[ > 17+34;
```

To perform a computation, click the mouse on the command you want to execute and push the normal carriage return key, labeled **Enter** or **Return**. Try it out on the above command. If everything works the way it should, then you will get the answer 51. If you don't, ask for help. Now, let's try it without the semicolon:

```
[ > 17+34
```

Hit **Enter** in the above group. Maple uses the semicolon as a sign that it has reached the end of this input. You can throw your hands in the air and cry out "They can put a man on the moon, but I will have to type these silly semicolons all semester!?!?" The simple answer is "yes". Now that you have that out of your system we can get back to work.

Try having Maple evaluate the following:

```
[ > 3+4;
[ > 13-47;
[ > 12*23;
[ > 284/16;
[ > 2^143;
[ >
```

Notice that you need to type a * for multiplication and a ^ to get a power.

Now that you have made some modifications to this worksheet, you should save your work. Doing so is easy: Click on "Save" in the "File" menu. There is also a keyboard "shortcut" for this and many other commands. To find the shortcuts, look at the right side of the menus. Be sure to save regularly to avoid losing your work if the machine malfunctions.

Everything in a Maple worksheet is organized into "groups". We saw how to open a group at the very beginning: by clicking the mouse on the plus sign in the square box next to the title. The scope of a group is indicated by the vertical line to the left.

▣ A group within a group - open me up

Thanks for opening me! Notice that there are now two lines on the left side of the screen: one for me and one for the larger group containing me. You already know how to open groups, so the big question is "How do you close a group?"

24

As you may have guessed, one way is to click on the minus sign in the square box at the beginning of the group. Try that now, and then open me back up to read the rest.

If a group is really long and you want to close it without scrolling to the top, then click once on the line at the left side of the screen for **this** group. It will change appearance to show that it is highlighted. Then, press the **space** bar and *shazam*, the group closes.

The two most important things you will find inside of a group is text (what you are reading now), and input groups (also referred to as executable groups). You can easily spot input groups by the square bracket at the left, the > symbol at the start, and the font used to display the group contents.

In this course, you will need to perform calculations that go beyond those which have been set up for you. To create a new input group, click the mouse between paragraphs at the spot you would like it to appear. Then, click the mouse on the symbol which resembles [> in the toolbar. You can also follow the menus:

Insert -> Execution Group -> After Cursor.

This will create a new input group. One word of warning:

> **After you hit return to execute a command in an input group, Maple will automatically jump to the *next* input group in the worksheet.**

If the next input group happens to be ten screenfuls down in the worksheet, so be it; that is where the cursor will go. An easy way to avoid having this feature surprise you is to make *two* input groups in a row, and use the first one. After your command has been evaluated, the cursor will land in the second one.

Create one (or two) input groups below this paragraph to compute 123!. (The "!" is not at the end of the sentence for emphasis; it's there to indicate the factorial operation. Factorials are entered just as shown).

You can obtain decimal approximations using the function evalf. For instance, when you executed the command to compute 284/16, Maple reduced the fraction but returned a fraction as output. To get a decimal approximation, run the command below:

```
[ > evalf(284/16);
```

The name of the command is an abreviation for evaluate floating point number. You can get more decimal places. For instance, here are 100 digits of π:

```
[ > evalf(Pi, 100);
[ >
```

To close this group, go back to the beginning and click on the minus sign in the box in the left margin. Alternatively, click once on the vertical line at the left and press the space bar.

■ Adept Algebra with Maple Click on the plus sign in the box in the left margin to open this group.

Here is $(x + 2\,y)^{25}$ multiplied out:

```
[ > expand((x+2*y)^25);
```

If you don't hang the command expand around the product, Maple will not multiply out the terms because it was not told to do so. Try it.

```
[ > (x+2*y)^25;
```

Get it? Try changing the numbers above and rerunning.

Maple can also factor polynomials:

```
[ > factor(1+x^15);
[ >
```

▓ *Exercise 1*

The following calculations suggest a pattern. What is the general formula suggested by these calculations?

```
[ > expand((1-x)*(1+x));
[ > expand((1-x)*(1+x+x^2));
[ > expand((1-x)*(1+x+x^2+x^3));
[ > expand((1-x)*(1+x+x^2+x^3+x^4));
[ >
```

                                  Save your work!

To close this group, click on the minus sign in the box at the beginning of the group. (Or, click on the leftmost vertical line, and press the space bar.)

▓ Defining Functions and Constants Click on the plus sign in the box in the left margin to open this group.

Defining functions in Maple is easy. Here's an example:

```
[ > f := x-> cos(x);
[ >
```

By executing the above command (which you should do now if you haven't already), you tell Maple that the function f is defined to be $\cos(x)$. If you think of a function as a mapping, this notation makes sense. Using ":=" to mean "this is a definition" is a convention used by some mathematicians in their writing, and some programming languages. So, the input statement can be read

f is defined to be the mapping which takes x to $\cos(x)$.

To evaluate a function, you just "plug in" a value. For instance, if we want to know $f(7\pi/12)$, we just type it in:

```
[ > f(7*Pi/12);
```

Note: For this command to evaluate properly, you must first have executed the definition for f above. All definitions must be executed prior to their use during each Maple session. This is important. When a function is provided in a worksheet, you need to evaluate it (click the mouse anywhere in the definition and press **Enter**).

Now, here's $f(1)$:

```
[ > f(1);
```

Remember that Maple gives exact answers unless told to do otherwise. Here's a decimal approximation to $f(1)$:

```
[ > evalf(f(1));
```

Defining constants is just like defining functions. Suppose that you wish to set the constant *a* to be equal to 1.4142135624, which happens to be the first ten decimal places of the square root of 2. All you do is the following:

```
[ > a:=1.4142135624;
[ >
```
                          Save your work!

■ Two Useful Commands Click on the . . . . . , well, you know the drill!

Two commands that you will use constantly are

  1) Copy

  2) Paste.

Here is an example of how to use these commands.

```
[ > evalf(3432902008176640000/121645100408832000);
[ >
```

1. Select the executable group above by double clicking on the bracket at the left. Alternatively, you can drag the mouse across the group while holding down the left mouse button.

2. Select the **Edit** menu and click on **Copy**. This copies the group to the computer's memory.

3. Click the mouse once at the beginning of a group. The new copy is going to go after that group.

4. Select the **Edit** menu and click on **Paste**. A copy of the group you chose should now appear on the screen. (This is the pasting part.)

5. Use the mouse to select the first digit in the numerator. Use the keyboard to change it from 3 to 2. Run your new command. The answer should be 20.

Some other things about copy-and-paste:

6. You can use copy-and-paste on words, lines, or practically anything you see on the screen. A good way to use copy-and-paste is to create a new input group first, and then paste stuff into it. To create an input group below the current group, remember that all you have to do is click on the [> button on the toolbar.

7. You can use copy-and-paste to copy materials from one part of a worksheet to another. You can also use copy-and-paste between worksheets. This can be very useful when you have an elaborate set of instructions in one worksheet that you want to copy to another worksheet.

                          Save your work!

■ Getting Help in Maple

Maple is a big program with lots of parts. However, it's not necessary to learn everything about Maple in order to use it. Feel free to experiment--look at the menus, try out different things, play with the program. Don't be afraid of breaking anything. You can't break the machine just by pushing keys!

You can get help from the Maple program itself. If you want to find out what a certain function does, you can type ?"FunctionName". Maple will open a Help window. To get rid of the Help window, select "Close Help Topic" from the File menu.

To find out what the functions divisors and ifactor do, you would type:
```
[ > ?divisors
[ > ?ifactor
[ >
```

The "Help" menu provides various means of accessing Maple's documentation. The "Full Text Search" can be especially useful for finding commands.

As you go along, you may find that something doesn't work the way you expect it to, or that there is something you would like Maple to do, but you don't know how to make it work. In those cases, you should seek help from the following sources:

1. Your fellow students.

2. The lab assistant (if there is one).

3. Your teacher.

4. The ? command.

5. The Help menu.

Exercise 2

(a) What does the Maple function floor do?

(b) What does the Maple function ceil do?

Save your work!

Mathematica Lab: Introduction to Using Mathematica

■ Mathematica Basics

Click twice on the outer bracket at the right to open the cell.

If you're reading this cell now, then you've mastered the second hardest thing about Mathematica — figuring out how to open a closed collection of cells.

The hardest thing about Mathematica is figuring out how to have Mathematica execute a computation. For example, to have Mathematica evaluate 17 + 34, we type

17 + 34

To perform a computation, click the mouse on the command you want to execute, hold down the **Shift** key, and press the normal carriage return key, labeled **Enter** or **Return**. (Hitting only the **Enter** key in the numeric keypad portion of your keyboard will also work, although we do not openly advocate hitting computers.) Try it out on the above command. The first time that you execute a command in a Mathematica session, you will be asked "Do you want to automatically evaluate all of the initialization cells in this notebook?" Your response to this question should always be "Yes".

If everything works the way it should, then you will get the answer 51. If you don't, ask for help.

Try having Mathematica evaluate the following:

3 + 4

13 – 47

12 23

12 * 23

284 / 16

2 ^ 143

Notice that **12 23** and **12*23** have the same effect. Multiplication can be indicated either with the symbol * or by leaving a space between the two quantities that are to be multiplied together.

Now that you have made some modifications to this notebook, you should save your work. Doing so is easy: Click on **Save** in the **File** menu. There is also a keyboard "shortcut" that you can use. On a Windows machine, hold down the **Ctrl** key and type **s**. On a Macintosh, hold down the "cloverleaf" key adjacent to the **Space** bar and type **s**. (In the future, we shall use the word *control* to refer to the **Ctrl** key on a Windows machine and the "cloverleaf" key on a Macintosh. If you are using something other than a Windows machine or a Macintosh, check with your instructor to determine your "control" key.) You should save frequently to avoid losing your work if the machine malfunctions.

Everything in a Mathematica notebook is contained in "cells". The scope of a given cell is indicated by the square bracket at the right. There are a variety of different kinds of cells. This one is called a "text" cell, and (you guessed it) is used for text. There are other kinds of cells, such as those shown below:

This is an input cell. (Do not execute.)

This is a text cell.

■ This is a subsection cell.

You can change the type of cell by clicking on the cell bracket (it will become shaded) and then by clicking on the **Format** menu and selecting **Style**. The different types of cells allow great freedom in customizing notebooks.

To create a new cell, move the cursor until it is between cells (the cursor will change from vertical to horizontal) and then click. A horizontal line should appear. Then start typing; you will get an **input** cell, which can be changed to a different type of cell using the **Style** menu within the **Format** menu.

Create a cell below this one to compute 123!. (The "!" is not at the end of the sentence for emphasis, it's there to indicate the factorial operation. Factorials are entered just as shown).

You can obtain decimal approximations using the command "**N**". For instance, when you executed the command to compute **284/16**, Mathematica reduced the fraction but returned a fraction as output. To get a decimal approximation, run the command below:

```
N[284 / 16]
```

You can get more decimal places. For instance, here are 100 digits of π:

```
N[Pi, 100]
```

```
To close this collection of cells, click twice on the outer cell bracket.
```

■ Adept Algebra with Mathematica
```
Click twice on the outer bracket at the right to open the cell.
```

Here is $(x + 2y)^{25}$ multiplied out:

```
Clear[x, y]
Expand[(x + 2 y) ^ 25]
```

If you don't hang the command **Expand** with the square brackets **[ ]** around the product, Mathematica will not multiply out the terms because it was not told to do so. Try it.

```
(x + 2 y) ^ 25
```

Get it? Try changing the numbers above and rerunning.

Mathematica can also factor polynomials:

```
Factor[1 + x ^ 15]
```

As mentioned in the preceding section, you don't have to use an asterisk between two variables when you multiply them, but you can if you want to.

```
a = 7;
b = 4;
a b
a * b
```

However, remember that if you don't use an asterisk then you must put a space between the variable names. To Mathematica, "**ab**" is not a product but a two-letter variable name.

```
a b
ab
```

Note that Mathematica remembers the values of **a** and **b**.

■ **Exercise 1**

The following calculations suggest a pattern.   What is the general formula suggested by these calculations?

```
Clear[x]
Expand[ (1 - x)  (1 + x) ]
Expand[ (1 - x)  (1 + x + x ^ 2) ]
Expand[ (1 - x)  (1 + x + x ^ 2 + x ^ 3) ]
Expand[ (1 - x)  (1 + x + x ^ 2 + x ^ 3 + x ^ 4) ]
```

Save your work!

To close this collection of cells, click twice on the outer cell bracket.

■ Defining Functions and Constants

Click twice on the outer bracket at the right to open the cell.

Defining functions in Mathematica is easy. Here's an example:

```
f[x_] := Cos[x];
```

By executing the above command (which you should do now if you haven't already), you tell Mathematica that the function **f** is defined to be **Cos[x]**. There are three important points that you need to remember when defining functions:

1. You must use the underscore "_" on the left side of the definition.

2. You must use square brackets "**[ ]**" rather then parentheses "**( )**".

3. You should use "**:=**" in the definition. (In most cases, "=" will also work.)

To evaluate a function, you just "plug in" a value. For instance, if we want to know **f[7π/12]**, we just type it in:

```
f[7 Pi / 12]
```

Note: For this command to evaluate properly, you must first have executed the definition for **f** above. All definitions must be executed prior to their use during **each** Mathematica session.

Here's **f[1]**:

```
f[1]
```

Remember that Mathematica gives exact answers unless told to do otherwise.   Here's a decimal approximation to **f[1]**:

```
N[f[1]]
```

Defining constants is even easier than defining functions. Suppose that you wish to set the constant **a** to be equal to 1.4142135624, which happens to be the first ten decimal places of the square root of 2. All you do is the following:

```
a = 1.4142135624
```

To close this collection of cells, click twice on the outer cell bracket.

■ Two Useful Commands
```
Click twice on the outer b....., well, you know the drill!
```

Two commands that you will use constantly are

1. Copy
2. Paste.

Here is an example of how to use these commands.

$$N[3432902008176640000 / 121645100408832000]$$

1. Select the cell above by clicking on the bracket at the right.

2. Hold down the "control" key and hit **c**. This copies the cell to the computer's memory. (This is the copy part.)

3. Move the cursor down below this cell. Put the cursor in a space between two of the brackets on the right. The cursor will switch from vertical to horizontal. Click the mouse once. A horizontal line will appear on the screen.

4. Hold down the "control" key, and hit **v**. A copy of the cell you chose should now appear on the screen. (This is the pasting part.)

5. Use the mouse to select the first digit in the numerator. Use the keyboard to change it from 3 to 2. Run your new command. The answer should be 20.

Some other things about copy-and-paste:

6. You can use copy-and-paste on whole cells. Click on the bracket defining the cell, hit "control"-**c** to copy. Later, use "control"-**v** to paste.

7. You can use copy-and-paste to copy materials from one part of a notebook to another. You can also use copy-and-paste between notebooks. This can be very useful when you have an elaborate set of instructions in one notebook that you want to copy to another notebook.

■ Getting Help in Mathematica

Mathematica is a big program with lots of parts. However, it's not necessary to learn everything about Mathematica in order to use it. Feel free to experiment —look at the menus, try out different things, play with the program. Don't be afraid of breaking anything. You can't break the machine just by pushing keys!

You can get help from the Mathematica program itself. If you want to find out what a certain function does, you can type ?"**FunctionName**". For example, to find out what the functions **Divisors** and **FactorInteger** do, you would type:

    ? Divisors

    ? FactorInteger

You get more information by using two question marks:

    ?? Divisors

    ?? FactorInteger

Selecting "Help . . ." in the "Help" menu provides a means of searching for commands.

As you go along, you may find that something doesn't work the way you expect it to, or that there is something you would like Mathematica to do, but you don't know how to make it work. In those cases, you should seek help from the following sources:

1. Your fellow students.
2. The lab assistant (if there is one).
3. Your teacher.
4. The ? command.
$4\frac{1}{2}$. The ?? command.
5. The function browser in the Help menu.
6. The Mathematica manual. (The entire Mathematica manual is built into the Help menu!)

■ **Exercise 2**

    (a) What does the Mathematica function **Floor** do?
    (b) What does the Mathematica function **Ceiling** do?

Save your work!

Web Lab: Introduction to Active Web Pages

0.1 Computing in a Web Page: Basics

If you're reading this text now, then you were able to follow a link with your web browser. You are off to a good start.

The main thing about these web pages which is different than most is that they are able to do computations. This is accomplished through what are called *Java applets*. They give you places to enter information, and buttons to click on to compute answers.

Our first example shows the basic form for many of our Java applets. It lets you input a positive integer n and computes the sum of the integers from 1 to n. When this page first loads, an initial value of n is chosen for you. Clicking on the button which says **Compute!** will perform the computation.

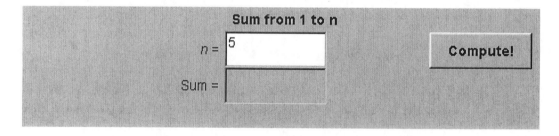

You can click the mouse in an input field (where the number 5 is located) and edit its value. Change the 5 above to a 7 and recompute the answer.

In general, your web browser should show input areas with a white background and output areas with a grey background so that you can easily tell what values can be changed.

A second type of applet we will see is one where the ouput is more complicated than one or two numbers. In that case, the output is put in a scrolling output area. For example, the following applet takes two integers a and b and computes $a+b$, $a*b$, and a^b.

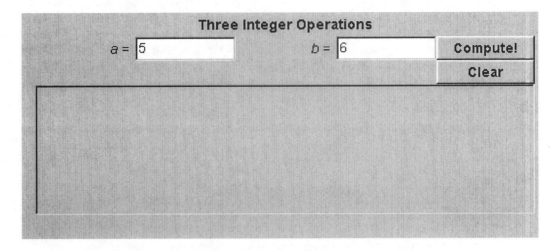

You can click on the **Compute!** button, change the values of a and b, and click on it again. Try that a few times. You should notice that the output area can be scrolled up and down to see all of the output, even if it is more than can be seen at once. If you want to erase the contents of the output area, click on the **Clear** button.

The output may get very wide, so that you would want to scroll from side to side. To see that, try using $a = 5$ and $b = 200$.

In this last computation, we could see something else of interest: the Java applet does computations using arbitrary precision integers. This will be true of all of the Java applets.

0.2   The Computation Line

A specific applet we will see from time to time is one which just does arithmetic computations. Here is an example. Again, you simply click on the **Compute!** button to have it compute.

Calculator	
17+34	Compute!
	Integers ▾

Of course, you can edit the entry field to do other computations. Here is how we express some basic operations.

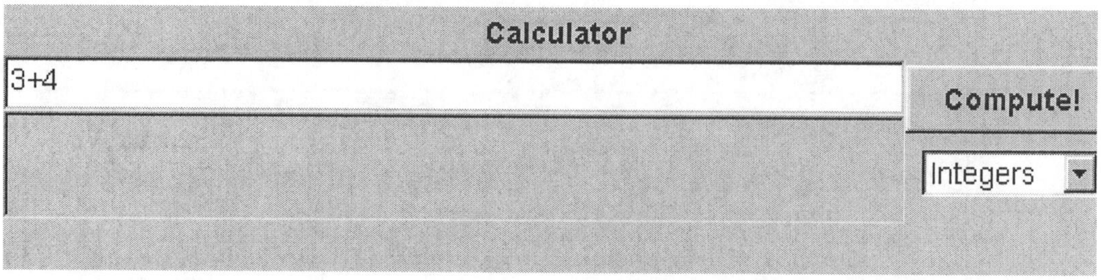

Calculator

3+4

Compute!

Integers ▼

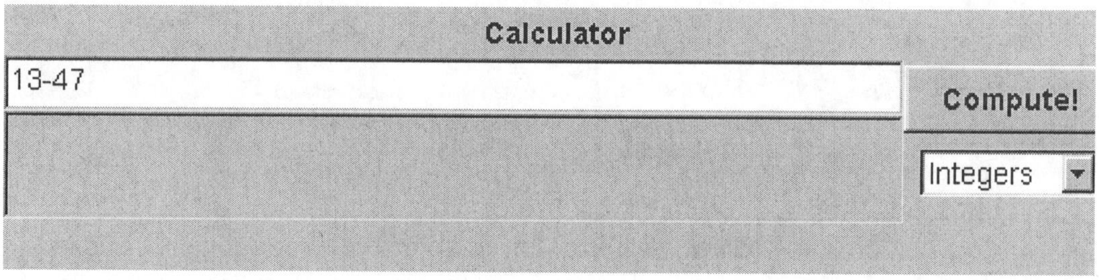

Calculator

13-47

Compute!

Integers ▼

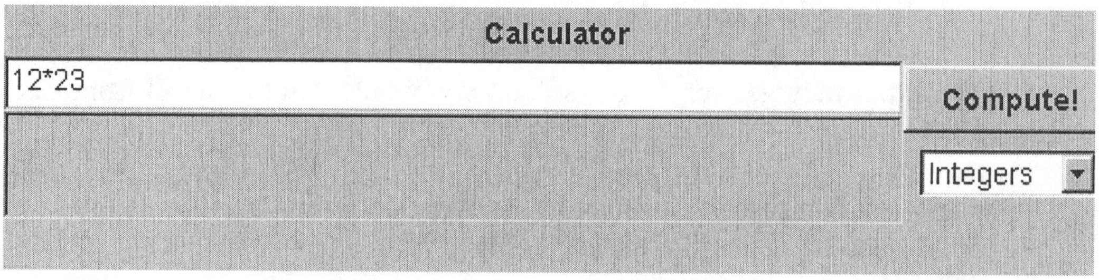

Calculator

12*23

Compute!

Integers ▼

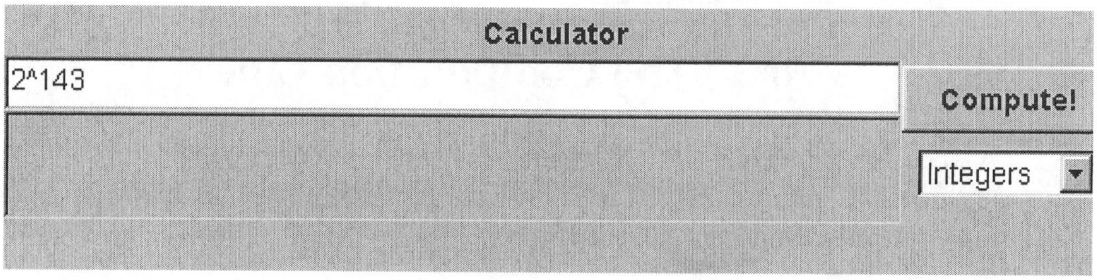

Calculator

2^143

Compute!

Integers ▼

Notice that you need to type a "*" for multiplication and a "^" to get a power. From the last example, we see that these applets will do computations with large integers. In fact, all applets will do arbitrary precision arithmetic with integers.

When doing division with integers, the default is typically set to truncate any values after the decimal point.

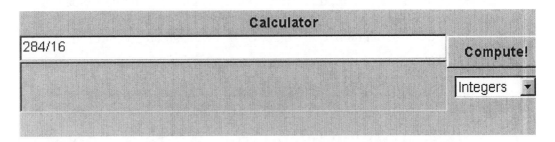

The basic computation applet gives you the option of doing floating point computations as well. To see the decimal value for this quotient, click on the place on the right side of the applet above where it says "Integers", and select "Floats" instead. Then click on the **Compute!** button again to see the result.

Most of the time we will simply want integer computations, but occasionally it is useful to do floating point computations as well.

It is possible to ask a Java applet to do a computation which will take it a very long time. The next example is set up to do just that. It might well take your computer years to finish the computation. Even so, click on the **Compute!** button, then keep reading.

The **Compute!** button should switch to saying **Stop It!**. All of the **Compute!** buttons do this. If the computation finishes, the button goes back to saying **Compute!**, but while it is computing, it switches so that you can use the same button to stop the computation. Click on the **Stop It!** button now to stop the computation. It may take a few moments for your computer to recover, so be patient.

Web browsers differ as to how gracefully they handle Java applets. If things start acting strangely,

1. First try clicking on the "Reload" button for your browser.
2. If that doesn't help, do a "full reload" by clicking on the Reload button while holding down the Shift key.
3. If that doesn't help, quit your web browser, and then start it up again.

Try doing a reload, and then a full reload now, whether you need it or not. It might help you to remember what to do if problems do arise later.

0.3   Arithmetic and the Arbitrary Applet

There will be times when you will be working with a specialized applet, but you may want to enter an arithmetic expression instead of a number. Well, go right ahead. Every applet which takes numbers as inputs can take arithmetic expressions and they will do integer arithmetic with them.

For example, we saw this next applet before. Notice how the entries for *a* and *b* are given as expressions. The applet does not mind.

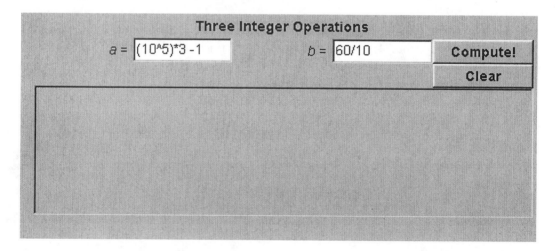

Try different expressions and recompute the result to see how this works.

0.4   Two Useful Commands

Two commands that you will use frequently are

1. Copy

    2. Paste.

Web browsers differ in how they handle these commands. Life is much easier if they work smoothly on your system. The idea is to copy-and-paste text to and from applets.

Here we describe an approach which often works. Experiment a little to see if it works for you.

The first thing you will need is someplace to copy-and-paste to. We recommend opening a text editor to run along with your web browser. For example, Windows users might want to use Notepad. Start your text editor now.

Let's try to copy information from a Java applet and paste it into your text editor. Click on the **Compute!** button on the next applet.

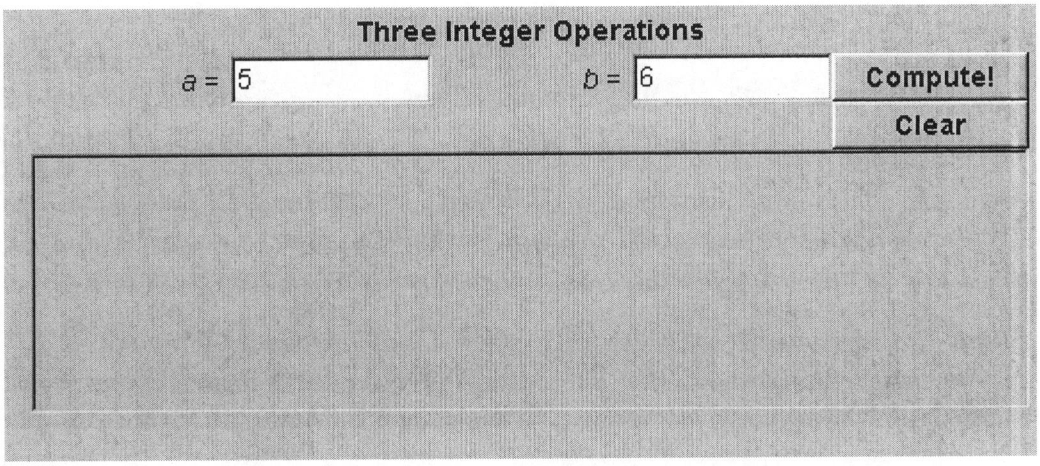

Three Integer Operations

a = 5   b = 6   Compute!

Clear

Select some of the output text you produced. Then pick "Copy" from the "Edit" menu on your web browser (all of the popular browsers have Edit menus). Now switch to your text editor. You should be able to just switch windows; you can leave the web browser going. Finally, select "Paste" from the Edit menu on your text editor.

If all went well, you now have a way to keep a lasting record of results from computations done in the web browser. Alternatively, some people like to simply take notes by hand on paper.

Whatever approach you use, the important thing is that you keep a record of the numerical experiments you perform. Being able to look back at your experiments is very useful in seeing relationships which lead to conjectures in this course.

Finally, bear in mind that the copy-and-paste process should also work in the other direction. You should be able to copy information from one place and paste it into the input fields of Java applets. In general, this is less important than going the other way, but it might be handy from time to time.

0.5   A Little Algebra

In general, Java applets do not do algebra unless they are specifically designed for it. Here we will illustrate with an example.

The following Java applet multiplies out the polynomials $1 - x$ and $1 + x + x^2 + \ldots + x^n$. All you need to supply is n.

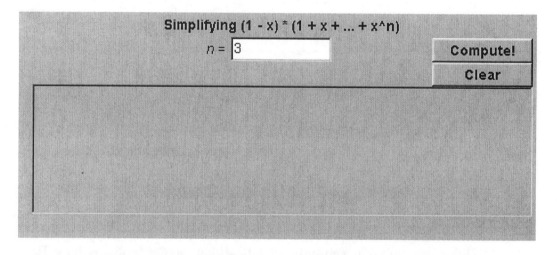

Exercise 1

Try different values of n in the applet above. What is the general formula suggested by these calculations?

---

1. Divisibility and Factorization

Prelab

In this chapter, we will be studying some basic and important properties of the integers. Before plunging ahead, recall that the integers are the set of positive and negative "counting numbers", together with 0 as shown:

$$\ldots, -3, -2, -1, 0, 1, 2, 3, \ldots$$

For ease of notation, we shall often refer to this set as $\mathbf{Z}$. Throughout this course, the lower case letters a, b, c, $\ldots$, x, y, z will denote integers unless stated otherwise.

Definition We say that a *divides* b (or a *is a divisor of* b) if there exists an integer d such that $ad = b$. In this case we write $a \mid b$. If a does not divide b, then we write $a \nmid b$.

As examples, we can easily see that $2 \mid 6$, $12 \mid 48$, and $13 \mid 13$. Moreover, $-7 \mid 21$ and $9 \mid -45$ since $-7(-3) = 21$ and $9(-5) = -45$ respectively. On the other hand, $2 \nmid 7$ and $0 \nmid 8$.[1] Here is our first prelab exercise:

1. **Compute all positive divisors for each of** 12, 18, **and** 30.

We next consider the following problem:

> Whoopdedoo Enterprises manages traveling amusement rides. Customers purchase tickets at a main booth and then spend the tickets on the rides. There are two levels of rides: the tame rides (e.g., the merry-go-round) which cost $0.85 each, and the good rides (e.g., the plunge-o-death) which cost $1.35 each. The management wants to sell only one denomination of ticket for simplicity, but they also want to minimize the number of tickets to be used for the rides. How much should the tickets be worth?

Note that in our problem, the solution turns out to be the largest integer which is a divisor of both 85 and 135. (Did you figure out the ticket value? If not, do so now!) In fact, if we generalize the problem so that the tame rides cost a cents and the good rides cost b cents, then the required ticket denomination is the largest integer that is a divisor of both a and b. This leads to the following:

---

[1] With this definition of *divides*, it is possible to have 0 divide another integer. Specifically, $0 \mid 0$ because we can satisfy the definition with $d = 2$. In fact, we satisfy the definition with *any* value of d!

Definition We say that d is the *greatest common divisor* of a and b, written $d = \gcd(a,b)$ or more briefly as $d = (a,b)$, if the following conditions are satisfied:

1. $d \mid a$ and $d \mid b$.

2. If $c \mid a$ and $c \mid b$, then $c \le d$.

As long as at least one of a and b is nonzero, there will be such an integer d.[2]

The most direct way to compute $\gcd(a,b)$ is to compute the set of all positive divisors of both a and b, compare the sets, and pick out the largest common factor. For example, we have

$$\text{Positive divisors of } 45 = \{1,3,5,9,15,45\}.$$
$$\text{Positive divisors of } 60 = \{1,2,3,4,5,6,10,12,15,20,30,60\}.$$

Comparing these sets, we see that $\gcd(45,60) = 15$.

2. **Use the method illustrated above to compute** $\gcd(12,18)$, $\gcd(6,22)$, **and** $\gcd(10,49)$.

As we shall soon see, the case where $\gcd(a,b) = 1$ is of particular interest, and is important enough to warrant special terminology:

Definition If $\gcd(a,b) = 1$, then we say that a and b are *relatively prime*.

The notion of two numbers being relatively prime is connected to their prime factorizations. Before we get to prime factorizations, let's back up a little bit. Although you probably already know what it means for a number to be prime, a definition is supplied below for the sake of completeness.

Definition An integer $p > 1$ is *prime* if the only positive divisors of p are 1 and p.

Prime numbers will play a central role in this course. Note that we have specifically excluded 1 from the definition of prime. This is done because 1 plays a special role with respect to factorization: it is the *identity for multiplication*. This means that $1 \cdot a = a$ for all integers a. (See footnote 3 for more information on why the definition of prime number excludes 1.) A positive integer that is neither prime nor equal to 1 is said to be *composite*. Thus, composite integers can be factored as a product of two smaller positive integers.

3. **Prove or disprove: if** $a \mid b$ **and** $a > 0$, **then** $\gcd(a,b) = a$.

We now turn to the question of the *prime-power factorization* (or just *factorization*) of positive integers.

---

[2]For every integer x, $0x = 0$ and so we find that $x \mid 0$. Thus, there is no greatest integer which divides 0 and 0.

Theorem 1.1 (Fundamental Theorem of Arithmetic) *Each integer $n > 1$ can be expressed as a product of powers of primes in the form*

$$n = p_1^{a_1} p_2^{a_2} \cdots p_k^{a_k},$$

where $p_1, \ldots, p_k$ are distinct, and each of the $a_1, \ldots, a_k$ are positive integers. Moreover, other than the order of the prime-power factors, this factorization is unique.

Thus, for example, we may write

$$300 = 2^2 \cdot 3^1 \cdot 5^2$$

and be guaranteed that this is the only way of expressing 300 as a product of prime powers.[3] This may not seem like a big deal, but as we shall see later, the uniqueness of factorization is very useful. For example, we can see immediately from this factorization that $7 \nmid 300$.[4]

4. **Compute the prime-power factorization for each of the following:** $n = 108$, $n = 315$, **and** $n = 1040$.

We will not give a proof of the Fundamental Theorem of Arithmetic here; a proof is sketched in the summary section of the next chapter. However, we present a tool which is typically used in proving the Fundamental Theorem, and which will arise in several places later in the course.[5]

Theorem 1.2 (Division Algorithm) *Let a be an integer and b be a positive integer. Then there exist unique integers q and r such that*

$$a = bq + r \quad and \quad 0 \le r < b.$$

The name of the theorem is misleading in that the statement is not an algorithm. However, the statement asserts the existence of a well-defined quotient and remainder typically produced by the algorithm known in grade school as *long division*.

5. **Find q and r for each pair of a and b.**

   (a) $a = 21$, $b = 6$ (c) $a = 6$, $b = 21$
   (b) $a = 143$, $b = 11$ (d) $a = -15$, $b = 4$

---

[3]If 1 were defined to be a prime number, then we would also be able to write 300 as a product of "prime" powers as $1^{17} \cdot 2^2 \cdot 3^1 \cdot 5^2$. Thus, the factorization would not be unique.

[4]If $7 \mid 300$, then $300 = 7k$ for some integer k. Factor k as a product of primes and we then get a factorization of 300 which includes 7, contradicting the uniqueness of the factorization of 300.

[5]You may recall seeing the Division Algorithm before. The statement and proof are given in the Course Guide.

We will use the following notation for the computation of remainders: the remainder of a when divided by b is written $a \% b$. You should think of $\%$ as an additional arithmetic operation (in addition to $+$, $-$, $/$, etc.).[6] So for example, $12 \% 5 = 2$. There will be no room for confusion of the symbol $\%$ with percentages in this text; we will *never* use $\%$ to signify percentage.

The notation $\%$ is fairly standard for this operation in computer programming languages (e.g., in C, C++, Java, and Pascal to name a few). Note that we define $a \% b$ to be the remainder from the Division Algorithm when applied to a and b. This means that $a \% b$ is only defined when a is an integer and b is a positive integer. When $a < 0$, our definition differs from what most programming languages give. By our definition, $-12 \% 5 = 3$ because $-12 = (-3) \cdot 5 + 3$ (and the remainder in the Division Algorithm must satisfy $0 \le r < b$). In the programming language C, $-12 \% 5$ would give you -2.

Finally, you may use the operation $\%$ like any other arithmetic operation in the Java applets for the web-based lab notebooks which accompany this course. It will function according to our definition.

Maple electronic notebook `01-divis.mws`

Mathematica electronic notebook `01-divis.nb`

Web electronic notebook Start with the web page `index.html`

---

[6]The placement of $\%$ in the order of operations is along with multiplication and division because of its connection to division. So after all exponentiations are done, multiplications, divisions, and remainders are done left to right. Then additions and subtractions are done left to right.

Maple Lab: Divisibility and Factorization

■ Divisors and GCDs

In the Prelab section of this chapter, you had the opportunity to compute the divisors of some integers by hand. Maple has a command called `divisors` that will automatically compute all of the positive divisors for a given integer. Before we can use it, you need to execute the following group:

```
[ > with(numtheory);
```

This loads a library full of commands related to number theory. The output shows the new functions at our disposal, including `divisors`. Now, here it is in action:

```
[ > divisors(12345);
```

The output is the set of all the positive divisors of 12345, given in the form of a Maple set. Sets in Maple are contained in curly braces, { }, just like in the rest of mathematics. Maple only lists the positive divisors because it is easy to compute all of the divisors from these. We would just have to take these numbers and their negations to get all of the divisors of 12345. In the Prelab section, an illustration was given that showed how to compute gcd(45, 60). The first step is to compute the positive divisors of both 45 and 60. Here's the Maple version:

```
[ > divisors(45);
[ > divisors(60);
```

The next step is to pick out the largest divisor that occurs in both lists. A handy way to do this is by using the Maple `intersect` command. This command is used just like the intersection symbol in mathematics: it produces a new set of all elements common to all of the input sets. In our case, we will get a list of all positive divisors of both 45 and 60. Here it is:

```
[ > divisors(45) intersect divisors(60);
```

With no work at all it is now clear that gcd(45, 60) = 15. The above example is cool, but after all, you could do that one by hand. Let's try some larger choices to really see the power of Maple:

```
[ > a:=223092870;
  > b:=6227020800;
  > divisors(a) intersect divisors(b);
```

Imagine trying to do this one by hand! By the way, what is gcd(a, b)? It may not be obvious because Maple does not sort the elements in a set. Take a close look! We can work around this feature of Maple by defining our own function. The next group defines a function which produces the common divisors of two positive integers, listed in order. You must execute the group below before you can use this new function:

```
[ > commondivisors := (n,m)-> sort(convert(
          divisors(n) intersect divisors(m),list));
[ >
```

45

Now let's try it:

```
[ > commondivisors(48, 600);
```

Try the previous example again:

```
[ > a:=223092870;
  > b:=6227020800;
  > commondivisors(a, b);
[ >
```

OK, now what is gcd(a, b)? In some cases, we may want to see the positive divisors of a single integer sorted as a list. Here is a function to do it:

```
[ > sortdivisors:= a -> sort(convert(divisors(a),list));
```

To see it in action, let's look at the divisors of gcd(223092870, 6227020800). You need to replace the ??? with the value of gcd(a, b) before executing the next line.

```
[ > sortdivisors(???);
[ >
```

■ Research Question 1

In the preceding computations, we have found the set of common positive divisors of $a = 223092870$ and $b = 6227020800$, and have found the set of positive divisors of gcd(a, b). Compare these two sets. Change the values of a and b, and repeat the experiment: compute the set of common positive divisors of a and b, compute the set of positive divisors of gcd(a, b), and compare the two sets. Once you think you see a pattern, state a conjecture and try to prove it!

Hints: Since your conjecture will involve sets, when working on the proof you should keep in mind the suggestions concerning proofs involving sets in the "Tips for Writing Proofs" section of the Course Guide. Also, you may find it helpful to use the results in the "GCDs and Factorization" section below when working on your proof.

■ Divisors and Factorization

How do we find the set of positive divisors for an integer? If you happen to have Mathematica at your disposal, you could just type in the `divisors` command, as you did above. However, it is important to understand how this works, especially if you ever stray more than a few feet from your computer.

The most basic method for computing divisors is exhaustive trial division. If we want to find the positive divisors for an integer n, we just take the integers $1, 2, 3, \ldots, n$, divide n by each, and those that divide evenly make up the set of positive divisors for n. Here is a function which does exactly that.

```
[ > getdivs1 := proc(n)
     local divlist, m;
     divlist := NULL;
     for m from 1 to n do
      if n/m = floor(n/m) then divlist := divlist, m; fi;
     od;
     divlist;
   end:
[
```

```
[ >
```

■ Maple Note: Why is there no semicolon?

A regular colon is also OK for ending Maple input. The difference is that a plain colon tells Maple to keep the output to itself. When doing computations, you probably want to see the result, so use a semicolon. When defining a function, the output is not terribly useful, so we usually suppress it.

The test for whether m is a divisor of n in this function is `n/m=floor(n/m)`. The = tests whether two quantities are equal. On the basis of what you learned about the `floor` command in the "Introduction to Using Maple", can you see why this works? Let's try our new function:

```
[ > getdivs1(12345);
```

We can compare it with the built in `divisors` command:

```
[ > divisors(12345);
[ >
```

Our method seems to work and is rather simple, but it is also quite inefficient.

One way to improve upon the above procedure is by applying the following observation: if m is a divisor of n, then $k = n/m$ is also a divisor of n, because $mk = n$. Thus, the positive divisors can be organized into pairs of the form $(m, n/m)$, where $m < n/m$. (The one exception to this is if n is a perfect square and $m = \sqrt{n}$, in which case $m = n/m$.) For example, if $n = 100$, then the positive divisors of n are given by

```
[ > divisors(100);
[ >
```

For this value of n, we have the pairs (1, 100), (2, 50), (4, 25), and (5, 20), and (because 100 is a perfect square) the single divisor $\sqrt{100} = 10$. Note that for each pair of positive divisors, the smaller divisor is less than $\sqrt{n}$. This is also the case in general: if $m \mid n$ and $m < n/m$, then $m < \sqrt{n}$. (The proof of this assertion is left as a homework exercise.) Thus, all positive divisors can be found by first using exhaustive trial division by all integers up to $\sqrt{n}$, and then generating the pairs to identify the remaining positive divisors.

For example, suppose that $n = 152$. Then we have

```
[ > evalf(sqrt(152));
[ >
```

Dividing each of 1, 2, 3, . . . , 12 into 152 reveals the divisors 1, 2, 4, and 8. The remaining positive divisors of 152 are then given by

$$152/1 = 152$$
$$152/2 = 76$$
$$152/4 = 38$$
$$152/8 = 19$$

Here's a check:

```
[ > divisors(152);
```

Here is a function which uses this second method for finding all positive divisors of an integer *n*:

```
> getdivs2 := proc(n)
    local divlist, m;
    divlist := NULL;
    for m from 1 to floor(sqrt(n)) do
     if n/m=floor(n/m) then
       divlist := divlist, m, n/m;
     fi;
    od;
    sort(convert({divlist},list));
  end:
```

Here it is in action.

```
> getdivs2(12345);
>
```

How much time do we save by using the second method instead of the first? We can measure how long it takes each method to compute the divisors of 10000 as shown:

```
> starttime := time():
  getdivs1(10000);
  printf('Time taken: %f seconds', time()-starttime);

> starttime := time():
  getdivs2(10000);
  printf('Time taken: %f seconds', time()-starttime);
>
```

The second method is much faster. The reason for the difference in speed is clear: `getdivs1` has to test *n* possible integers, while `getdivs2` checks only $\sqrt{n}$ integers. So the running time for `getdivs1` should be roughly proportional to *n* whereas the running time for `getdivs2` should be roughly proportional to $\sqrt{n}$. (There are other things going on inside the computer, so one should not expect timing estimates like this to be exact. However, they can provide good ballpark estimates.)

Experiment a little with each function. What do you think will happen to the running times if we multiply *n* by 10 for each function? Then check the result by changing 10000 in the commands above.

Now try to answer the following question.

■ *Exercise 1*

On the basis of the timings above, extrapolate the amount of time (in years) required to find the positive divisors for 10^{18} using `getdivs1` and `getdivs2`. For simplicity, you may assume that all years have 365 days.

A problem related to finding the set of divisors for an integer is that of finding the prime-power factorization of an integer. Indeed, we shall see later in the chapter how to use the factorization of an integer to easily compute all of the divisors for the integer.

A number of different methods have been developed for finding the factorization of an integer. The method that we shall describe here is fairly simple, and borrows from the ideas used above to find divisors; that is, it uses trial division.

Recall that finding the prime factorization of an integer requires us to express an integer *n* as

$$n = p_1^{m_1} p_2^{m_2} \bullet \bullet \bullet p_k^{m_k}.$$

The basic procedure for finding the factorization is as follows:

1. Start with the first prime, $p = 2$, and check to see if $2 \mid n$. If so, then replace *n* with *n*/2. Repeat until 2 will no longer divide in evenly, keeping track of the number of factors of 2.

2. Repeat the above step with the next prime, $p = 3$, and then with the next prime, $p = 5$, and so on. As above, keep track of the number of factors along the way.

3. Stop when you are left with 1 or with a number you know is prime.

As an example, let us find the prime-power factorization for $n = 571450$. We start by dividing *n* by 2:

```
[ > 571450/2;
```

Although it is pretty obvious from the output, let's see if 2 will divide evenly into the above quotient:

```
[ > 285725/2;
```

Since we get a fraction for output, we know that 2 won't divide in again. Thus we know that 2^1 is the correct power of 2 in the factorization. Now let's move to the prime 3:

```
[ > 285725/3;
```

This doesn't divide in evenly, so 3 is not included in the factorization. Let's try 5:

```
[ > 285725/5;
```

Thus 5 divides in evenly, and as we can see from the output, 5 will divide in evenly again:

```
[ > 57145/5;
```

That's it for the 5s. Hence we know that 5^2 is the correct power of 5 in the factorization. Here's the test for the prime 7:

```
[ > 11429/7;
```

No luck. Here's the prime 11:

```
[ > 11429/11;
```

Continuing in this manner, we find that 1039 is not divisible by 11, 13, 17, 19, 23, 29, or 31. We already know that 1039 is not divisible by 2, 3, or 5 because we divided away all factors of 2, 3, and 5 already. If we are clever, we can stop now since $\sqrt{1039}$ is approximately . . .

```
[ > evalf(sqrt(1039));
[ >
```

Because 1039 has no prime divisors less than or equal $\sqrt{1039}$, it must be prime. Thus, we now have the entire factorization:

$$571450 = 2 \cdot 5^2 \cdot 11 \cdot 1039.$$

The method described and illustrated above will work well for fairly small integers and can be used to factor numbers on a hand calculator. However, this technique is inefficient for larger integers, and more-advanced methods must be used in these cases. But even the more-sophisticated factoring techniques have their limits. The general problem of factoring large integers (upwards of 200 digits!) is a difficult one, and is currently an area of intense mathematical research. The problem is not only interesting in its own right, but also has some important applications. We'll see how factoring is related to cryptography in a later chapter.

GCDs and Factorization

Here's a computation we saw earlier:

```
> a:=223092870;
> b:=6227020800;
> commondivisors(a,b);
```

On the basis of the output, we see that $\gcd(a, b) = 30030$.

Another way to compute gcd(a, b) is to look at the factorizations of a and b. The Maple command `ifactor` computes factorizations of integers. To illustrate, let's take $a = 7920$ and $b = 4536$.

```
> a:=7920;
  b:=4536;
  ifactor(a);
  ifactor(b);
>
```

The last two lines give us the prime factorizations of $a = 7920$ and of $b = 4536$. Now if d divides both a and b, then all of the prime divisors of d have to divide a and all of the prime divisors of d have to divide b. The only primes that divide both a and b are 2 and 3, so any positive common divisor must have the form

$$d = 2^m \, 3^n.$$

What are the largest choices of m and n we can take? Well, on the basis of the factorizations,

if $d \mid a$ then $m \le 4$, and if $d \mid b$ then $m \le 3$.

So, we've got to take $m \le 3$. Similarly,

if $d \mid a$ then $n \le 2$, and if $d \mid b$ then $n \le 4$.

This time, we've got to take $n \le 2$. Thus, taking the biggest possible values for n and m, we see that the greatest common divisor is

$$2^3 \, 3^2 = 72.$$

In general, suppose we have the prime-power factorizations

$$a = p_1^{m_1} p_2^{m_2} \cdots p_k^{m_k}$$

and

$$b = p_1^{n_1} p_2^{n_2} \cdots p_k^{n_k}.$$

Note: It is possible that some of the $m_1, \ldots, m_k$ and $n_1, \ldots, n_k$ are equal to 0. Writing two integers as products of prime powers over the same set of primes is a useful trick when writing proofs involving prime-power factorizations. You can use it yourself, as long as you advise the reader of what you are doing.

The greatest common divisor of a and b is

$$\gcd(a, b) = p_1^{\min(m_1, n_1)} p_2^{\min(m_2, n_2)} \cdots p_k^{\min(m_k, n_k)}.$$

To see how this works in practice, let's look at another example. Suppose that we have $a = 165620000$ and $b = 65984625$. Here are the factorizations:

```
> a:=165620000:
  b:=65984625:
  ifactor(a);
  ifactor(b);
>
```

To compute $\gcd(a, b)$, we just start comparing prime factors. Since 2 is a factor of a but not b, 2 is not a factor of $\gcd(a, b)$. Similarly, since 3 is a factor of b but not a, 3 is not a factor of $\gcd(a, b)$. Now 5 is a factor of both a and b, and the smallest exponent on 5 is 3, so that 5^3 is a factor of $\gcd(a, b)$. Also, 7 is a factor of both a and b, where the smallest exponent is 2, so that 7^2 is a factor of $\gcd(a, b)$. The factors 13 and 19 are eliminated in the same manner as 2 and 3, and we arrive at

$$\gcd(a, b) = 5^3 \, 7^2 = 6125.$$

Maple has a built-in command to compute greatest common divisors. The name of the command is `gcd`. We can use it to check our computation:

```
> gcd(165620000, 65984625);
>
```

Looks good.

▨ Maple Note: What about igcd?

There is also a function in Maple called `igcd`. It is for taking the gcd of integers. Since that is exactly what we are doing, why are we using plain old `gcd`? Because it is fast enough for our purposes (`igcd` is slightly faster), and because the name of the function is more natural. Feel free to use `igcd` for your own calculations if you wish. (The "i" stands for integer. The general `gcd` function can take greatest common divisors of polynomials too.)

If you play around with Maple's `gcd` command, you will find that it computes gcds very quickly, even for huge numbers. In fact, we can use it on numbers which are too large for Maple to factor:

```
> gcd(10^300+1, 10^500+1);
>
```

How does it find gcds without factoring the numbers? We will learn the secret in the next chapter.

▨ Counting Divisors

By using a combination of the `divisors` and `nops` commands, it's easy to get Maple to compute the number of positive divisors of a given integer. Here's an example:

```
[ > nops(divisors(7920));
[ >
```

All that this code does is compute the list of positive divisors, and then report the number of elements in the list.

In this section, we shall perform some experiments to learn how the number of positive divisors of a given integer can be determined from the factorization of the integer. We'll do this by starting with fairly simple cases and building up to more complicated situations. Along the way, a pattern should (hopefully!) emerge.

Let us begin with the simplest case: Suppose that $n = p$, where p is a prime number. For instance, let $n = 13$. Here's the positive divisors:

```
[ > divisors(13);
```

And here's $d(n)$, the number of positive divisors of n:

```
[ > nops(divisors(13));
[ >
```

▨ Research Question 2

Select different primes, and repeat the above experiment: Compute the divisor list, and then compute $d(n)$. When you have enough data, form a conjecture that states $d(n)$ for $n = p$, where p is prime. (It probably won't take much data for you to form a conjecture!) Then prove your conjecture.

▨ Research Question 3

Repeat Research Question 2, but this time for $n = p^a$, where p is a prime and a is a positive integer.

Hints: When doing computational experiments, select a small prime p and then increment the value of a by 1. Also, note that when $a = 1$, you revert to the case covered by Research Question 2. Thus, whatever your conjecture, it must be consistent with your results from the preceding Research Question.

Now suppose that $n = p\,q$, where p and q are distinct primes. (If $p = q$, then we would be in the situation covered in Research Question 3.) What is $d(n)$ in this case? Let's try an experiment, taking $n = 3 \cdot 7$:

```
[ > n:=3*7:
    divisors(n);
    nops(divisors(n));
[ >
```

▨ Research Question 4

Select different primes p and q, and repeat the above experiment: Compute the list of positive divisors for $n = p\,q$, and then compute $d(n)$. When you have enough data, form a conjecture that states $d(n)$ for $n = p\,q$, where p and q are distinct primes.

Note: We will no longer state that you should prove your conjecture, since you should always prove (or try to prove) all conjectures made in this course.

Research Question 5

Repeat Research Question 4, but this time take $n = p^a q^b$, where p and q are distinct primes and a and b are positive integers.

Hints: When doing your numerical experiments, it will make life easier if you are systematic. (In fact, systematic experimentation is the hallmark of good scientific work.) You might try starting with fixed values of p and q, and make simple variations in the exponents a and b. Note also that the previous Research Questions are all special cases of this one. Therefore, your conjecture in this case must be consistent with your previous results.

All of the earlier work has been leading up to the most general case of all. Suppose that

$$n = p_1^{a_1} p_2^{a_2} \cdots p_k^{a_k},$$

where $p_1, p_2, \ldots, p_k$ are distinct primes and $a_1, a_2, \ldots, a_k$ are positive integers. What is $d(n)$ in this case? On the basis of your earlier investigations, you may not need any additional experimentation to form a conjecture. On the other hand, there is no harm in further experimentation, so feel free to do more if you wish.

Research Question 6

Form a conjecture that states $d(n)$ for

$$n = p_1^{a_1} p_2^{a_2} \cdots p_k^{a_k},$$

where $p_1, p_2, \ldots, p_k$ are distinct primes and $a_1, a_2, \ldots, a_k$ are positive integers.

It may seem that we led to this formula in an inefficient manner; why not go for the whole formula at once? As you may have discovered along the way, it can help a great deal to work up to the general case through various special cases. In number theory, starting with cases such as n prime, and then n a prime power, and so on, is a standard progression of study. Keep this in mind for later investigations in the course.

An important part of proving your conjectures for Research Questions 2 through 6 is being able to list the positive divisors of an integer of the forms $n = p$, $n = pq$, $n = p^a$, and so on. The basic idea of listing these divisors was implicit in the section above where we derived a formula for the greatest common divisor of two integers in terms of their factorizations. In the next exercise, you should formulate the statement precisely.

Exercise 2

Suppose $n = p_1^{m_1} p_2^{m_2} \cdots p_k^{m_k}$. What are the positive divisors of n in terms of their prime-power factorizations?

The Division Algorithm

We begin this section with a statement of the Division Algorithm, which you saw at the end of the Prelab section of this chapter:

Theorem 1.2 (Division Algorithm)
Let a be an integer and b be a positive integer. Then there exist unique integers q and r such that

$$a = b\,q + r \quad \text{and} \quad 0 \le r < b.$$

Remember learning long division in grade school? (If not, pretend that you do.) Long division is a procedure for dividing a number a by another number b, and coming up with a quotient q and a remainder r. The Division Algorithm is really nothing more than a guarantee that good old long division really works. Although this result doesn't seem too profound, it is nonetheless quite handy. For instance, it is used in proving the Fundamental Theorem of Arithmetic, and will also appear in the next chapter. A proof of the Division Algorithm is given at the end of the "Tips for Writing Proofs" section of the Course Guide.

Now, suppose that you have a pair of integers a and b, and would like to find the corresponding q and r. If a and b are small, then you could find q and r by trial and error. However, suppose that $a = 124389001$ and $b = 593$. In this case, trial and error would be extremely tedious. What happens if we try the division using Maple? Since q is the quotient, it should be close to a/b.

```
> a:=124389001;
  b:=593;
  a/b;
```

Maybe a decimal approximation would be more useful:

```
> evalf(a/b);
>
```

Now we're getting somewhere. Since we know that $a = b\,q + r$, it follows that

$$\frac{a}{b} = q + \frac{r}{b}.$$

Thus it follows that $q \le a/b$. (Remember that $0 \le r < b$.) So, in our above example, it makes sense to take $q = 209762$, because this is the biggest integer that is less than (or equal to) a/b. (Recall that this is the idea used repeatedly when performing long division by hand.) With this choice of q, what is r? That's easy enough, because we must have

$$a - b\,q = r.$$

Here's the computation:

```
> q:=209762:
  r:=a-b*q;
>
```

Since $r < b$, we have fulfilled the requirements of the Division Algorithm, and we're done.

Let's return to the determination of q. The procedure used in the above example will work in general: Find the largest integer that is less than or equal to a/b. This process can be automated by using the Maple `floor` command, which does exactly what is required. So, we could have computed q this way:

```
```

```
[ > q:=floor(a/b);
```

The whole process is automated in the command `divalg` defined below:

```
[ > divalg := proc(a,b)
     local q, r;
     q:= floor(a/b);
     r:= a-b*q;
     [q,r];
   end:
```

Here it is in action on our sample values of *a* and *b*:

```
[ > divalg(124389001,593);
[ >
```

◼ *Exercise 3*

Use `divalg` to repeat Exercise 5 in the Prelab section of this chapter.

Mathematica Lab: Divisibility and Factorization

■ Divisors and GCDs

In the Prelab section of this chapter, you had the opportunity to compute the divisors of some integers by hand. Mathematica has a built-in command called **Divisors** that will automatically compute all of the positive divisors for a given integer. Here it is in action:

```
Divisors[12345]
```

The output is the set of all the positive divisors of 12345, given in the form of a Mathematica *list*. Lists in Mathematica are contained in curly braces, **{ }**. Mathematica only lists the positive divisors because it is easy to compute all of the divisors from these. We would just have to take these numbers and their negations to get all of the divisors of 12345.

In the Prelab section, an illustration was given that showed how to compute gcd(45, 60). The first step is to compute the positive divisors of both 45 and 60. Here's the Mathematica version:

```
Divisors[45]
Divisors[60]
```

The next step is to pick out the largest divisor that occurs in both lists. A handy way to do this is by using the Mathematica **Intersection** command. This command takes two (or more) lists as input, and produces a new sorted list of all elements common to all of the input lists. In our case, we will get a list of all positive divisors of both 45 and 60. Here it is:

```
Intersection[Divisors[45], Divisors[60]]
```

With no work at all it is now clear that gcd(45, 60) = 15. The above example is cool, but after all, you could do that one by hand. Let's try some larger choices to really see the power of Mathematica:

```
a = 223092870;
b = 6227020800;
Intersection[Divisors[a], Divisors[b]]
```

Imagine trying to do this one by hand! By the way, what is gcd(a, b)? Then use Mathematica to compute the positive divisors of gcd(a, b). (You have to replace **???** with gcd(a, b) in the next cell before executing it.)

```
Divisors[???]
```

■ **Research Question 1**

In the preceding computations, we have found the set of common positive divisors of $a = 223092870$ and $b = 6227020800$, and have found the set of positive divisors of gcd(a, b). Compare these two sets. Change the values of a and b, and repeat the experiment: compute the set of common positive divisors of a and b, compute the set of positive divisors of gcd(a, b), and compare the two sets. Once you think you see a pattern, state a conjecture and try to prove it!

Hints: Since your conjecture will involve sets, when working on the proof you should keep in mind the suggestions concerning proofs involving sets in the "Tips for Writing Proofs" section of the Course Guide. Also, you may find it helpful to use the results in the "GCDs and Factorization" section below when working on your proof.

■ Divisors and Factorization

How do we find the set of positive divisors for an integer? If you happen to have Mathematica at your disposal, you could just type in the **Divisors** command, as you did above. However, it is important to understand how this works, especially if you ever stray more than a few feet from your computer.

The most basic method for computing divisors is exhaustive trial division. If we want to find the positive divisors for an integer n, we just take the integers 1, 2, 3, ..., n, divide n by each, and those that divide evenly make up the set of positive divisors for n. Here is a function which does exactly that.

```
getdivs1[n_] := Module[{divlist},
    divlist = { };
    Do[If[n / m == Floor[n / m],
        AppendTo[divlist, m]], {m, 1, n}];
    Return[divlist]];
```

The test for whether **m** is a divisor of **n** in this function is **n/m == Floor[n/m]**. The **==** tests whether two quantities are equal. On the basis of what you learned about the **Floor** command in the "Introduction to Using Mathematica", can you see why this works? Let's try our new function:

```
getdivs1[12345]
```

We can compare it with the built-in **Divisors** command:

```
Divisors[12345]
```

Our method seems to work and is rather simple, but it is also quite inefficient.

One way to improve upon the above procedure is by applying the following observation: if m is a divisor of n, then $k = n/m$ is also a divisor of n, because $mk = n$. Thus, the positive divisors can be organized into pairs of the form $(m, n/m)$, where $m < n/m$. (The one exception to this is if n is a perfect square and $m = \sqrt{n}$, in which case $m = n/m$.) For example, if $n = 100$, then the positive divisors of n are given by

```
Divisors[100]
```

For this value of n, we have the pairs (1, 100), (2, 50), (4, 25), and (5, 20), and (because 100 is a perfect square) the single divisor $\sqrt{100} = 10$. Note that for each pair of positive divisors, the smaller divisor is less than $\sqrt{n}$. This is also the case in general: if $m \mid n$ and $m < n/m$, then $m < \sqrt{n}$. (The proof of this assertion is left as a homework exercise.) Thus, all positive divisors can be found by first using exhaustive trial division by all integers up to $\sqrt{n}$, and then generating the pairs to identify the remaining positive divisors.

For example, suppose that $n = 152$. Then we have

```
N[Sqrt[152]]
```

Dividing each of 1, 2, 3, . . . , 12 into 152 reveals the divisors 1, 2, 4, and 8. The remaining positive divisors of 152 are then given by

$$
\begin{aligned}
152/1 &= 152 \\
152/2 &= 76 \\
152/4 &= 38 \\
152/8 &= 19
\end{aligned}
$$

Here's a check:

```
Divisors[152]
```

Here is a function which uses this second method for finding all positive divisors of an integer n:

```
getdivs2[n_] := Module[{divlist},
    divlist = { };
    Do[If[n / m == Floor[n / m],
        AppendTo[divlist, {m, n / m}]],
    {m, 1, Sqrt[n]}];
    Return[Union[Flatten[divlist]]]];
```

Here it is in action.

```
getdivs2[12345]
```

How much time do we save by using the second method instead of the first? We can measure how long it takes each method to compute the divisors of 10000 using the **Timing** command, as shown:

```
Timing[getdivs1[10000]]
```

```
Timing[getdivs2[10000]]
```

The second method is much faster. The reason for the difference in speed is clear: **getdivs1** has to test n possible integers, while **getdivs2** checks only $\sqrt{n}$ integers. So the running time for **getdivs1** should be roughly proportional to n whereas the running time for **getdivs2** should be roughly proportional to $\sqrt{n}$. (There are other things going on inside the computer, so one should not expect timing estimates like this to be exact. However, they can provide good ballpark estimates.)

Experiment a little with each function. What do you think will happen to the running times if we multiply n by 10 for each function? Then check the result by changing 10000 in the commands above.

Now try to answer the following question.

■ Exercise 1

On the basis of the timings above, extrapolate the amount of time (in years) required to find the positive divisors for 10^{18} using **getdivs1** and **getdivs2**. For simplicity, you may assume that all years have 365 days.

A problem related to finding the set of divisors for an integer is that of finding the prime-power factorization of an integer. Indeed, we shall see later in the chapter how to use the factorization of an integer to easily compute all of the divisors for the integer.

A number of different methods have been developed for finding the factorization of an integer. The method that we shall describe here is fairly simple, and borrows from the ideas used above to find divisors; that is, it uses trial division.

Recall that finding the prime factorization of an integer requires us to express an integer n as

$$n = p_1^{n_1}\, p_2^{n_2} \cdots p_k^{n_k}.$$

The basic procedure for finding the factorization is as follows:

1. Start with the first prime, $p = 2$, and check to see if $2 \mid n$. If so, then replace n with $n/2$. Repeat until 2 will no longer divide in evenly, keeping track of the number of factors of 2.

2. Repeat the above step with the next prime, $p = 3$, and then with the next prime, $p = 5$, and so on. As above, keep track of the number of factors along the way.

3. Stop when you are left with 1 or with a number you know is prime.

As an example, let us find the prime-power factorization for $n = 571450$. We start by dividing n by 2:

> **571450 / 2**

Although it is pretty obvious from the output, let's see if 2 will divide evenly into the above quotient:

> **285725 / 2**

Since we get a fraction for output, we know that 2 won't divide in again. Thus, we know that 2^1 is the correct power of 2 in the factorization. Now let's move to the prime 3:

> **285725 / 3**

This doesn't divide in evenly, so 3 is not included in the factorization. Let's try 5:

> **285725 / 5**

Thus 5 divides in evenly, and as we can see from the output, 5 will divide in evenly again:

> **57145 / 5**

That's it for the 5s. Hence we know that 5^2 is the correct power of 5 in the factorization. Here's the test for the prime 7:

> **11429 / 7**

No luck. Here's the prime 11:

> **11429 / 11**

Continuing in this manner, we find that 1039 is not divisible by 11, 13, 17, 19, 23, 29, or 31. We already know that 1039 is not divisible by 2, 3, or 5 because we divided away all factors of 2, 3, and 5 already. If we are clever, we can stop now since $\sqrt{1039}$ is approximately . . .

> **N[Sqrt[1039]]**

Because 1039 has no prime divisors less than or equal $\sqrt{1039}$, it must be prime. Thus we now have the entire factorization:

$$571450 = 2^1 \cdot 5^2 \cdot 11^1 \cdot 1039.$$

The method described and illustrated above will work well for fairly small integers and can be used to factor numbers on a hand calculator. However, this technique is inefficient for larger integers, and more-advanced methods must be used in these cases. But even the more-sophisticated factoring techniques have their limits. The general problem of factoring large integers (upwards of 200 digits!) is a difficult one, and is currently an area of intense mathematical research. The problem is not only interesting in its own right, but also has some important applications. We'll see how factoring is related to cryptography in a later chapter.

■ GCDs and Factorization

Here's a computation we saw earlier:

```
a = 223092870;
b = 6227020800;
Intersection[Divisors[a], Divisors[b]]
```

On the basis of the output, we see that $\gcd(a, b) = 30030$.

Another way to compute $\gcd(a, b)$ is to look at the factorizations of a and b. The Mathematica command **FactorInteger** computes factorizations of integers. To illustrate, let's take $a = 7920$ and $b = 4536$.

```
a = 7920;
b = 4536;
FactorInteger[a]
FactorInteger[b]
```

The output tells us that

$$a = 7920 = 2^4\, 3^2\, 5^1\, 11^1$$

and

$$b = 4536 = 2^3\, 3^4\, 7^1.$$

Now if d divides both a and b, then all of the prime divisors of d have to divide a and all of the prime divisors of d have to divide b. The only primes that divide both a and b are 2 and 3, so any positive common divisor must have the form

$$d = 2^m\, 3^n.$$

What are the largest choices of m and n we can take? Well, on the basis of the factorizations,

$$\text{if } d \mid a \text{ then } m \le 4, \text{ and if } d \mid b \text{ then } m \le 3.$$

So, we've got to take $m \le 3$. Similarly,

$$\text{if } d \mid a \text{ then } n \le 2, \text{ and if } d \mid b \text{ then } n \le 4.$$

This time, we've got to take $n \le 2$. Thus, taking the biggest possible values for n and m, we see that the greatest common divisor is

$$2^3\, 3^2 = 72.$$

In general, suppose we have the prime-power factorizations

$$a = p_1^{m_1}\, p_2^{m_2} \cdots p_k^{m_k}$$

and

$$b = p_1^{n_1}\, p_2^{n_2} \cdots p_k^{n_k}.$$

Note: It is possible that some of the $m_1, \ldots, m_k$ and $n_1, \ldots, n_k$ are equal to 0. Writing two integers as products of prime powers over the same set of primes is a useful trick when writing proofs involving prime power factorizations. You can use it yourself, as long as you advise the reader of what you are doing.

The greatest common divisor of *a* and *b* is

$$\gcd(a,\ b)\ =\ p_1^{\min(m_1,\,n_1)}\ p_2^{\min(m_2,\,n_2)}\ \cdots\ p_k^{\min(m_k,\,n_k)}.$$

To see how this works in practice, let's look at another example. Suppose that we have $a = 165620000$ and $b = 65984625$. Here are the factorizations:

```
a = 165620000;
b = 65984625;
FactorInteger[a]
FactorInteger[b]
```

To compute gcd(*a*, *b*), we just start comparing prime factors. Since 2 is a factor of *a* but not *b*, 2 is not a factor of gcd(*a*, *b*). Similarly, since 3 is a factor of *b* but not *a*, 3 is not a factor of gcd(*a*, *b*). Now 5 is a factor of both *a* and *b*, and the smallest exponent on 5 is 3, so that 5^3 is a factor of gcd(*a*, *b*). Also, 7 is a factor of both *a* and *b*, where the smallest exponent is 2, so that 7^2 is a factor of gcd(*a*, *b*). The factors 13 and 19 are eliminated in the same manner as 2 and 3, and we arrive at

$$\gcd(a,\ b) = 5^3 7^2 = 6125.$$

Mathematica has a built-in command to compute greatest common divisors. The name of this command is **GCD**. We can use it to check our computation:

```
GCD[165620000, 65984625]
```

Looks good.

If you play around with Mathematica's **GCD** command, you will find that it computes gcds very quickly, even for huge numbers. In fact, we can use it on numbers which are too large for Mathematica to factor:

```
GCD[10^300 + 1, 10^500 + 1]
```

How does it find gcds without factoring the numbers? We will learn the secret in the next chapter.

■ Counting Divisors

By using a combination of the **Divisors** and **Length** commands, it's easy to get Mathematica to compute the number of positive divisors of a given integer. Here's an example:

```
Length[Divisors[7920]]
```

All that this code does is compute the list of positive divisors, and then report the number of elements in the list.

In this section, we shall perform some experiments to learn how the number of positive divisors of a given integer can be determined from the factorization of the integer. We'll do this by starting with fairly simple cases and building up to more complicated situations. Along the way, a pattern should (hopefully!) emerge.

Let us begin with the simplest case: Suppose that $n = p$, where *p* is a prime number. For instance, let $n = 13$. Here's the positive divisors:

```
Divisors[13]
```

And here's *d*(*n*), the number of positive divisors of *n*:

```
Length[Divisors[13]]
```

■ Research Question 2

Select different primes, and repeat the above experiment: Compute the divisor list, and then compute $d(n)$. When you have enough data, form a conjecture that states $d(n)$ for $n = p$, where p is prime. (It probably won't take much data for you to form a conjecture!) Then prove your conjecture.

■ Research Question 3

Repeat Research Question 2, but this time for $n = p^a$, where p is a prime and a is a positive integer.

Hints: When doing computational experiments, select a small prime p and then increment the value of a by 1. Also, note that when $a = 1$, you revert to the case covered by Research Question 2. Thus, whatever your conjecture, it must be consistent with your results from the preceding Research Question.

Now suppose that $n = pq$, where p and q are distinct primes. (If $p = q$, then we would be in the situation covered in Research Question 3.) What is $d(n)$ in this case? Let's try an experiment, taking $n = 3 \cdot 7$:

```
n = 3 * 7;
Divisors[n]
Length[Divisors[n]]
```

■ Research Question 4

Select different primes p and q, and repeat the above experiment: Compute the list of positive divisors for $n = p\,q$, and then compute $d(n)$. When you have enough data, form a conjecture that states $d(n)$ for $n = p\,q$, where p and q are distinct primes.

Note: We will no longer state that you should prove your conjecture, since you should always prove (or try to prove) all conjectures made in this course.

■ Research Question 5

Repeat Research Question 4, but this time take $n = p^a q^b$, where p and q are primes and a and b are positive integers.

Hints: When doing your numerical experiments, it will make life easier if you are systematic. (In fact, systematic experimentation is the hallmark of good scientific work.) You might try starting with fixed values of p and q, and make simple variations in the exponents a and b. Note also that the previous Research Questions are all special cases of this one. Therefore, your conjecture in this case must be consistent with your previous results.

All of the earlier work has been leading up to the most general case of all. Suppose that

$$n = p_1^{a_1}\, p_2^{a_2} \cdots p_k^{a_k},$$

where $p_1, p_2, \ldots, p_k$ are distinct primes and $a_1, a_2, \ldots, a_k$ are positive integers. What is $d(n)$ in this case? On the basis of your earlier investigations, you may not need any additional experimentation to form a conjecture. On the other hand, there is no harm in further experimentation, so feel free to do more if you wish.

■ Research Question 6

> Form a conjecture that states $d(n)$ for
>
> $$n = p_1^{a_1}\, p_2^{a_2} \cdots p_k^{a_k},$$
>
> where $p_1, p_2, \ldots, p_k$ are distinct primes and $a_1, a_2, \ldots, a_k$ are positive integers.

It may seem that we led to this formula in an inefficient manner; why not go for the whole formula at once? As you may have discovered along the way, it can help a great deal to work up to the general case through various special cases. In number theory, starting with cases such as n prime, and then n a prime power, and so on, is a standard progression of study. Keep this in mind for later investigations in the course.

An important part of proving your conjectures for Research Questions 2 through 6 is being able to list the positive divisors of an integer of the forms $n = p$, $n = p\,q$, $n = p^a$, and so on. The basic idea of listing these divisors was implicit in the section above where we derived a formula for the greatest common divisor of two integers in terms of their factorizations. In the next exercise, you should formulate the statement precisely.

■ Exercise 2

Suppose $n = p_1^{m_1}\, p_2^{m_2} \cdots p_k^{m_k}$. What are the positive divisors of n in terms of their prime-power factorizations?

■ The Division Algorithm

We begin this section with a statement of the Division Algorithm, which you saw at the end of the Prelab section of this chapter:

Theorem 1.2 (Division Algorithm) Let a be an integer and b be a positive integer. Then there exist unique integers q and r such that

$$a = b\,q + r \quad \text{and} \quad 0 \le r < b.$$

Remember learning long division in grade school? (If not, pretend that you do.) Long division is a procedure for dividing a number a by another number b, and coming up with a quotient q and a remainder r. The Division Algorithm is really nothing more than a guarantee that good old long division really works. Although this result doesn't seem too profound, it is nonetheless quite handy. For instance, it is used in proving the Fundamental Theorem of Arithmetic, and will also appear in the next chapter. A proof of the Division Algorithm is given at the end of the "Tips for Writing Proofs" section of the Course Guide.

Now, suppose that you have a pair of integers a and b, and would like to find the corresponding q and r. If a and b are small, then you could find q and r by trial and error. However, suppose that $a = 124389001$ and $b = 593$. In this case, trial and error would be extremely tedious. What happens if we try the division using Mathematica? Since q is the quotient, it should be close to a/b.

```
a = 124389001;
b = 593;
a / b
```

Maybe a decimal approximation would be more useful:

```
N[a / b, 10]
```

Now we're getting somewhere. Since we know that $a = b\,q + r$, it follows that

$$\frac{a}{b} = q + \frac{r}{b}.$$

Thus it follows that $q \le a/b$. (Remember that $0 \le r < b$.) So, in our above example, it makes sense to take $q = 209762$, because this is the biggest integer that is less than (or equal to) a/b. (Recall that this is the idea used repeatedly when performing long division by hand.) With this choice of q, what is r? That's easy enough, because we must have

$$a - b\,q = r.$$

Here's the computation:

```
q = 209762;
r = a - b q
```

Since $r < b$, we have fulfilled the requirements of the Division Algorithm, and we're done.

Let's return to the determination of q. The procedure used in the above example will work in general: Find the largest integer that is less than or equal to a/b. This process can be automated by using the Mathematica **Floor** command, which does exactly what is required. So, we could have computed q this way:

```
q = Floor[a / b]
```

The whole process is automated in the command **divalg** defined below:

```
divalg[a_, b_] := Module[{q, r},
        q = Floor[a / b];
        r = a - b q;
        Return[{q, r}]]
```

Note that this function is made up of two operations, separated by semicolons. Here it is in action on our sample values of a and b:

```
divalg[124389001, 593]
```

■ Exercise 3

Use **divalg** to repeat Exercise 5 in the Prelab section of this chapter.

Web Lab: Divisibility and Factorization

1.1 Divisors and GCDs

In the Prelab section of this chapter, you had the opportunity to compute the divisors of some integers by hand. The applet below will automatically compute all of the positive divisors for a given integer. Give it a try:

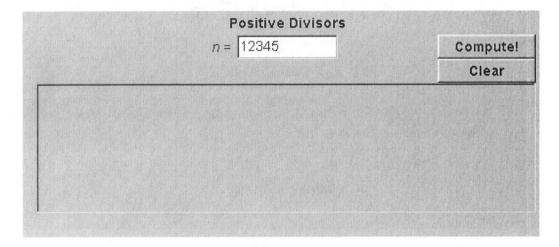

The output is the set of all the positive divisors of 12345. We only list the positive divisors because it is easy to compute all of the divisors from these. We would just have to take these numbers and their negations to get all of the divisors of 12345.

In the Prelab section, an illustration was given that showed how to compute gcd(45, 60). The first step is to compute the positive divisors of both 45 and 60. Here are the divisors of 45:

And here are the divisors of 60:

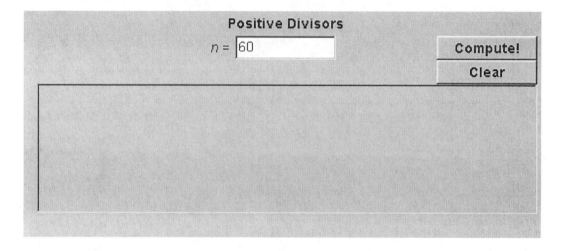

The last step is to pick out the largest number that occurs in both lists. A handy way to do this in one shot is to use the applet below. This applet takes two integers as input, and produces the set of positive divisors shared by both input values.

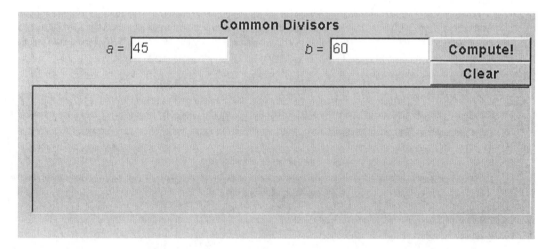

With no work at all it is clear that gcd(45, 60) = 15. The above example is cool, but after all, you could do that one by hand. Let's try some larger choices:

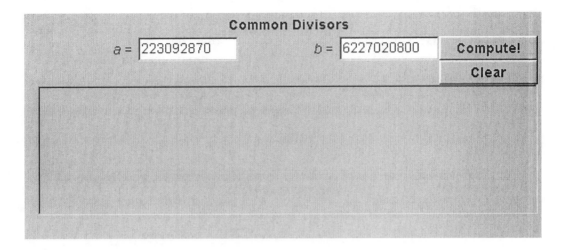

Imagine trying to do this one by hand! By the way, what is gcd(a, b)? Once you figure it out, use the applet below to compute the positive divisors of gcd(a, b). (Replace ??? with the value of gcd(a, b).)

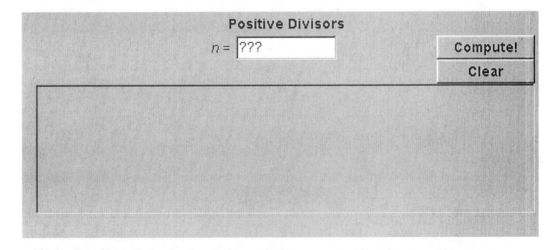

Research Question 1

In the preceding computations, we have found the set of common positive divisors of a = 223092870 and b = 6227020800, and have found the set of positive divisors of gcd (a, b). Compare these two sets. Change the values of a and b, and repeat the experiment: compute the set of common positive divisors of a and b, compute the set of positive divisors of gcd (a, b), and compare the two sets. Once you think you see a pattern, state a conjecture and try to prove it!

Hints: Since your conjecture will involve sets, when working on the proof you should keep in mind the suggestions concerning proofs involving sets in the "Tips for Writing Proofs" section of the Course Guide. Also, you may find it helpful to use the results in the "GCDs and Factorization" section below when working on your proof.

1.2 Divisors and Factorization

How do we find the set of positive divisors for an integer? If you happen to have a computer at your disposal, you could just use the Java applet from the previous section. However, it is important to understand how this applet works, especially if you ever stray more than a few feet from your computer.

The most basic method for computing divisors is exhaustive trial division. If we want to find the positive divisors for an integer n, we just take the integers $1, 2, 3, \ldots, n$, divide n by each, and those that divide evenly make up the set of positive divisors for n. This method works well and is rather simple, but it is also quite inefficient.

One way to improve upon the above procedure is by applying the following observation: if m is a divisor of n, then $k = n/m$ is also a divisor of n, because $mk = n$. Thus, the positive divisors can be organized into pairs of the form $(m, n/m)$, where $m < n/m$. The one exception to this is if n is a perfect square and $m = \sqrt{n}$, in which case $m = n/m$. For example, if $n = 100$, then the positive divisors of n are given by

Positive Divisors

$n =$ | 100 | Compute!

Clear

For this value of n, we have the pairs $(1, 100)$, $(2, 50)$, $(4, 25)$, and $(5, 20)$, and because 100 is a perfect square, the single divisor $\sqrt{100} = 10$. Note that for each pair of positive divisors, the smaller divisor is less than $\sqrt{n}$. This is also the case in general: if $m \mid n$ and $m < n/m$, then $m < \sqrt{n}$. (The proof of this assertion is left as a homework exercise.) Thus, all positive divisors can be found by first using exhaustive trial division by all integers up to $\sqrt{n}$, and then generating the pairs to identify the remaining positive divisors.

For example, suppose that $n = 152$. Then a decimal approximation for $\sqrt{n}$ is given by

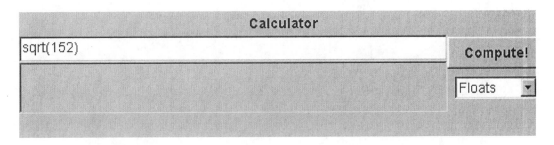

Dividing each of $1, 2, 3, \ldots, 12$ into 152 reveals the divisors 1, 2, 4, and 8. The remaining positive divisors of 152 are then given by

$$152/1 = 152$$
$$152/2 = 76$$
$$152/4 = 38$$
$$152/8 = 19$$

Here's a check:

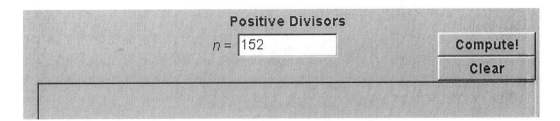

How much time do we save by using the second method instead of the first? We can measure how long it takes each method to compute the divisors of 10000 using the applet below. "Trial n" finds divisors by using trial division up to n, and "Trial sqrt(n)" finds divisors by using trial division up to $\sqrt{n}$:

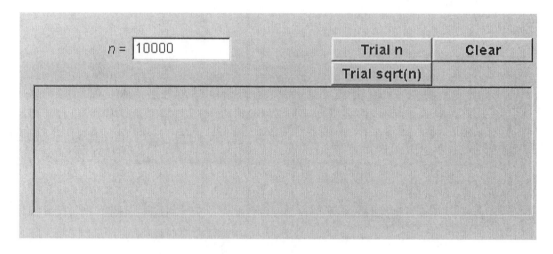

"Trial sqrt(n)" is significantly faster. The reason for the difference in speed is clear: "Trial n" has to test n possible integers, while "Trial sqrt(n)" checks only $\sqrt{n}$ integers. So the running time for "Trial n" should be roughly proportional to n whereas the running time for "Trial sqrt(n)" should be roughly proportional to $\sqrt{n}$.

There are other things going on inside the computer, so one should not expect timing estimates like this to be exact. However, they can provide good ballpark estimates.

What do you think will happen to the running times for each method if we multiply n by 10? Then check the result by changing 10000 in the above applet.

Now try to answer the following question.

Exercise 1

On the basis of the timings above, extrapolate the amount of time (in years) required to find the positive divisors for 10^{18} using both methods discussed above. For simplicity, you may assume that all years have 365 days.

A problem related to finding the set of divisors for an integer is that of finding the prime-power factorization of an integer. Indeed, we shall see later in the chapter how to use the factorization of an integer to easily compute all of the divisors for the integer.

A number of different methods have been developed for finding the factorization of an integer. The method that we shall describe here is fairly simple, and borrows from the ideas used above to find divisors; that is, it uses trial division.

Recall that finding the prime factorization of an integer requires us to express an integer n as

$$n = p_1^{a_1} p_2^{a_2} \cdots p_k^{a_k}.$$

The basic procedure for finding the factorization is as follows:

1. Start with the first prime, $p = 2$, and check to see if $2 \mid n$. If so, then replace n with $n/2$. Repeat until 2 will no longer divide in evenly, keeping track of the number of factors of 2.

2. Repeat the above step with the next prime, $p = 3$, and then with the next prime, $p = 5$, and so on. As above, keep track of the number of factors along the way.

3. Stop when you are left with 1 or with a number you know is prime.

As an example, let us find the prime-power factorization for $n = 571450$. We start by dividing n by 2:

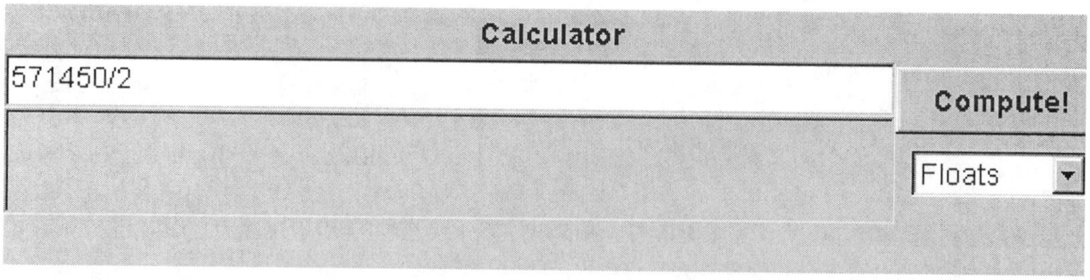

Calculator

571450/2

Compute!

Floats ▾

Although it is pretty obvious from the output, let's see if 2 will divide evenly into the above quotient:

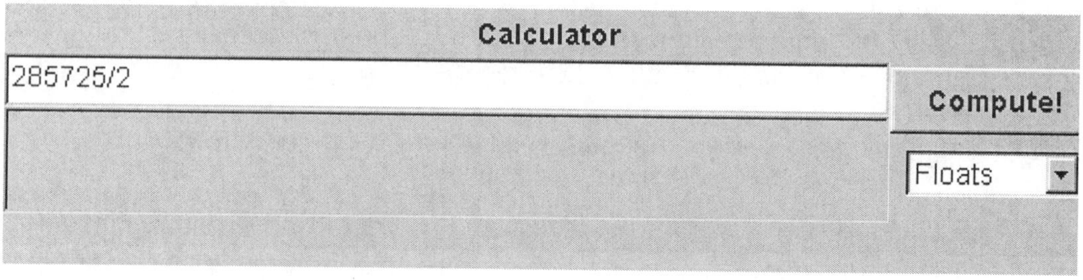

Calculator

285725/2

Compute!

Floats ▾

Since we get a noninteger for output, we know that 2 won't divide in again. Thus, we know that 2^1 is the correct power of 2 in the factorization. Now let's move to the prime 3:

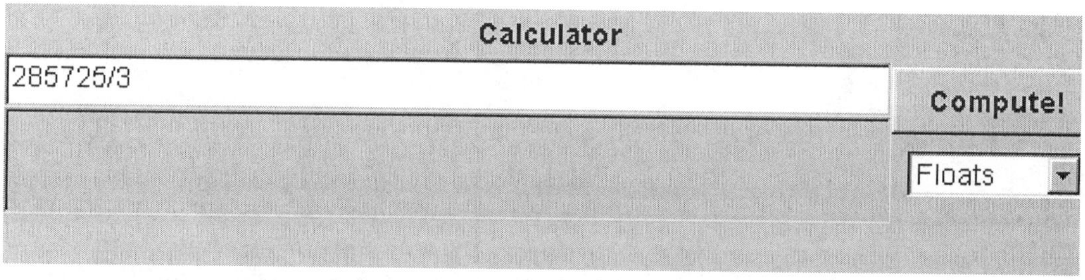

Calculator

285725/3

Compute!

Floats ▾

This doesn't divide in evenly, so 3 is not included in the factorization. Let's try 5:

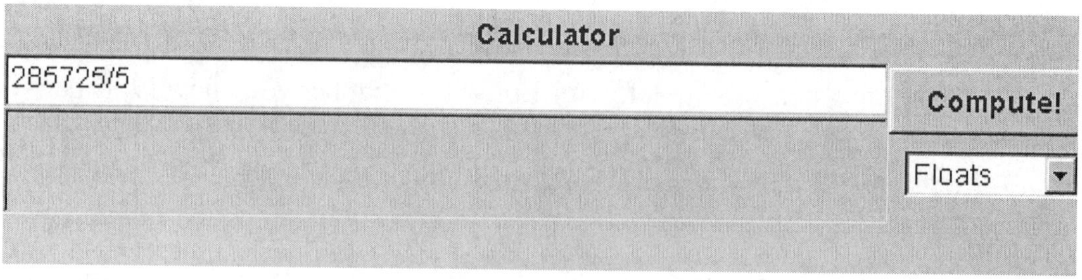

Calculator

285725/5

Compute!

Floats ▾

Thus 5 divides in evenly, and as we can see from the output, 5 will divide in evenly again:

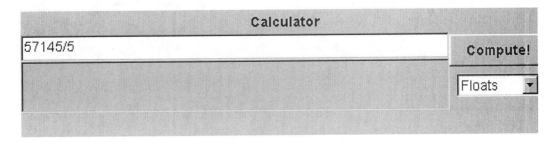

That's it for the 5s. Hence we know that 5^2 is the correct power of 5 in the factorization. Here's the test for the prime 7:

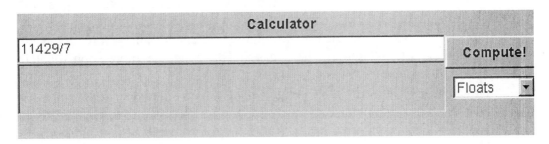

No luck. Here's the prime 11:

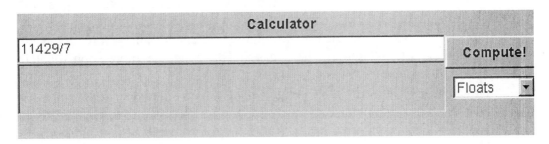

Continuing in this manner, we find that 1039 is not divisible by 11, 13, 17, 19, 23, 29, or 31. We already know that 1039 is not divisible by 2, 3, or 5 because we divided away all factors of 2, 3, and 5 already. If we are clever, we can stop now since $\sqrt{1039}$ is approximately . . .

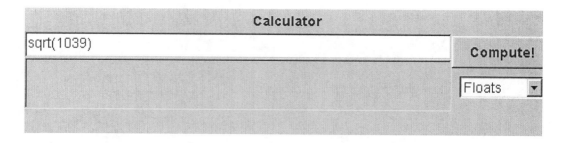

Because 1039 has no prime divisors less than or equal $\sqrt{1039}$, it must be prime. Thus we now have the entire factorization:

$$571450 = 2^1 \cdot 5^2 \cdot 11^1 \cdot 1039$$

The method described and illustrated above will work well for fairly small integers and can be used to factor numbers on a hand calculator. However, this technique is inefficient for larger integers, and more-advanced methods must be used in these cases. But even the more-sophisticated factoring techniques have their limits. The general problem of factoring large integers (upwards of 200 digits!) is a difficult one, and is currently an area of intense mathematical research. The problem is not only interesting in its own right, but also has some important applications. We'll see how factoring is related to cryptography in a later chapter.

1.3 GCDs and Factorization

Here's a computation we saw earlier:

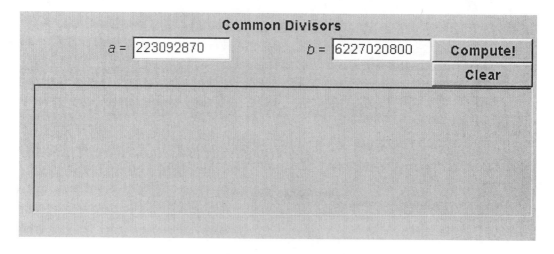

On the basis of the output, we see that gcd(223092870, 6227020800) = 30030.

Given two integers a and b, another way to compute gcd(a, b) is to look at the factorizations of a and b. The applet below computes factorizations of integers. For example, here is the factorization of $n = 272$:

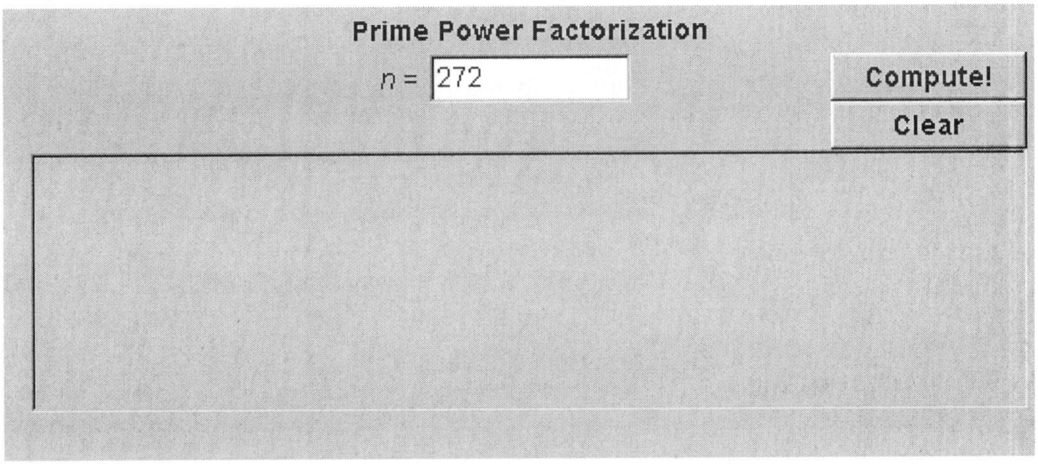

Prime Power Factorization

$n =$ `272` | **Compute!** |
| **Clear** |

From the output we see that
$$272 = 2^4 \cdot 17^1.$$

When computing gcds, it will be handy to be able to easily compute the factorization of two integers. The applet below allows us to do just that. To illustrate, let's take $a = 7920$ and $b = 4536$.

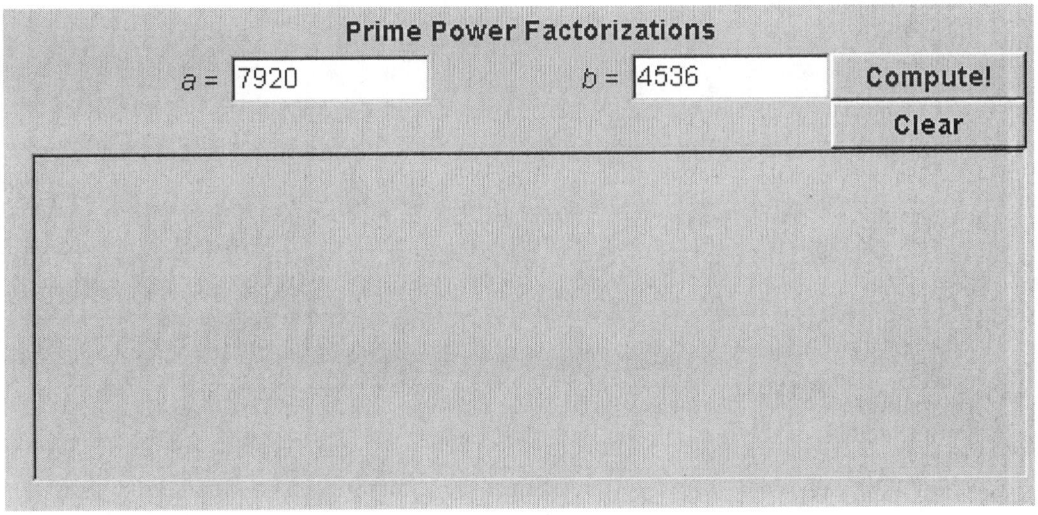

Prime Power Factorizations

$a =$ `7920` $b =$ `4536` | **Compute!** |
| **Clear** |

The output tells us that

$$a = 7920 = 2^4 \cdot 3^2 \cdot 5^1 \cdot 11^1$$

and

$$b = 4536 = 2^3 \cdot 3^4 \cdot 7^1.$$

Now if d divides both a and b, then all of the prime divisors of d have to divide a and all of the prime divisors of d have to divide b. The only primes that divide both a and b are 2 and 3, so any positive common divisor must have the form

$$d = 2^m \cdot 3^n.$$

What are the largest choices of m and n we can take? Well, on the basis of the factorizations,

if $d \mid a$, then $m \leq 4$, and if $d \mid b$ then $m \leq 3$.

So, we've got to take $m \leq 3$. Similarly,

if $d \mid a$ then $n \leq 2$, and if $d \mid b$ then $n \leq 4$.

This time, we've got to take $n \leq 2$. Thus, taking the biggest possible values for n and m, we see that the greatest common divisor is

$$2^3 \cdot 3^2 = 72.$$

In general, suppose we have the prime-power factorizations

$$a = p_1^{m_1} p_2^{m_2} \cdots p_k^{m_k}$$

and

$$b = p_1^{n_1} p_2^{n_2} \cdots p_k^{n_k}$$

Note: It is possible that some of the $m_1, \ldots, m_k$ and $n_1, \ldots, n_k$ are equal to 0. Writing two integers as products of prime powers over the same set of primes is a useful trick when writing proofs involving prime-power factorizations. You can use it yourself, as long as you advise the reader of what you are doing.

The greatest common divisor of a and b is

$$\gcd(a, b) = p_1^{\min(m_1, n_1)} p_2^{\min(m_2, n_2)} \cdots p_k^{\min(m_k, n_k)} .$$

To see how this works in practice, let's look at another example. Suppose that we have $a = 165620000$ and $b = 65984625$. Here are the factorizations:

To compute $\gcd(a, b)$, we just start comparing prime factors. Since 2 is a factor of a but not b, 2 is not a factor of $\gcd(a, b)$. Similarly, since 3 is a factor of b but not a, 3 is not a factor of $\gcd(a, b)$. Now 5 is a factor of both a and b, and the smallest exponent on 5 is 3, so 5^3 is a factor of $\gcd(a, b)$. Also, 7 is a factor of both a and b, where the smallest exponent is 2, so that 7^2 is a factor of $\gcd(a, b)$. The factors 13 and 19 are eliminated in the same manner as 2 and 3, and we arrive at

$$\gcd(a, b) = 5^3 \cdot 7^2 = 6125.$$

The Java applet below will compute greatest common divisors. We can use it to check our computation:

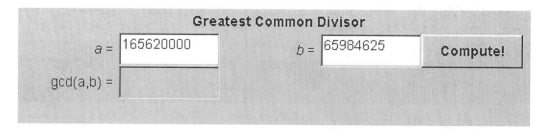

Looks good.

If you play around with the gcd applet, you will find that it computes gcds very quickly, even for huge numbers. In fact, we can use it on numbers that are too large for us to factor:

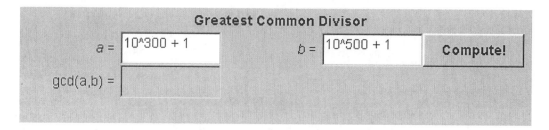

How does it find gcds without factoring the numbers? We will learn the secret in the next chapter.

1.4 Counting Divisors

The applet below takes an integer n as input, and computes $d(n)$, the number of positive divisors of n. Here's an example:

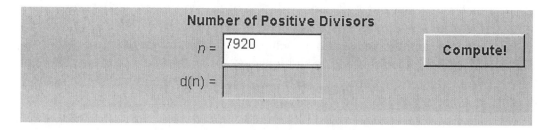

In this section, we shall perform some experiments to learn how the number of positive divisors of a given integer can be determined from the factorization of the integer. We'll do this by starting with fairly simple cases and building up to more complicated situations. Along the way, a pattern should (hopefully!) emerge.

Let us begin with the simplest case: Suppose that $n = p$, where p is a prime number. For instance, let $n = 13$. The applet below will generate a list of positive divisors and compute $d(n)$:

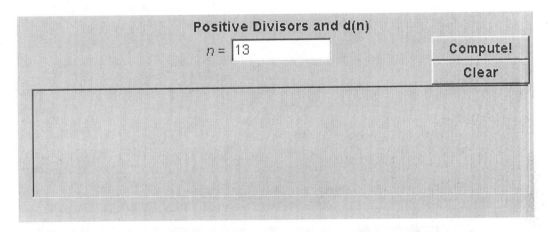

Research Question 2

Select different primes, and repeat the above experiment: Compute the divisor list, and then compute $d(n)$. When you have enough data, form a conjecture that states $d(n)$ for $n = p$, where p is prime. (It probably won't take much data for you to form a conjecture!) Then prove your conjecture.

Research Question 3

Repeat Research Question 2, but this time for $n = p^a$, where p is a prime and a is a positive integer.

Hints: When doing computational experiments, select a small prime p and then increment the value of a by 1. Also, note that when $a = 1$, you revert to the case covered by Research Question 2. Thus, whatever your conjecture, it must be

consistent with your results from the preceding Research
Question.

Now suppose that $n = pq$, where p and q are distinct primes. (If $p = q$, then
we would be in the situation covered in Research Question 3.) What is $d(n)$
in this case? Let's try an experiment, taking $n = 3 \cdot 7$:

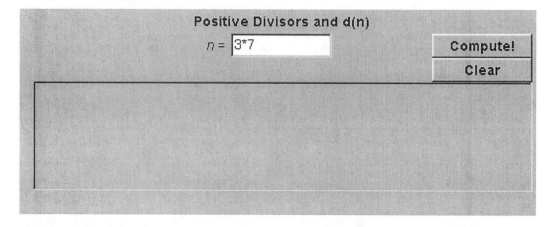

Research Question 4

Select different primes p and q, and repeat the above
experiment: Compute the list of positive divisors for $n = pq$,
and compute $d(n)$. When you have enough data, form a
conjecture that states $d(n)$ for $n = pq$, where p and q are distinct
primes.

Note: We will no longer state that you should prove your
conjecture, since you should always prove (or try to prove) all
conjectures made in this course.

Research Question 5

Repeat Research Question 4, but this time take $n = p^a q^b$, where
p and q are primes and a and b are positive integers.

Hints: When doing your numerical experiments, it will make life easier if you are systematic. (In fact, systematic experimentation is the hallmark of good scientific work.) You might try starting with fixed values of p and q, and make simple variations in the exponents a and b. Note also that the previous Research Questions are all special cases of this one. Therefore, your conjecture in this case must be consistent with your previous results.

All of the earlier work has been leading up to the most general case of all. Suppose that

$$n = p_1^{a_1} p_2^{a_2} \cdots p_k^{a_k},$$

where $p_1, p_2, \ldots, p_k$ are distinct primes and $a_1, a_2, \ldots, a_k$ are positive integers. What is $d(n)$ in this case? On the basis of your earlier investigations, you may not need any additional experimentation to form a conjecture. On the other hand, there is no harm in further experimentation, so feel free to do more if you wish.

Research Question 6

Form a conjecture that states $d(n)$ for

$$n = p_1^{a_1} p_2^{a_2} \cdots p_k^{a_k},$$

where $p_1, p_2, \ldots, p_k$ are distinct primes and $a_1, a_2, \ldots, a_k$ are positive integers.

It may seem that we led to this formula in an inefficient manner; why not go for the whole formula at once? As you may have discovered along the way, it can help a great deal to work up to the general case through various special cases. In number theory, starting with cases such as n prime, and then n a prime power, and so on, is a standard progression of study. Keep this in mind for later investigations in the course.

An important part of proving your conjectures for Research Questions 2 through 6 is being able to list the positive divisors of an integer of the forms $n = p$, $n = pq$, $n = p^a$, and so on. The basic idea of listing these divisors was implicit in the section above where we derived a formula for the greatest common divisor of two integers in terms of their factorizations. In the next exercise, you should formulate the statement precisely.

Exercise 2

Suppose that $n = p_1^{a_1} p_2^{a_2} \cdots p_k^{a_k}$. What are the positive divisors of n in terms of their prime-power factorizations?

1.5   The Division Algorithm

We begin this section with a statement of the Division Algorithm, which you saw at the end of the Prelab section of this chapter:

Theorem 1.2 (Division Algorithm) Let a be an integer and b be a positive integer. Then there exist unique integers q and r such that

$$a = bq + r \quad \text{and} \quad 0 \leq r < b.$$

Remember learning long division in grade school? (If not, pretend that you do.) Long division is a procedure for dividing a number a by another number b, and coming up with a quotient q and a remainder r. The Division Algorithm is really nothing more than a guarantee that good old long division really works. Although this result doesn't seem too profound, it is nonetheless quite handy. For instance, it is used in proving the Fundamental Theorem of Arithmetic, and will also appear in the next chapter. A proof of the Division Algorithm is given at the end of the "Tips for Writing Proofs" section of the Course Guide.

Now, suppose that you have a pair of integers a and b, and would like to find the corresponding q and r. If a and b are small, then you could find q and r by trial and error. However, suppose that $a = 124389001$ and $b =$

593. In this case, trial and error would be extremely tedious. What happens if we try the division using our on-line calculator? Since q is the quotient, it should be close to a/b.

Calculator	
124389001/593	**Compute!**
	Floats ▾

Now we're getting somewhere. We know that $a = bq + r$. Dividing on both sides of the equation by b yields

$$a/b = q + r/b.$$

Thus it follows that $q \leq a/b$. (Remember that $0 \leq r < b$.) So, in our above example, it makes sense to take $q = 209762$, because this is the biggest integer that is less than (or equal to) a/b. (Recall that this is the idea used repeatedly when performing long division by hand.) Setting the mode to "Integers" in the on-line calculator will automatically give you the "integer part" of a/b, which is the required value for q:

Calculator	
124389001/593	**Compute!**
	Integers ▾

With this choice of q, what is r? That's easy enough, because we must have

$$a - bq = r.$$

Here's the computation:

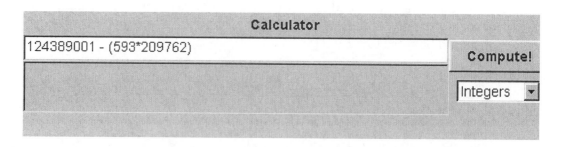

Since $r < b$, we have fulfilled the requirements of the Division Algorithm, and we're done. Let's summarize the procedure:

1. Begin with values of a and b;
2. Set q equal to the integer part of a/b;
3. Set $r = a - bq$.

The applet below automates this procedure. Enter values for a and b, and it will return the corresponding values of the quotient q and the remainder r. Try it out.

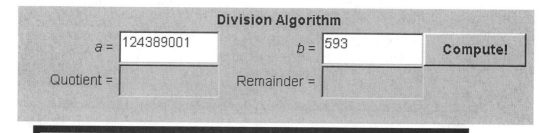

Exercise 3

Use the Division Algorithm applet to repeat Exercise 5 in the Prelab section of this chapter.

Homework

1. For each of $n = 24$, $n = 105$, and $n = 594$:

   (a) Find the prime factorization of n.

   (b) Use the prime factorization from part (a) to compute $d(n)$.

   (c) Use the prime factorization from part (a) to compute $\sigma(n)$.

   (d) Give a list of the positive divisors of n.

2. Find the smallest positive integer n with the given value of $d(n)$.

   (a) $d(n) = 4$

   (b) $d(n) = 6$

   (c) $d(n) = 12$

   (d) $d(n) = p$, where p is a prime number. Prove your answer.

3. Suppose that $n = p_1^{a_1} p_2^{a_2} \cdots p_k^{a_k}$, where $p_1, p_2, \ldots, p_k$ are distinct primes. Prove that n is a perfect square if and only if each of $a_1, a_2, \ldots, a_k$ is even.

4. Show that if $a \mid b$ and $b \mid c$, then $a \mid c$.

5. Prove that if $a \mid c$ and $b \mid d$, then $ab \mid cd$.

6. Prove or disprove: If $m \mid n$, then $m^2 \mid n^2$.

7. Prove or disprove: If $m^2 \mid n^2$, then $m \mid n$.

8. Prove or disprove: If $a < b$ and $m^a \mid n^b$, then $m \mid n$.

9. Prove or disprove: If $0 < b \le a$ and $m^a \mid n^b$, then $m \mid n$.

10. Prove that if $m \mid n$ and $m < n/m$, then $m < \sqrt{n}$.

11. Suppose that $n > 1$.

    (a) Suppose that m is the smallest integer that is both a divisor of n and is greater than 1. Prove that m is prime.

    (b) Show that if n is composite, then n has a *prime* divisor p with $p \le \sqrt{n}$.

12. Prove or disprove: If $n \mid ab$ and $n \nmid a$, then $n \mid b$.

13. Prove or disprove: If $n \mid ab$ and $\gcd(n, a) = 1$, then $n \mid b$.

14. Prove or disprove: If $\gcd(a, b) = d$, then $\gcd(a/d, b/d) = 1$.

15. Show that $\gcd(a, b) = \gcd(a, b + na)$ for all integers n.

16. Prove or disprove: If $\gcd(m, n) = 1$, then $\gcd(a, mn) = \gcd(a, m) \cdot \gcd(a, n)$.

17. Prove or disprove: If $\gcd(m, n) = d$, then $\gcd(a, mn) = \gcd(a, m) \cdot \gcd(a, n)/d$.

18. Prove or disprove: $\gcd(ma, mb) = m \cdot \gcd(a, b)$.

19. Prove or disprove: $\gcd(a^n, b^n) = \gcd(a, b)^n$.

20. Suppose that a is any integer.

    (a) Prove or disprove: If $n > 0$, then for any integer a, either $n \mid a$ or $\gcd(n, a) = 1$.

    (b) Repeat part (a) with "$n > 0$" replaced by "n is prime".

21. What are the possible values of $\gcd(a, a + 1)$?

22. What are the possible values of $\gcd(a, a + 2)$?

23. What are the possible values of $\gcd(a, a + 3)$?

24. What are the possible values of $\gcd(a, a + 4)$?

25. Let $\gcd(a, b) = 1$. What are the possible values of $\gcd(a + b, a - b)$?

26. Suppose that a is an even integer. For what values of b is $\gcd(a/2, b) = \gcd(a, b)$?

27. Suppose that a and b are any integers.

    (a) Suppose that p is prime.

        i. Prove or disprove: If $p \mid a$ or $p \mid b$, then $p \mid ab$.
        ii. Prove or disprove: If $p \mid ab$, then $p \mid a$ or $p \mid b$.

    (b) Repeat part (a) with the prime "p" replaced with the composite "n".

28. Suppose that a is any integer.

    (a) Let p and q be distinct primes such that $p \mid a$ and $q \mid a$. Prove or disprove: The product pq is a divisor of a.

    (b) Repeat part (a), but with "p" and "q" replaced with "m" and "n", respectively, where m and n are any integers satisfying $\gcd(m, n) = 1$.

    (c) Repeat part (b) without the condition that $\gcd(m, n) = 1$.

29. Let n be a positive integer.

    (a) Prove or disprove: $2 \mid (n^2 - n)$.

    (b) Prove or disprove: $2 \mid (n^a - n)$ for any $a \geq 1$.

30. The *least common multiple* of two positive integers m and n is denoted by $\operatorname{lcm}(m, n)$, and is defined to be the smallest positive integer ℓ such that $m \mid \ell$ and $n \mid \ell$.

    (a) Compute $\operatorname{lcm}(12, 20)$, $\operatorname{lcm}(15, 25)$, and $\operatorname{lcm}(7, 13)$.

    (b) Suppose that $m = p_1^{a_1} p_2^{a_2} \cdots p_k^{a_k}$ and $n = p_1^{b_1} p_2^{b_2} \cdots p_k^{b_k}$ are the prime factorizations of n and m, respectively. (It is possible that some of the a_i's and b_i's are equal to zero.)

        i. Show that $\operatorname{lcm}(m, n) = p_1^{\max(a_1, b_1)} p_2^{\max(a_2, b_2)} \cdots p_k^{\max(a_k, b_k)}$.
        ii. Show that $\operatorname{lcm}(m, n) = mn / \gcd(m, n)$.

31. For $n = p_1^{a_1} p_2^{a_2} \cdots p_k^{a_k}$, show that

$$\sigma(n) = \left(\frac{p_1^{a_1+1} - 1}{p_1 - 1} \right) \left(\frac{p_2^{a_2+1} - 1}{p_2 - 1} \right) \cdots \left(\frac{p_k^{a_k+1} - 1}{p_k - 1} \right).$$

32. A function f defined on the positive integers is said to be *multiplicative* if $f(mn) = f(m)f(n)$ for all pairs m and n of relatively prime integers.

    (a) Show that $d(n)$ is multiplicative.

    (b) Show that $\sigma(n)$ is multiplicative.

33. Suppose that $n = p_1^{a_1} p_2^{a_2} \cdots p_k^{a_k}$. Find a general formula for:

    (a) $\sigma_2(n) = \sum_{d \mid n} d^2$.

    (b) $\sigma_m(n) = \sum_{d \mid n} d^m \ (m > 0)$.

34. An integer n is *perfect* if $\sigma(n) = 2n$.

    (a) Show that 6, 28, and 496 are all perfect.

    (b) Show that if $2^k - 1$ is prime, then $n = 2^{k-1}(2^k - 1)$ is perfect.

35. Two integers m and n are called an *amicable pair* if $\sigma(m) - m = n$ and $\sigma(n) - n = m$. Show that 220 and 284 are an amicable pair.

36. A sequence of integers $a_1, a_2, \ldots, a_k$ is called an *aliquot cycle of length k* if

$$\sigma(a_1) - a_1 = a_2,\ \sigma(a_2) - a_2 = a_3,\ \ldots\ \sigma(a_{k-1}) - a_{k-1} = a_k,\ \text{and}\ \sigma(a_k) - a_k = a_1.$$

Show that the integers given below form an aliquot cycle.

$$a_1 = 12496 = 2^4 \cdot 11 \cdot 71$$
$$a_2 = 14288 = 2^4 \cdot 19 \cdot 47$$
$$a_3 = 15472 = 2^2 \cdot 967$$
$$a_4 = 14536 = 2^3 \cdot 23 \cdot 79$$
$$a_5 = 14264 = 2^3 \cdot 1783$$

2. The Euclidean Algorithm and Linear Diophantine Equations

Prelab

One goal of this chapter is to develop a more efficient method for computing greatest common divisors, or gcds. The starting point happens to be the Division Algorithm, Theorem 1.2.

1. **Find the quotient q and the remainder r from the Division Algorithm for each pair a and b.**

   (a) $a = 29$, $b = 17$

   (b) $a = 126$, $b = 21$

   (c) $a = 241$, $b = 38$

2. **For each pair a and b and remainder r (computed by you) in the preceding problem, determine $\gcd(a, b)$ and $\gcd(b, r)$.**

Our second goal for this chapter is to develop a systematic method for finding all of the integer solutions x and y (if there are any) to the equation

$$ax + by = c,$$

where a, b, and c are integer constants. As you might guess, the nature of the solutions to this equation depends on the values of a, b, and c. Equations of this form are called *linear diophantine equations.*[1]

3. **By inspection, find a few integer solutions (if there are any) to the equation $6x + 2y = 4$.**

4. **By inspection, find a few integer solutions (if there are any) to the equation $6x + 2y = 3$.**

---

[1]The equation is linear because the variables x and y appear to the first power. It is called *diophantine* after the Greek mathematician Diophantus (thought to have lived in the third or fourth century A.D.), who first systematically studied integer solutions to equations of this form.

Finding integer solutions to linear diophantine equations is not only interesting in its own right, but is a useful tool as well. Below is a problem that can be solved by first translating to an appropriate linear diophantine equation, and then finding the solutions to the equation. We'll be returning to this problem later in the chapter.

> You have purchased a new 49-liter fish tank, and are setting it up. When it comes time to fill the tank, you realize that your delicate South American cichlids will never tolerate the regular tap water. So, you go off to the store to purchase pure Artesian spring water, which is sold in 5- and 8-liter bottles. How many bottles of each size will you need to exactly fill the tank?

Maple electronic notebook `02-euclid.mws`

Mathematica electronic notebook `02-euclid.nb`

Web electronic notebook Start with the web page `index.html`

Maple Lab: The Euclidean Algorithm and Linear Diophantine Equations

■ The Division Algorithm Revisited

Below is a statement of the division algorithm, as given in the Prelab section of this chapter:

Theorem (Division Algorithm)
Let a be an integer and b be a positive integer. Then there exist unique integers q and r such that
$$a = b\,q + r \quad \text{and} \quad 0 \le r < b.$$

Recall that at the end of the last chapter, you used a Maple function called `divalg` to compute q and r for a given a and b. Here is the definition for `divalg`:

```
> divalg := proc(a,b)
    local q, r;
    q:= floor(a/b);
    r:= a-b*q;
    [q,r];
  end:
```

Here's how it works for $a = 123456789$ and $b = 369$:

```
> divalg(123456789,369);
```

Thus, for this choice of a and b, we have $q = 334571$ and $r = 90$. In Prelab exercise 2, you were asked to compute $\gcd(a, b)$ and $\gcd(b, r)$ for different choices of a and b. If we do this computation in this case, we get

```
> a := 123456789;
  b := 369;
  r := 90;
  [gcd(a,b), gcd(b,r)];
>
```

We can save ourselves a little typing by combining the computation required to compute r with the computation to compute $\gcd(a, b)$ and $\gcd(b, r)$. We do this by using the fact that `divalg(a,b)[2]` will return just r.

```
> a:=123456789:
  b:=369:
  [gcd(a,b),gcd(b,divalg(a,b)[2])];
>
```

■ `Maple Note`
In general, if you have a list such as mylist = [1, 2, 4, 8, 16], you can select a particular element from the list using brackets:

`mylist[4] = 8`

97

For those with computer programming experience, this is the analogue of working with an array, where we would specify the *n*th entry of the array **mylist** with **mylist[*n*]**.

■ Research Question 1

Compute the values of gcd(a, b) and gcd(b, r) as above for different pairs a and b of your choosing until you have enough data to form a conjecture concerning the relationship between gcd(a, b) and gcd(b, r). As always, once you have formed your conjecture, try to prove it!

■ The Euclidean Algorithm 1

Now that you have completed Research Question 1, you know that for a pair of integers a and b, we have

$$\gcd(a, b) = \gcd(b, r),$$

where r is the remainder when a is divided by b. (Sure, this gives away the conjecture for Research Question 1, but you already knew the conjecture anyway, right?) Thus, for example, if $a = 123456789$ and $b = 369$ (as in the preceding section), then $r = 90$, and we see that

$$\gcd(123456789, 369) = \gcd(369, 90).$$

It's clear that computing gcd($369, 90$) will be easier than computing gcd($123456789, 369$). (This will be true using any of the methods discussed in the previous chapter.) But rather than trying to compute gcd($369, 90$), why don't we make the problem even easier by repeating what we did above? If now we think of $a = 369$ and $b = 90$, then we have r equal to

```
[ > divalg(369, 90)[2];
```

Thus it follows that gcd($369, 90$) = gcd($90, 9$). Repeating the above, we compute

```
[ > divalg(90, 9)[2];
[ >
```

It shouldn't be surprising that the remainder in this case is equal to 0. After all, it's easy to see that 90/9 = 10. Thus 9 is a divisor of 90, so that we have gcd($90, 9$) = 9. Finally, we just string all of this together to arrive at

$$\gcd(123456789, 369) = \gcd(90, 9) = 9.$$

The process illustrated is called the *Euclidean Algorithm*, and it is very efficient for computing gcds. In fact, this is probably the method that the Maple command gcd uses for computing greatest common divisors.

■ *Exercise 1*

Repeat the above procedure to compute gcd($7920, 4536$).

■ Linear Diophantine Equations 1

As stated in the Prelab section, one goal of this chapter is to develop a systematic method for finding all integer solutions x and y (when there are any) to the linear diophantine equation

(1) $ax + by = c,$

where a, b, and c are integer constants. As you discovered when working on the Prelab exercises, for a given choice of a and b, equation (1) may have several solutions or possibly no solutions; it depends on the value of c. If we have specific values for a and b, how can we figure out the values of c for which equation (1) will have a solution? As a first step, let's look at the situation for a specific choice of a and b, say $a = 6$ and $b = 4$. In this case, the above equation becomes

(2) $6\,x + 4\,y = c,$

and our question is: For what values of c is there a solution to this equation? One way to approach this is to try plugging a bunch of different values for x and y into the left-hand side of (2), and see what we get out. After all, any value that comes out must be a suitable value for c. (Do you see why?) Below is the Maple code to compute the values of $6\,x + 4\,y$ for each choice of x and y satisfying $-3 \le x \le 5$ and $-3 \le y \le 5$:

```
[ > seq(seq(6*x+4*y, x=-3..5), y=-3..5);
[ >
```

On the basis of this list, it looks like c must be an even integer in order for equation (2) to have a solution. We can also see that some values appear several times; this indicates that for a given value of c there may be lots of solutions.

If we think about it, if $x = m$ and $y = n$ are solutions to $6\,x + 4\,y = c$, then $x = -m$ and $y = -n$ are solutions to $6\,x + 4\,y = -c$. Therefore, if we know the positive values of c for which there are solutions to equation (2), then we also know the story for negative values of c. (Of course, $c = 0$ is easy. Quick, name choices for x and y such that $6\,x + 4\,y = 0$.) Thus, we can restrict ourselves to $c > 0$.

Below is a Maple function that will compute $a\,x + b\,y$ for specified values of a and b by plugging in x and y satisfying $-bound \le x \le bound$ and $-bound \le y \le bound$ and then removing the repeated and nonpositive values:

```
[ > findc := (a,b,bound) -> select((u)-> u>0, sort( convert( {seq( seq(
[   abs(a*x+b*y), x=-bound..bound), y=-bound..bound)}, list))):
[ >
```

█ Maple Note

> The seq command creates a sequence. We use two of them since there are two parameters, x and y. The function abs takes the absolute value of the linear combination.
>
> From there, it gets a bit technical Maple-wise. We use set braces so that Maple will treat our numbers as a set (so it will automatically remove duplicates). Next, we convert the set to a list so that Maple will think that the order of elements in our set/sequence is important. Then we can sort the result. Finally, we select those elements which are positive.

Here is what we get when we try findc out on our equation, which corresponds to $a = 6$ and $b = 4$:

```
[ > findc(6,4,10);
[ >
```

As we observed above, it looks like c must be a multiple of 2 in order for equation (2) to have a solution.

Let's try a different equation, say $9x + 12y = c$. What do we get in this case?

```
[ > findc(9,12,10);
[ >
```

Hmm, this time it looks like we get solutions only if c is a multiple of 3. How about $5x + 8y = c$?

```
[ > findc(5,8,10);
[ >
```

■ Research Question 2

Evaluate `findc(a,b)` for values of a and b of your choosing until you have enough data to fill in the blank at the end of the following conjecture:

"In order for $ax + by = c$ to have solutions, c must be of the form _____."

■ The Euclidean Algorithm 2

■ Finding One Solution

Your conjecture in Research Question 2 gives conditions that c must satisfy in order for

$$a x + b y = c$$

to have solutions. Suppose that c does have the form given in your conjecture; in this case, are we guaranteed that there will be solutions to our equation? This is a pretty general question. Let's look at one special case. Suppose that $a = 408$, $b = 126$, and $c = \gcd(408, 126) = 6$. Are there any solutions to the equation

$$408 x + 126 y = 6 ?$$

It turns out that the answer to this question is yes. To find a solution, we start by going through the steps of the Euclidean Algorithm to show that $\gcd(408, 126) = 6$:

```
[ > divalg(408,126);
[ >
```

Therefore, $408 = 3 \cdot 126 + 30$. Remember that

$$\gcd(408, 126) = \gcd(126, 30).$$

Now we just repeat:

```
[ > divalg(126,30);
[ >
```

Therefore, $126 = 4 \cdot 30 + 6$.

```
[ > divalg(30,6);
[ >
```

Therefore, $30 = 5 \cdot 6 + 0$. Since clearly 30 is divisible by 6, it follows that $\gcd(30, 6) = 6$. Thus it must be that $\gcd(408, 126) = 6$. Now, what does this have to do with solutions to

$$408\, x + 126\, y = 6?$$

Good question. Let's take a look at the steps in the Euclidean Algorithm again:

(a) $408 = 3 \cdot 126 + 30$.

(b) $126 = 4 \cdot 30 + 6$.

(c) $30 = 5 \cdot 6 + 0$.

Reorganizing the equation in step (b), we have

(3) $6 = 126 - 4 \cdot 30$.

From step (a), we see that $30 = 408 - 3 \cdot 126$. Substituting into equation (3), we get

$$\begin{aligned}
6 &= 126 - 4 \cdot 30 \\
&= 126 - 4 \cdot (408 - 3 \cdot 126) \\
&= -4 \cdot 408 + (1 + 12) \cdot 126 \\
&= -4 \cdot 408 + 13 \cdot 126
\end{aligned}$$

Therefore, we see that $x = -4$, $y = 13$ is a solution to $408\, x + 126\, y = 6$. Just to be on the safe side, let's check using Maple by executing the code below.

```
[ > evalb(408*(-4) + 126*13 = 6);
[ >
```

■ Maple Note

The function `evalb` stands for "evaluate Boolean expression". When you execute this code, Maple compares the values on both sides of this equation and returns true if both sides are equal and false otherwise.

The *true* means that the left and right sides of the equation are equal. Let's look at another example. Suppose that $a = 1232$ and $b = 573$, and we want to find a solution to $1232\, x + 573\, y = d$, where $d = \gcd(1232, 573)$. First we compute d using the Euclidean Algorithm:

```
[ > divalg(1232,573);
```

1. $1232 = 2 \cdot 573 + 86$.

```
[ > divalg(573,86);
```

2. $573 = 6 \cdot 86 + 57$.

```
[ > divalg(86,57);
```

3. $86 = 1 \cdot 57 + 29$.

```
[ > divalg(57,29);
```

    4. $57 = 1 \cdot 29 + 28.$

```
[ > divalg(29,28);
```

    5. $29 = 1 \cdot 28 + 1.$

```
[ > divalg(28,1);
[ >
```

    6. $28 = 28 \cdot 1 + 0.$

We see that $d = \gcd(1232, 573) = 1$, and so we are looking for a solution to

$$1232\, x + 573\, y = 1.$$

Here's a summary of the above steps:

    1. $1232 = 2 \cdot 573 + 86.$

    2. $573 = 6 \cdot 86 + 57.$

    3. $86 = 1 \cdot 57 + 29.$

    4. $57 = 1 \cdot 29 + 28.$

    5. $29 = 1 \cdot 28 + 1.$

    6. $28 = 28 \cdot 1 + 0.$

Now we work our way back up the chain:

$$
\begin{aligned}
1 &= 29 - 1 \cdot 28 \\
  &= 29 - 1 \cdot (57 - 1 \cdot 29) \\
  &= -1 \cdot 57 + 2 \cdot 29 \\
  &= -1 \cdot 57 + 2 \cdot (86 - 1 \cdot 57) \\
  &= 2 \cdot 86 - 3 \cdot 57 \\
  &= 2 \cdot 86 - 3 \cdot (573 - 6 \cdot 86) \\
  &= -3 \cdot 573 + 20 \cdot 86 \\
  &= -3 \cdot 573 + 20 \cdot (1232 - 2 \cdot 573) \\
  &= 20 \cdot 1232 - 43 \cdot 573.
\end{aligned}
$$

So, $x = 20$ and $y = -43$ should be a solution to $1232\, x + 573\, y = 1$. As before, let's check:

```
[ > evalb(20*1232 - 43*573  = 1);
[ >
```

Solving the linear equation

$$a\,x + b\,y = \gcd(a, b)$$

is useful in a variety of places in number theory. Indeed, the fact that this equation always has a solution is so handy that we give it a special name: *The GCD Trick*. We'll have several opportunities to exploit the GCD Trick as the course progresses. Maple has a built-in command that implements the procedure illustrated above. Given a and b, the command `igcdex(a,b,'x','y')` computes both $\gcd(a, b)$ and the numbers x and y which satisfy the above equation. For example, if we start with a = 1232 and b = 573 as above, we get:

```
[ > igcdex(5,8, 'x', 'y');
```

The output is simply the gcd. The x and y values have been stored by Maple in the variables called x and y. Let's take a look at them:

```
[ > x;
[ > y;
```

For example, if we start with $a = 1232$ and $b = 573$ as above,

```
[ > igcdex(1232, 573, 'x', 'y');
```

So, gcd(1232, 573) = 1, and we can find the x and y values:

```
[ > [x, y];
[ >
```

Note that it gives the same answer as was found above.

Finding All Solutions

So far, so good. The method that we used in the two previous examples is quite general, and will work to find a solution to any equation of the form

$$(4) \qquad a\,x + b\,y = d,$$

where $d = \gcd(a, b)$. Although we have made some progress towards our original goal of finding all solutions to $a\,x + b\,y = c$, we still have a long way to go. The next question that we shall consider is: Are there any solutions to equation (4) other than the one guaranteed by the GCD Trick and found using the *reverse Euclidean Algorithm* method? Let's go with a simple example,

$$7\,x + 2\,y = 1.$$

It's clear that gcd(7, 2) = 1, and we get the solution $x_0 - 1$, $y_0 = -3$ by reversing the steps of the Euclidean Algorithm. One way to look for other solutions is to use Maple to do a simple search. Solving for y in terms of x, we get the equation

$$y = \frac{1 - 7\,x}{2}.$$

If we plug in an integer value of x, and the corresponding value of y is also an integer, then the pair (x, y) is a solution to our equation. (Be sure that you understand why before moving forward.) Here's the Maple code to do the work. In this case, we are just using values of x satisfying $-10 \leq x \leq 10$.

```
> for x from -10 to 10 do
    if type((1-7*x)/2, integer) then printf('{%d, %d}\n', x,
  (1-7*x)/2); fi; od;
[ >
```

■ Maple Note

type(n, integer) returns true if n is an integer, and false otherwise.

As you can see, our solution $x_0 = 1$, $y_0 = -3$ is among those found. Let's try a different equation, say

$$8\,x + 3\,y = 1.$$

In this case, it's easy to see that $\gcd(8, 3) = 1$. The Euclidean Algorithm yields the solution $x_0 = -1$, $y_0 = 3$. Solving for y in terms of x, we have $y = (1 - 8\,x)/3$. Here's the test:

```
> for x from -10 to 10 do
    if type((1-8*x)/3, integer) then printf('{%d, %d}\n', x,
  (1-8*x)/3); fi; od;
[ >
```

As you might guess, we are working towards forming a conjecture, and it will be handy to have a function that will automatically look for solutions. Here's one provided for your convenience:

```
> sols := proc(a,b,d,n)
    local x;
    for x from -n to n do
     if type((d-a*x)/b, integer) then
       printf('{%d, %d}\n', x,(d-a*x)/b);
     fi;
    od;
  end:
```

The function `sols` will search for solutions to $a\,x + b\,y = d$, trying values of x satisfying $-n \leq x \leq n$. Let's check this function against the preceding output:

```
[ > sols(8,3,1,10);
[ >
```

Looks fine. Here's one more equation: $11\,x + 4\,y = 1$. As we can see, $\gcd(11, 4) = 1$, and the Euclidean Algorithm produces the solution $x_0 = -1$, $y_0 = 3$. Here's the results of a search:

```
[ > sols(11,4,1,20);
[ >
```

■ **Research Question 3**

Suppose that gcd(a, b) = 1, and that (x_0, y_0) is the solution to

$$a\,x + b\,y = 1$$

produced by the Euclidean Algorithm. Find a general formula for all solutions (x, y) to the above equation, giving x in terms of x_0 and y in terms of y_0. Be sure to prove that your formula works and that it accounts for *all* solutions.

Research Question 4

Now suppose that $\gcd(a, b) = d > 1$, and that (x_0, y_0) is the solution to

$$a\,x + b\,y = d$$

produced by the Euclidean Algorithm. Find a general formula for all solutions (x, y) to the above equation, giving x in terms of x_0 and y in terms of y_0.

Hint: The equations $a\,x + b\,y = d$ and $(a/d)\,x + (b/d)\,y = 1$ have the same set of solutions.

Linear Diophantine Equations 2

Now we're getting somewhere. With the research questions in the preceding section complete, we now know all of the solutions to the equation

$$a\,x + b\,y = d,$$

where $d = \gcd(a, b)$. Therefore, it remains to determine all of the solutions to the equation

$$a\,x + b\,y = k\,d,$$

where k is an integer. To get a feel for what is going on, let's look at an example. In the preceding section, we looked at the equation $7\,x + 2\,y = 1$. Let's change this a bit, say to

$$7\,x + 2\,y = 5.$$

Using the Euclidean Algorithm, we found that $x = 1$, $y = -3$ is a solution to $7\,x + 2\,y = 1$. Thus, it is easy to see that $x = 1 \cdot 5 = 5$ and $y = -3 \cdot 5 = -15$ is a solution to the equation $7\,x + 2\,y = 5$. What about the other solutions? Let's put `sols` to work to find some others:

```
[ > sols(7,2,5,10);
[ >
```

Research Question 5

Suppose that $\gcd(a, b) = d$, and that (x_0, y_0) is a solution to $a\,x + b\,y = d$. Find the general form of all solutions (x, y) to

$$a\,x + b\,y = k\,d,$$

giving x in terms of x_0 and y in terms of y_0.

Mathematica Lab: The Euclidean Algorithm and Linear Diophantine Equations

■ The Division Algorithm Revisited

Below is a statement of the division algorithm, as given in the Prelab section of this chapter:

Theorem (Division Algorithm) Let a be an integer and b be a positive integer. Then there exist unique integers q and r such that

$$a = bq + r \quad \text{and} \quad 0 \le r < b.$$

Recall that at the end of the last chapter, you used a Mathematica function called **divalg** to compute q and r for a given a and b. Here is the definition for **divalg**:

```
divalg[a_, b_] :=
  Module[{q, r}, q = Floor[a / b]; r = a - b q; Return[{q, r}]];
```

Here's how it works for $a = 123456789$ and $b = 369$:

```
divalg[123456789, 369]
```

Thus, for this choice of a and b, we have $q = 334571$ and $r = 90$. In Prelab exercise 2, you were asked to compute $\gcd(a, b)$ and $\gcd(b, r)$ for different choices of a and b. If we do this computation in this case, we get

```
a = 123456789;
b = 369;
r = 90;
{GCD[a, b], GCD[b, r]}
```

We can save ourselves a little typing by combining the computation required to compute r with the computation to compute $\gcd(a, b)$ and $\gcd(b, r)$. We do this by using the fact that **divalg[a,b][[2]]** will return just r.

```
a = 123456789;
b = 369;
{GCD[a, b], GCD[b, divalg[a, b][[2]]]}
```

♀ Mathematica Note

In general, if you have a list such as **mylist = {1, 2, 4, 8, 16}**, you can select a particular element from the list by using double brackets:

```
mylist[[4]] = 8
```

For those with computer programming experience, this is the analogue of working with an array, where we would specify the nth entry of the array **mylist** with **mylist[[n]]**.

106

■ Research Question 1

Compute the values of gcd(a, b) and gcd(b, r) as above for different pairs a and b of your choosing until you have enough data to form a conjecture concerning the relationship between gcd(a, b) and gcd(b, r). As always, once you have formed your conjecture, try to prove it!

■ The Euclidean Algorithm 1

Now that you have completed Research Question 1, you know that for a pair of integers a and b, we have

$$\gcd(a, b) = \gcd(b, r),$$

where r is the remainder when a is divided by b. (Sure, this gives away the conjecture for Research Question 1, but you already knew the conjecture anyway, right?) Thus, for example, if $a = 123456789$ and $b = 369$ (as in the preceding section), then $r = 90$, and we see that

$$\gcd(123456789, 369) = \gcd(369, 90).$$

It's clear that computing gcd(369, 90) will be easier than computing gcd(123456789, 369). (This will be true using any of the methods discussed in the previous chapter.) But rather than trying to compute gcd(369, 90), why don't we make the problem even easier by repeating what we did above? If now we think of $a = 369$ and $b = 90$, then we have r equal to

```
divalg[369, 90][[2]]
```

Thus it follows that gcd(369, 90) = gcd(90, 9). Repeating the above, we compute

```
divalg[90, 9][[2]]
```

It shouldn't be surprising that the remainder in this case is equal to 0. After all, it's easy to see that 90/9 = 10. Thus 9 is a divisor of 90, so that we have gcd(90, 9) = 9. Finally, we just string all of this together to arrive at

$$\gcd(123456789, \ 369) \ = \ \gcd(369, \ 90) \ = \ \gcd(90, \ 9) \ = \ 9.$$

The process illustrated is called the *Euclidean Algorithm,* and it is very efficient for computing gcds. In fact, this is probably the method that the Mathematica command **GCD** uses for computing greatest common divisors.

■ Exercise 1

Repeat the above procedure to compute gcd(7920, 4536).

■ Linear Diophantine Equations 1

As stated in the Prelab section, one goal of this chapter is to develop a systematic method for finding all integer solutions x and y (when there are any) to the linear diophantine equation

$$a x + b y = c, \tag{1}$$

where a, b, and c are integer constants. As you discovered when working on the Prelab exercises, for a given choice of a and b, equation (1) may have several solutions or possibly no solutions; it depends on the value of c. If we have specific values for a and b, how can we figure out the values of c for which equation (1) will have a solution? As a first step, let's look at the situation for a specific choice of a and b, say $a = 6$ and $b = 4$. In this case, the above equation becomes

$$6x + 4y = c, \tag{2}$$

and our question is: For what values of c is there a solution to this equation? One way to approach this is to try plugging a bunch of different values for x and y into the left-hand side of (2), and see what we get out. After all, any value that comes out must be a suitable value for c. (Do you see why?) Below is the Mathematica code to compute the values of $6x + 4y$ for each choice of x and y satisfying $-3 \le x \le 5$ and $-3 \le y \le 5$:

```
Flatten[Table[6 x + 4 y, {x, -3, 5}, {y, -3, 5}]]
```

▽ Mathematica Note

The **Flatten** is there to remove extra sets of unnecessary braces. If you really want to see what it does, remove the **Flatten** command and re-execute the cell.

On the basis of this list, it looks like c must be an even integer in order for equation (2) to have a solution. We can also see that some values appear several times; this indicates that for a given value of c there may be lots of solutions.

If we think about it, if $x = m$ and $y = n$ are solutions to $6x + 4y = c$ then $x = -m$ and $y = -n$ are solutions to $6x + 4y = -c$. Therefore, if we know the positive values of c for which there are solutions to equation (2), then we also know the story for negative values of c. (Of course, $c = 0$ is easy. Quick, name choices for x and y such that $6x + 4y = 0$.) Thus, we can restrict ourselves to $c > 0$.

Below is a Mathematica function that will compute $ax + by$ for specified values of a and b by plugging in x and y satisfying $-n \le x \le n$ and $-n \le y \le n$ and then removing the repeated and nonpositive values:

```
findc[a_, b_, n_] :=
    Module[{clist},
                            clist =
        Table[a x + b y, {x, -n, n}, {y, -n, n}];
                            clist =
        Rest[Union[Abs[Flatten[clist]]]];
                            Return[clist]];
```

▽ Mathematica Note

The **Union** command behaves just as you would expect with sets: it removes duplicate entries. Along the way, it also sorts the list.

Here we also see **Abs**, which takes the absolute value of a number. In this case, it is applied to a list. The end result is to take the absolute value of every entry in the list. Mathematica tries to do something sensible whenever an operation is applied to a list. The best way to understand what it will do is by experimenting.

Finally, we encounter the **Rest** command. It removes the first entry of a list. Can you guess why we wanted to do that? If not, make a copy of **findc** in a new cell, give it a new name, and remove **Rest** from the definition. See if you can spot the difference between the original function **findc** and the new one you created.

Here is what we get when we try **findc** out on our equation, which corresponds to $a = 6$ and $b = 4$:

```
findc[6, 4, 10]
```

As we observed above, it looks like c must be a multiple of 2 in order for equation (2) to have a solution. Let's try a different equation, say $9x + 12y = c$. What do we get in this case?

```
findc[9, 12, 10]
```

Hmm, this time it looks like we get solutions only if c is a multiple of 3. How about $5x + 8y = c$?

```
findc[5, 8, 10]
```

■ Research Question 2

> Evaluate **findc[a,b]** for values of a and b of your choosing until you have enough data to fill in the blank at the end of the following conjecture:
>
> "In order for $ax + by = c$ to have solutions, c must be of the form _____."

■ The Euclidean Algorithm 2

■ Finding One Solution

Your conjecture in Research Question 2 gives conditions that c must satisfy in order for

$$ax + by = c$$

to have solutions. Suppose that c does have the form given in your conjecture; in this case, are we guaranteed that there will be solutions to our equation? This is a pretty general question. Let's look at one special case. Suppose that $a = 408$, $b = 126$, and $c = \gcd(408, 126) = 6$. Are there any solutions to the equation

$$408x + 126y = 6?$$

It turns out that the answer to this question is yes. To find a solution, we start by going through the steps of the Euclidean Algorithm to show that $\gcd(408, 126) = 6$:

```
divalg[408, 126]
```

Therefore, $408 = 3 \cdot 126 + 30$. Remember that

$$\gcd(408, 126) = \gcd(126, 30).$$

Now we just repeat:

```
divalg[126, 30]
```

Therefore, $126 = 4 \cdot 30 + 6$.

```
divalg[30, 6]
```

Therefore, $30 = 5 \cdot 6 + 0$. Since clearly 30 is divisible by 6, it follows that $\gcd(30, 6) = 6$. Thus it must be that $\gcd(408, 126) = 6$. Now, what does this have to do with solutions to

$$408x + 126y = 6?$$

Good question. Let's take a look at the steps in the Euclidean Algorithm again:

(a) $408 = 3 \cdot 126 + 30$.

(b) $126 = 4 \cdot 30 + 6$.

(c) $30 = 5 \cdot 6 + 0$.

Reorganizing the equation in step (b), we have

$$6 = 126 - 4 \cdot 30. \tag{3}$$

From step (a), we see that $30 = 408 - 3 \cdot 126$. Substituting into equation (3), we get

$$\begin{aligned}
6 &= 126 - 4 \cdot 30 \\
&= 126 - 4 \cdot (408 - 3 \cdot 126) \\
&= -4 \cdot 408 + (1 + 12) \cdot 126 \\
&= -4 \cdot 408 + 13 \cdot 126
\end{aligned}$$

Therefore, we see that $x = -4$, $y = 13$ is a solution to

$$408\,x + 126\,y = 6.$$

Just to be on the safe side, let's check using Mathematica by executing the code below.

```
408 * (-4) + 126 * 13 == 6
```

♡ Mathematica Note

The "`==`" means that this code is treated as an equation rather than a definition. When you execute this code, Mathematica compares the values on both sides of the equation, and returns **True** if both sides are equal, and **False** otherwise.

The **True** means that the left and right sides of the equation are equal. Let's look at another example. Suppose that $a = 1232$ and $b = 573$, and we want to find a solution to

$$1232\,x + 573\,y = d,$$

where $d = \gcd(1232, 573)$. First we compute d using the Euclidean Algorithm:

```
divalg[1232, 573]
```

1. $1232 = 2 \cdot 573 + 86$.

```
divalg[573, 86]
```

2. $573 = 6 \cdot 86 + 57$.

```
divalg[86, 57]
```

3. $86 = 1 \cdot 57 + 29$.

```
divalg[57, 29]
```

4. $57 = 1 \cdot 29 + 28$.

```
divalg[29, 28]
```

5. $29 = 1 \cdot 28 + 1$.

```
divalg[28, 1]
```

6. $28 = 28 \cdot 1 + 0$.

We see that $d = \gcd(1232, 573) = 1$, and so we are looking for a solution to

$$1232\,x + 573\,y = 1.$$

Here's a summary of the above steps:

1. $1232 = 2 \cdot 573 + 86$.

2. $573 = 6 \cdot 86 + 57$.

3. $86 = 1 \cdot 57 + 29$.

4. $57 = 1 \cdot 29 + 28$.

5. $29 = 1 \cdot 28 + 1$.

6. $28 = 28 \cdot 1 + 0$.

Now we work our way back up the chain:

$$\begin{aligned}
1 &= 29 - 1 \cdot 28 \\
  &= 29 - 1 \cdot (57 - 1 \cdot 29) \\
  &= -1 \cdot 57 + 2 \cdot 29 \\
  &= -1 \cdot 57 + 2 \cdot (86 - 1 \cdot 57) \\
  &= 2 \cdot 86 - 3 \cdot 57 \\
  &= 2 \cdot 86 - 3 \cdot (573 - 6 \cdot 86) \\
  &= -3 \cdot 573 + 20 \cdot 86 \\
  &= -3 \cdot 573 + 20 \cdot (1232 - 2 \cdot 573) \\
  &= 20 \cdot 1232 - 43 \cdot 573.
\end{aligned}$$

So, $x = 20$ and $y = -43$ should be a solution to $1232\,x + 573\,y = 1$. As before, let's check:

```
20 * 1232 - 43 * 573  == 1
```

Solving the linear equation

$$a\,x + b\,y = \gcd(a,\, b)$$

is useful in a variety of places in number theory. Indeed, the fact that this equation always has a solution is so handy that we give it a special name: *The GCD Trick*. We'll have several opportunities to exploit the GCD Trick as the course progresses. Mathematica has a built-in command that implements the procedure illustrated above. Given a and b, the command **ExtendedGCD[a, b]** computes both $\gcd(a, b)$ and a pair (x, y) which satisfies the above equation. For example, if we start with $a = 1232$ and $b = 573$ as above, we get:

```
ExtendedGCD[1232, 573]
```

The output comes in the form $\{\gcd(a, b), \{x, y\}\}$. Note that it gives the same answer as was found above.

■ Finding All Solutions

So far, so good. The method that we used in the two previous examples is quite general, and will work to find a solution to any equation of the form

$$a\,x + b\,y = d, \tag{4}$$

where $d = \gcd(a, b)$. Although we have made some progress towards our original goal of finding all solutions to $a\,x + b\,y = c$, we still have a long way to go. The next question that we shall consider is the following: Are there any solutions to equation (4) other than the one guaranteed by the GCD Trick and found using the *reverse Euclidean Algorithm* method? Let's go with a simple example,

$$7\,x + 2\,y = 1.$$

It's clear that $\gcd(7, 2) = 1$, and we get the solution $x_0 = 1$, $y_0 = -3$ by reversing the steps of the Euclidean Algorithm. One way to look for other solutions is to use Mathematica to do a simple search. Solving for y in terms of x, we get the equation

$$y = (1 - 7x)/2.$$

If we plug in an integer value of x, and the corresponding value of y is also an integer, then the pair (x, y) is a solution to our equation. (Be sure that you understand why before moving forward.) Here's the Mathematica code to do the work. In this case, we are just using values of x satisfying $-10 \le x \le 10$.

```
Do[If[IntegerQ[(1 - 7 x) / 2], Print[{x, (1 - 7 x) / 2}]], {x, -10, 10}]
```

♡ Mathematica Note

`IntegerQ[t]` returns **True** if **t** is an integer, and **False** otherwise.

As you can see, our solution $x_0 = 1$, $y_0 = -3$ is among those found. Let's try a different equation, say

$$8x + 3y = 1.$$

In this case, it's easy to see that $\gcd(8, 3) = 1$. The Euclidean Algorithm yields the solution $x_0 = -1$, $y_0 = 3$. Solving for y in terms of x, we have $y = (1 - 8x)/3$. Here's the test:

```
Do[If[IntegerQ[ (1 - 8 x) / 3], Print[{x, (1 - 8 x) / 3}]], {x, -10, 10}]
```

As you might guess, we are working towards forming a conjecture, and it will be handy to have a function that will automatically look for solutions. Here's one provided for your convenience:

```
sols[a_, b_, d_, n_] :=
    Do[If[IntegerQ[(d - a x) / b], Print[{x, (d - a x) / b}]], {x, -n, n}]
```

The function **sols** will search for solutions to $ax + by = d$, trying values of x satisfying $-n \le x \le n$. Let's check this function against the preceding output:

```
sols[8, 3, 1, 10]
```

Looks fine. Here's one more equation: $11x + 4y = 1$. As we can see, $\gcd(11, 4) = 1$, and the Euclidean Algorithm produces the solution $x_0 = -1$, $y_0 = 3$. Here's the result of a search:

```
sols[11, 4, 1, 20]
```

■ Research Question 3

Suppose that $\gcd(a, b) = 1$, and that (x_0, y_0) is the solution to

$$ax + by = 1$$

produced by the Euclidean Algorithm. Find a general formula for all solutions (x, y) to the above equation, giving x in terms of x_0 and y in terms of y_0. Be sure to prove that your formula works and that it accounts for *all* solutions.

■ Research Question 4

Now suppose that $\gcd(a, b) = d > 1$, and that (x_0, y_0) is the solution to

$$a x + b y = d$$

produced by the Euclidean Algorithm. Find a general formula for all solutions (x, y) to the above equation, giving x in terms of x_0 and y in terms of y_0.

Hint: The equations $a x + b y = d$ and $(a/d) x + (b/d) y = 1$ have the same set of solutions.

■ Linear Diophantine Equations 2

Now we're getting somewhere. With the research questions in the preceding section complete, we now know all of the solutions to the equation

$$a x + b y = d,$$

where $d = \gcd(a, b)$. Therefore, it remains to determine all of the solutions to the equation

$$a x + b y = k d,$$

where k is an integer. To get a feel for what is going on, let's look at an example. In the preceding section, we looked at the equation $7 x + 2 y = 1$. Let's change this a bit, say to

$$7 x + 2 y = 5.$$

Using the Euclidean Algorithm, we found that $x = 1$, $y = -3$ is a solution to $7 x + 2 y = 1$. Thus, it is easy to see that $x = 1 \cdot 5 = 5$ and $y = -3 \cdot 5 = -15$ is a solution to the equation $7 x + 2 y = 5$. What about the other solutions? Let's put **sols** to work to find some others:

```
sols[7, 2, 5, 10]
```

■ Research Question 5

Suppose that $\gcd(a, b) = d$ and that (x_0, y_0) is a solution to $a x + b y = d$. Find the general form of all solutions (x, y) to

$$a x + b y = k d,$$

giving x in terms of x_0 and y in terms of y_0.

Web Lab: The Euclidean Algorithm and Linear Diophantine Equations

2.1 The Division Algorithm Revisited

Below is a statement of the Division Algorithm, as given in the Prelab section of this chapter:

> **Theorem (Division Algorithm)** Let a be an integer and b be a positive integer. Then there exist unique integers q and r such that
>
> $$a = bq + r \quad \text{and} \quad 0 \le r < b.$$

Recall that at the end of the last chapter, you computed q and r for a given a and b. Here's the computation for $a = 123456789$ and $b = 369$:

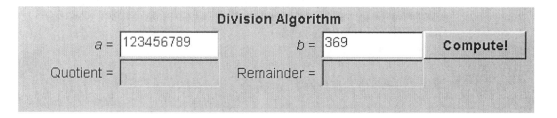

Thus, for this choice of a and b, we have $q = 334571$ and $r = 90$. In Exercise 2 of the Prelab section, you were asked to compute gcd(a, b) and gcd(b, r) for different choices of a and b. If we perform this computation for the values of a and b given above, we get

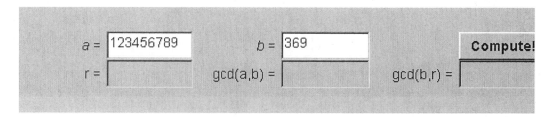

114

2.2 The Euclidean Algorithm 1

Now that you have completed Research Question 1, you know that for a pair of integers *a* and *b*, we have

$$\gcd(a, b) = \gcd(b, r),$$

where *r* is the remainder when *a* is divided by *b*. (Sure, this gives away the conjecture for Research Question 1, but you already knew the conjecture anyway, right?) Thus, for example, if *a* = 123456789 and *b* = 369 (as in the preceding section), then *r* = 90, and we see that

$$\gcd(123456789, 369) = \gcd(369, 90).$$

It's clear that computing gcd(369, 90) will be easier than computing gcd (123456789, 369). (This will be true using any of the methods discussed in the previous chapter.) But rather than trying to compute gcd(369, 90), why don't we make the problem even easier by repeating what we did above? If now we think of *a* = 369 and *b* = 90, then we may compute the new value of *r* below.

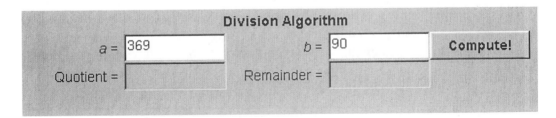

Thus we have gcd(369, 90) = gcd(90, 9). Repeating once again, we have

Division Algorithm

$a =$ `90` $b =$ `9`   **Compute!**

Quotient = ⎢   Remainder = ⎢

It shouldn't be surprising that the remainder in this case is equal to 0. After all, it's easy to see that $90/9 = 10$. Thus 9 is a divisor of 90, so that we have $\gcd(90, 9) = 9$. Finally, we just string all of this together to arrive at

$$\gcd(123456789, 369) = \gcd(369, 90) = \gcd(90, 9) = 9.$$

The process illustrated is called the *Euclidean Algorithm,* and it is very efficient for computing gcds. In fact, this is the method that this web page uses for computing greatest common divisors.

Exercise 1

Repeat the above procedure to compute $\gcd(7920, 4536)$.

2.3   Linear Diophantine Equations 1

As stated in the Prelab section, one goal of this chapter is to develop a systematic method for finding all integer solutions x and y (when there are any) to the linear diophantine equation

$$ax + by = c, \qquad\qquad (1)$$

where a, b, and c are integer constants. As you discovered when working on the Prelab exercises, for a given choice of a and b, equation (1) may have several solutions or possibly no solutions; it depends on the value of c. If we have specific values for a and b, how can we figure out the values of c for which equation (1) will have a solution? As a first step, let's look at the situation for a specific choice of a and b, say $a = 6$ and $b = 4$. In this case, the above equation becomes

$$6x + 4y = c, \qquad\qquad (2)$$

and our question is: For what values of c is there a solution to this equation? One way to approach this is to try plugging a bunch of different values for x and y into the left-hand side of (2), and see what we get out. After all, any value that comes out must be a suitable value for c. (Do you see why?) Below is an applet programmed to automatically do the "plugging in." It is initially set to compute the values of $6x + 4y$ for each choice of x and y satisfying $-3 \le x \le 5$ and $-3 \le y \le 5$:

On the basis of this list, it looks like c must be an even integer in order for equation (2) to have a solution. We can also see that some values appear several times; this indicates that for a given value of c there may be lots of solutions.

If we think about it, if $x = m$ and $y = n$ are solutions to $6x + 4y = c$, then $x = -m$ and $y = -n$ are solutions to $6x + 4y = -c$. Therefore, if we know the positive values of c for which there are solutions to equation (2), then we also know the story for negative values of c. (Of course, $c = 0$ is easy. Quick, name choices for x and y such that $6x + 4y = 0$.) Thus, we can restrict ourselves to $c > 0$.

Below is an applet that will compute $ax + by$ for specified values of a and b

(plugging in x and y satisfying $-n \leq x \leq n$ and $-n \leq y \leq n$), and then removes the repeated and nonpositive values. Here is what we get when we try it out on our equation, which corresponds to $a = 6$ and $b = 4$:

Positive Linear Combinations

$a =$ 6 $b =$ 4   Compute!

$n =$ 10   Clear

As we observed above, it looks like c must be a multiple of 2 in order for equation (2) to have a solution. Let's try a different equation, say $9x + 12y = c$. What do we get in this case?

Positive Linear Combinations

$a =$ 9 $b =$ 12   Compute!

$n =$ 10   Clear

Hmm, this time it looks like we get solutions only if c is a multiple of 3. How about $5x + 8y = c$?

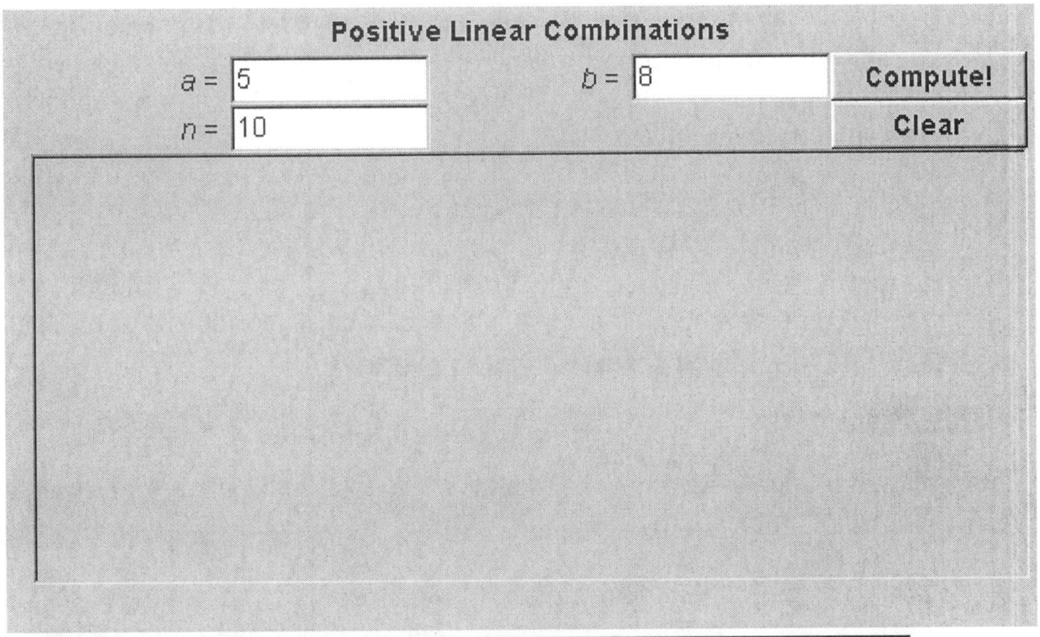

Positive Linear Combinations

$a = \boxed{5}$ $b = \boxed{8}$ **Compute!**

$n = \boxed{10}$ **Clear**

Research Question 2

Execute the above applet again using values of a and b of your choosing until you have enough data to fill in the blank at the end of the following conjecture:

"In order for $ax + by = c$ to have solutions, c must be of the form _____."

2.4 The Euclidean Algorithm 2

2.4.1 Finding One Solution

Your conjecture in Research Question 2 gives conditions that c must satisfy in order for

$$a x + b y = c$$

to have solutions. Suppose that c does have the form given in your conjecture; in this case, are we guaranteed that there will be solutions to our equation? This is a pretty general question. Let's look at one special case: suppose that $a = 408$, $b = 126$, and $c = \gcd(408, 126) = 6$. Are there any solutions to the equation

$$408x + 126y = 6?$$

It turns out that the answer to this question is yes. To find a solution, we start by going through the steps of the Euclidean Algorithm to show that $\gcd(408, 126) = 6$:

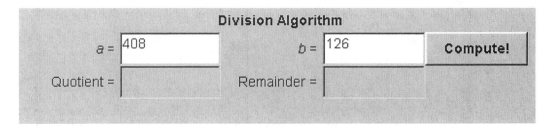

Therefore, $408 = 3 \cdot 126 + 30$. Remember that

$$\gcd(408, 126) = \gcd(126, 30).$$

Now we just repeat:

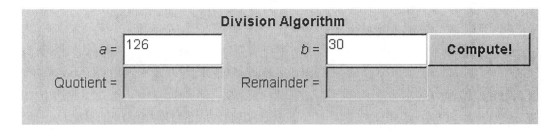

Therefore, $126 = 4 \cdot 30 + 6$.

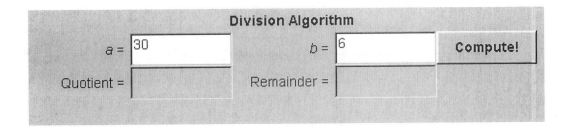

Therefore, $30 = 5 \cdot 6 + 0$. Since clearly 30 is divisible by 6, then it follows that $\gcd(30, 6) = 6$. Thus is must be that $\gcd(408, 126) = 6$. Now, what does this have to do with solutions to

$$408x + 126y = 6?$$

Good question. Let's take a look at the steps in the Euclidean Algorithm again:

(a) $408 = 3 \cdot 126 + 30$.

(b) $126 = 4 \cdot 30 + 6$.

(c) $30 = 5 \cdot 6 + 0$.

Reorganizing the equation in step (b), we have

$$6 = 126 - (4 \cdot 30). \qquad\qquad (3)$$

From step (a), we see that $30 = 408 - (3 \cdot 126)$. Substituting into equation (3), we get

$$\begin{aligned}
6 &= 126 - (4 \cdot 30) \\
&= 126 - 4 \cdot (408 - 3 \cdot 126) \\
&= (-4) \cdot 408 + (1 + 12) \cdot 126 \\
&= (-4) \cdot 408 + 13 \cdot 126
\end{aligned}$$

Therefore, we see that $x = -4$, $y = 13$ is a solution to

$$408x + 126y = 6.$$

Just to be on the safe side, let's check using the on-line calculator.

Calculator

408*(-4)+126*13   **Compute!**

Integers ▼

Let's look at another example. Suppose that $a = 1232$ and $b = 573$, and we want to find a solution to

$$1232x + 573y = d,$$

where $d = \gcd(1232, 573)$. First we compute d using the Euclidean Algorithm:

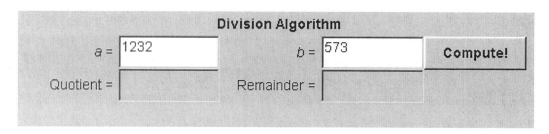

1. $1232 = 2 \cdot 573 + 86.$

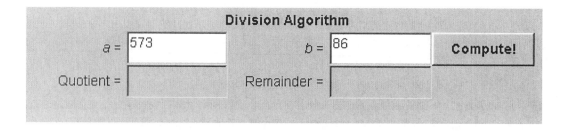

2. $573 = 6 \cdot 86 + 57.$

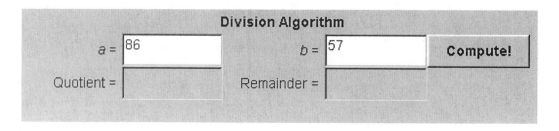

3. $86 = 1 \cdot 57 + 29$.

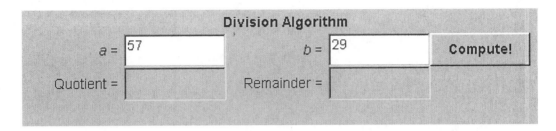

4. $57 = 1 \cdot 29 + 28$.

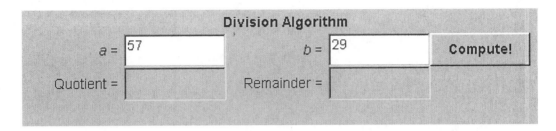

5. $29 = 1 \cdot 28 + 1$.

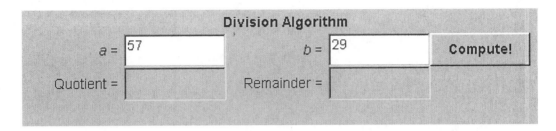

6. $28 = 28 \cdot 1 + 0$.

We see that $d = \gcd(1232, 573) = 1$, and so we are looking for a solution to

$$1232x + 573y = 1.$$

Here's a summary of the above steps:

1. $1232 = 2 \cdot 573 + 86.$
2. $573 = 6 \cdot 86 + 57.$
3. $86 = 1 \cdot 57 + 29.$
4. $57 = 1 \cdot 29 + 28.$
5. $29 = 1 \cdot 28 + 1.$
6. $28 = 28 \cdot 1 + 0.$

Now we work our way back up the chain:

$$
\begin{aligned}
1 &= 29 - 1 \cdot 28 \\
&= 29 - 1 \cdot (57 - 1 \cdot 29) \\
&= -1 \cdot 57 + 2 \cdot 29 \\
&= -1 \cdot 57 + 2 \cdot (86 - 1 \cdot 57) \\
&= 2 \cdot 86 - 3 \cdot 57 \\
&= 2 \cdot 86 - 3 \cdot (573 - 6 \cdot 86) \\
&= -3 \cdot 573 + 20 \cdot 86 \\
&= -3 \cdot 573 + 20 \cdot (1232 - 2 \cdot 573) \\
&= 20 \cdot 1232 - 43 \cdot 573.
\end{aligned}
$$

So, $x = 20$ and $y = -43$ should be a solution to $1232x + 573y = 1$. As before, let's check:

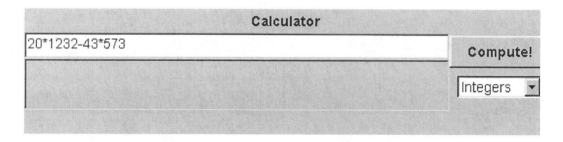

Solving the linear equation

$$ax + by = \gcd(a, b)$$

is useful in a variety of places in number theory. Indeed, the fact that this equation always has a solution is so handy that we give it a special name: *The GCD Trick*. We'll have several opportunities to exploit the GCD Trick as the course progresses.

2.4.2 Finding All Solutions

So far, so good. The method that we used in the two previous examples is quite general, and will work to find a solution to any equation of the form

$$ax + by = d, \qquad (4)$$

where $d = \gcd(a, b)$. Although we have made some progress towards our original goal of finding all solutions to $ax + by = c$, we still have a long way to go. The next question that we shall consider is the following: Are there any solutions to equation (4) other than the one guaranteed by the GCD Trick and found using the *reverse Euclidean Algorithm* method? Let's go with a simple example,

$$7x + 2y = 1.$$

It's clear that $\gcd(7, 2) = 1$, and we get the solution $x_0 = 1$, $y_0 = -3$ by reversing the steps of the Euclidean Algorithm. One way to look for other solutions is to perform a simple search. Solving for y in terms of x, we get the equation

$$y = (1 - 7x)/2.$$

If we plug in an integer value of x, and the corresponding value of y is also an integer, then the pair (x, y) is a solution to our equation. (Be sure that you understand why before moving forward.) The applet below will do the computing. It searches for solutions to $ax + by = c$, using values of x satisfying $-n \le x \le n$. Try it out on our example equation.

Solving a Linear Diophantine Equation			
$a =$ 7		$b =$ 2	Compute!
$n =$ 10		$d =$ 1	Clear

As you can see, our solution $x_0 = 1$, $y_0 = -3$ is among those found. Let's try a different equation, say

$$8x + 3y = 1.$$

In this case, it's easy to see that $\gcd(8, 3) = 1$, and the Euclidean Algorithm yields the solution $x_0 = -1$, $y_0 = 3$. Solving for y in terms of x, we have $y = (1 - 8x)/3$. Here's the computation:

Looks fine. As you might guess, we are working towards forming a conjecture. Here's one more sample equation: $11x + 3y = 1$. As we can see, $\gcd(11, 3) = 1$, and the Euclidean Algorithm produces the solution $x_0 = -1$, $y_0 = 4$. Here's the computation:

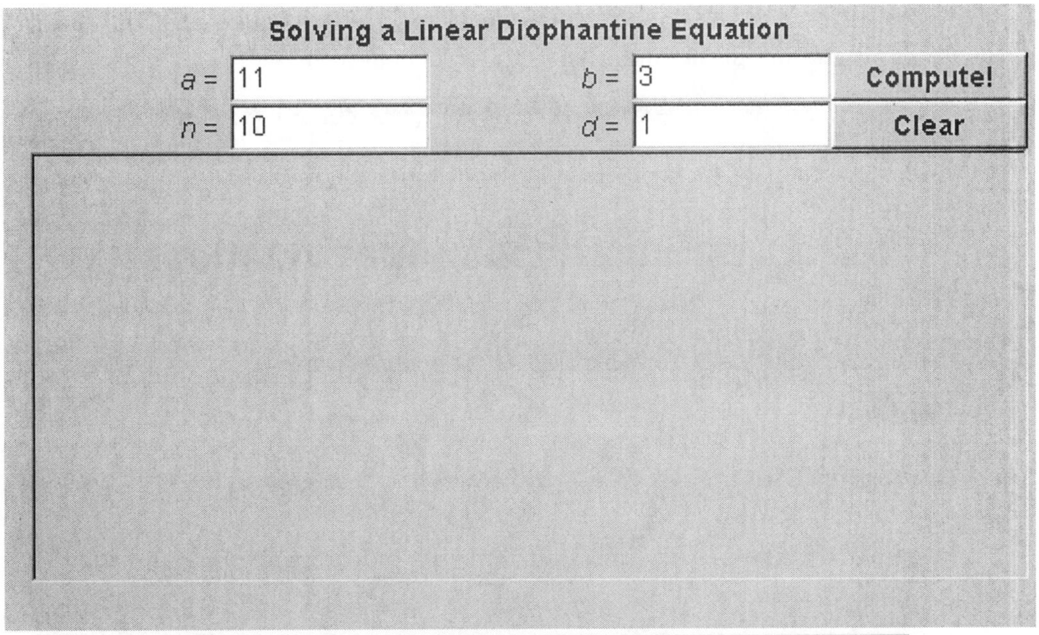

Solving a Linear Diophantine Equation

a =	11	b =	3	Compute!
n =	10	d =	1	Clear

Research Question 3

Suppose that $\gcd(a, b) = 1$, and that (x_0, y_0) is the solution to

$$ax + by = 1$$

produced by the Euclidean Algorithm. Find a general formula for all solutions (x, y) to the above equation, giving x in terms of x_0 and y in terms of y_0. Be sure to prove that your formula works and that it accounts for *all* solutions.

Research Question 4

Now suppose that $\gcd(a, b) = d > 1$, and that (x_0, y_0) is the solution to

$$ax + by = d$$

produced by the Euclidean Algorithm. Find a general formula for all solutions (x, y) to the above equation, giving x in terms of x_0 and y in terms of y_0.

Hint: The equations $ax + by = d$ and $(a/d)x + (b/d)y = 1$ have the same set of solutions.

2.5 Linear Diophantine Equations 2

Now we're getting somewhere. With the research questions in the preceding section complete, we now know all of the solutions to the equation

$$ax + by = d,$$

where $d = \gcd(a, b)$. Therefore, it remains to determine all of the solutions to the equation

$$ax + by = c,$$

where $c = kd$ and k is an integer. To get a feel for what is going on, let's look at an example. In the preceding section, we looked at the equation $7x + 2y = 1$. Let's change this a bit, say to

$$7x + 2y = 5.$$

Using the Euclidean Algorithm, we found that $x = 1$, $y = -3$ is a solution to $7x + 2y = 1$. Thus, it is easy to see that $x = 1 \cdot 5 = 5$ and $y = -3 \cdot 5 = -15$ is a solution to the equation $7x + 2y = 5$. What about the other solutions? The applet from the previous section can be used to search for additional solutions:

Solving a Linear Diophantine Equation

$a = $ 7 $b = $ 2   Compute!
$n = $ 10 $d = $ 5   Clear

Research Question 5

Suppose that $\gcd(a, b) = d$ and that (x_0, y_0) is a solution to $ax + by = d$. Find the general form of all solutions (x, y) to

$$ax + by = k\,d,$$

giving x in terms of x_0 and y in terms of y_0.

Homework

1. For each pair of integers a and b given below, compute $\gcd(a, b)$ using the Euclidean Algorithm, and then find a pair of integers x and y that satisfy $ax + by = \gcd(a, b)$. (Note: You should show each step of the Euclidean Algorithm in your solutions.)

    (a) $a = 387$, $b = 192$

    (b) $a = 487$, $b = 172$

    (c) $a = 100027$, $b = 100097$

2. Find the general form of all solutions (if there are any) for the following equations.

    (a) $57x + 289y = 1$

    (b) $246x + 372y = 20$

    (c) $154x + 91y = 42$

3. Find the solution to the problem posed in the Prelab section by setting up and solving an appropriate linear diophantine equation. For your convenience, the problem is stated below:

    > You have purchased a new 49-liter fish tank, and are setting it up. When it comes time to fill the tank, you realize that your delicate South American cichlids will never tolerate the regular tap water. So, you go off to the store to purchase pure Artesian spring water, which is sold in 5- and 8-liter bottles. How many bottles of each size will you need to exactly fill the tank?

4. You have a supply of 32¢ stamps and 21¢ stamps. You need to mail a package which requires \$1.48 in postage. How many of each type of stamp should you use?

5. When will there exist solutions to $ax + (a + 2)y = c$?

6. Write 3 as a linear combination of 15 and 21.

7. Write 1 as a linear combination of 3 and 37.

8. Your favorite brand of mulch can be purchased at a nearby store in bags containing 4 cubic feet, or from a gardening supply company by the cubic yard. Your project requires 220 cubic feet of mulch.

(a) List all possible ways to purchase exactly the required amount of mulch.

(b) If each bag costs \$3 from the nearby store and the gardening supply company charges \$14 per cubic yard plus a flat \$10 for delivery, what is the minimum cost for the required amount of mulch?

9. Show that if $ax + by = c$ has solutions, then there exists a solution (x_0, y_0) such that $0 \le y_0 < |a|$.

10. Find all solutions (x, y) such that $x \ge 0$ and $y \ge 0$ for the following equations.

   (a) $3x + 2y = 9$

   (b) $5x + 7y = 23$

   (c) $4x + 6y = 16$

11. Suppose that $\gcd(a, b) = 1$. Find additional conditions on a and b such that $ax + by = c$ will have infinitely many solutions with both $x > 0$ and $y > 0$.

12. Find the general form of all solutions (if there are any) for the following equations.

   (a) $22x + 48y + 4z = 18$

   (b) $14x + 21y + 91z = 34$

   (c) $15x + 31y + 35z = 1$

13. By setting up and solving an appropriate linear diophantine equation, find the number of ways to make

   (a) \$1.15 from dimes and quarters.

   (b) \$.55 from nickels, dimes, and quarters.

14. Compute the least common multiple of each of the following pairs. (Hint: Recall that $\mathrm{lcm}(a, b) = ab/\gcd(a, b)$.)

   (a) $a = 23$, $b = 17$

   (b) $a = 21$, $b = 91$

   (c) $a = 12$, $b = 30$

15. (a) Show that $\gcd(a, b, c) = \gcd(\gcd(a, b), c)$.

   (b) Use part (a) and the Euclidean Algorithm to find $\gcd(4488, 4675, 11543)$.

16. Show that if $r \mid a$, $r \mid b$, and $r \mid c$, then $r \mid \gcd(a, b, c)$.

3. Congruences

Prelab

What time is it 19 hours after 12:00? For most people, the answer is easy: 7:00. To arrive at the answer, we use a variation of normal arithmetic in which 1 comes after 12. (This is sometimes referred to as "clock arithmetic".) We can work in the negative direction as well: 19 hours before 12:00 is 5:00. If h is any integer, then the time h hours after 12:00 (or $-h$ hours before 12:00 if h is negative) must clearly be one of 12:00, 1:00, 2:00, ... , 11:00. Thus each integer h can be grouped together with all other integers that correspond to the same time, as shown below:

⋯	−48	−36	−24	−12	0	12	24	36	48	60	72	84	96 ⋯
⋯	−47	−35	−23	−11	1	13	25	37	49	61	73	85	97 ⋯
⋯	−46	−34	−22	−10	2	14	26	38	50	62	74	86	98 ⋯
⋯	−45	−33	−21	−9	3	15	27	39	51	63	75	87	99 ⋯
⋯	−44	−32	−20	−8	4	16	28	40	52	64	76	88	100 ⋯
⋯	−43	−31	−19	−7	5	17	29	41	53	65	77	89	101 ⋯
⋯	−42	−30	−18	−6	6	18	30	42	54	66	78	90	102 ⋯
⋯	−41	−29	−17	−5	7	19	31	43	55	67	79	91	103 ⋯
⋯	−40	−28	−16	−4	8	20	32	44	56	68	80	92	104 ⋯
⋯	−39	−27	−15	−3	9	21	33	45	57	69	81	93	105 ⋯
⋯	−38	−26	−14	−2	10	22	34	46	58	70	82	94	106 ⋯
⋯	−37	−25	−13	−1	11	23	35	47	59	71	83	95	107 ⋯

The first row consists of the set of all integers h corresponding to 12:00, the second row consists of the set of all integers h corresponding to 1:00, and so on.

Although most people use a 12-hour clock, there is no fundamental reason why a clock with a different number of hours cannot be used. For example, military time is based on a 24-hour clock.[1] In general, we will be working with n-hour clocks, where $n \geq 2$ is a fixed integer.[2] In this case, the integers divide naturally into n distinct groups in a manner analogous to that given above. Here are the groupings corresponding to a 4-hour clock:

⋯	−16	−12	−8	−4	0	4	8	12	16	20	24	28	32	36	40 ⋯
⋯	−15	−11	−7	−3	1	5	9	13	17	21	25	29	33	37	41 ⋯
⋯	−14	−10	−6	−2	2	6	10	14	18	22	26	30	34	38	42 ⋯
⋯	−13	−9	−5	−1	3	7	11	15	19	23	27	31	35	39	43 ⋯

---

[1] Make that a 24-hour clock, Sir!

[2] One could take $n = 1$, but it is not very useful. It would be like having a clock which always reads 0 o'clock

As you can see, the integers are listed in order going down. After listing an integer in the bottom row, we jump to the top of the next column. Of particular interest is the relationship among the integers in a given row. Refer to the list of groupings for the 4-hour clock to answer each of the following questions:

1. **What do the entries in the row containing 0 have in common?**

2. **How is any entry (in any row) related to the entries immediately to its right and to its left?**

3. **How is any entry related to *any* other entry in the same row (in terms of their difference)?**

4. **How are entries in the same row related (in terms of remainders upon division by 4)?**

The set of integers in a given row (for the 4-hour clock) are called a *congruence class modulo* 4 (or *congruence class mod* 4, for short). More generally, if the integers are similarly divided into n distinct rows, then each set from a given row is called a *congruence class modulo* n.

5. **Make a listing similar to those above that shows the 3 congruence classes modulo 3.**

6. **Answer Questions 1–4, using your set of congruence classes mod 3. Treat Questions 1–4 as parts (a), (b), (c), and (d) of this question.**

When working with two elements a and b from the same congruence class, it is a bit clumsy (not to mention tiring) to repeatedly be saying "a and b are in the same congruence class modulo n". This statement is shortened by using the following notation:[3]

Notation If two integers a and b are in the same congruence class modulo n, we write $a \equiv b \pmod{n}$, and say that "a is congruent to b mod n",

One of the wonderful things about congruence classes is that we can do arithmetic with them. For example, referring to the congruence classes modulo 12, note that if we add any number from the third row to any number in the seventh row, we always get a number in the ninth row. Similarly, if we add any number from the first row to a number from the sixth row, we get a number in the sixth row. The same principle holds for subtraction and multiplication.

---

[3]This notation was invented by Carl Friedrich Gauss (1777–1855), who is one of the greatest mathematicians in history. Although he made significant contributions to many areas of mathematics, he had a special fondness and esteem for number theory. One of his most enduring quotes is "Mathematics is the queen of the sciences, and the theory of numbers is the queen of mathematics."

7. **Check the claim made above by doing the following: select five pairs of numbers (a, b) from the congruence classes modulo 12, with a from the sixth row and b from the tenth row. For each pair, determine the row containing $a + b$.**

8. **For each of your five pairs in the preceding question, determine the row containing $a - b$. Then do the same for the product ab.**

Your answers to the two questions above serve to illustrate the principle that arithmetic of congruence classes modulo n is well defined. By *well-defined* we mean that, when doing arithmetic mod n, it does not matter which element of a congruence class we use. There is more on this subject in the electronic notebook, so don't be overly concerned if things seem vague at the moment.

Maple electronic notebook `03-cong.mws`

Mathematica electronic notebook `03-cong.nb`

Web electronic notebook Start with the web page `index.html`

Maple Lab: Congruences

■ Congruence Classes

We begin this section by reviewing the three different ways of thinking about congruence classes that were discussed in the Pre-lab section.

■ By Counting

Portions of the congruence classes modulo n can be viewed using the command congclasses(n), which is defined as follows:

```
> congclasses := proc(n)
    local i, j;
      for i from 0 to n-1 do
        for j from -3 to 10 do
          printf('%4d', i+n*j)
      od;
      printf('\n');
    od;
  end:
```

Here's what we get when $n = 7$:

```
> congclasses(7);
>
```

Keep in mind that the output only shows part of each congruence class. Try changing the 7 to some other positive integers n to see the congruence classes modulo n.

■ By Differences Within Rows

Referring to the above description of congruence classes, we see that each number is exactly n more than the number to its left (where n is the number of rows). By extension, if we pick any number r and look c columns to the right, we will have to add n to r a total of c times. Thus, c columns directly to the right of r is the number $r + c\,n$. Therefore it follows that two numbers are in the same row if they differ by a multiple of n. Using the congruence notation introduced in the Pre-lab section, our observation may be expressed as follows:

$$a \equiv b \ (\text{mod } n) \text{ if and only if } n \mid (b - a).$$

This characterization of congruence is extremely useful in proofs, since it brings everything known about divisibility into the picture and frequently reduces congruence problems to simple algebra.

■ By Remainders

In this section, we consider a third way to think about congruence classes. Recall that the Division Algorithm states that if any integer a is divided by a positive integer n, then the remainder r is always between 0 and $n - 1$. (Recall also our notation for the remainder: $r = a \% n$.) Maple has a built-in command for computing remainders: modp(a, n) computes $a \% n$. For example, modp(27, 4) gives the remainder when 27 is divided by 4:

```
[ > modp(27, 4);
[ >
```

■ Maple Note

Maple can also compute remainders with a different notation: a mod n computes *a* % *n*. The only difference between mod and modp is the syntax.

```
[ > 27 mod 4;
[ > modp(27,4);
[ >
```

Can you predict the result of each of the following before executing it?

```
[ > modp(4, 27);
[ > modp(12, 12);
[ > modp(-1, 6);
[ > modp(-1, 176);
[ >
```

The command modcongclasses(n) produces output similar to congclasses(n), but with each entry replaced by its remainder when divided by *n*. Here is the definition for modcongclasses:

```
[ > modcongclasses := proc(n)
     local i, j;
       for i from 0 to n-1 do
         for j from -3 to 10 do
           printf('%4d', modp(i+n*j, n))
       od;
       printf('\n');
     od;
   end:
```

Here's congclasses(7):

```
[ > congclasses(7);
```

And here's modcongclasses(7):

```
[ > modcongclasses(7);
[ >
```

Quite a striking pattern, isn't it? Try changing 7 to some other positive integers.

■ **Research Question 1**

Form a conjecture that explains the output of modcongclasses(n).

■ **Summing Up**

Above we have considered 3 ways of looking at congruence modulo *n*. Each is useful in its own way. The first description is somewhat visual and gives a good intuitive feel for congruence classes. The description in terms of differences frequently works the best in proofs. The characterization in terms of remainders makes it easy to compute modulo *n* since there are only *n* integers (the remainders 0, 1, ..., *n* − 1) to keep track of. (There will be more on computing modulo *n* later in the lab.)

■ A Tip

Mathematicians often think about and work with a concept in one way, but write their proofs in a different way. When reading proofs in a book, all one typically sees is the proof written in the most efficient manner, while the author may prefer to think about the problem in other terms. When there are several ways of describing the same thing, do not feel limited to work with only one of them. Learn how to use them all, so you can move back and forth between them depending on the situation.

■ Well-Defined Arithmetic

As mentioned in the Pre-lab section, it is possible to do arithmetic with congruence classes. To add two congruence classes modulo n, we just select any element a from the first class and any element b from the second class, and then compute $a + b$ as we would for normal integers. The sum of the two congruence classes is then defined to be equal to the congruence class containing the "usual" sum $a + b$. For example, here are the congruence classes modulo 9:

```
[ > congclasses(9);
[ >
```

Suppose that we wish to add the congruence class containing 3 to the congruence class containing 52. According to the recipe described above, we just select an element from each class, add them together, and see which class the resulting sum is in. Clearly 3 and 52 are elements of our classes, so let's try them first:

```
[ > 3+52;
[ >
```

As we can see, the sum is in the congruence class containing 55, which corresponds to the second row above. To simplify discussions involving congruence classes, it is helpful to specify the congruence classes by identifying them with the unique integer from the set $\{0, 1, 2, ..., n - 1\}$ contained in a given class. Of course, from your work above you know that the correct integer is equal to $a \bmod n$, where a is *any* element of the congruence class. The command modp can be used as shown to determine the congruence class in this way:

```
[ > modp(3+52, 9);
[ >
```

Things are fine so far, but suppose that we selected elements other than 3 and 52 from their respective congruence classes? After all, the recipe for addition says that we can use any elements from each class to do the addition. Will this work? Let's try some examples. From our table above, we can see that −15 and 66 are in the same class as 3 and that 7 and 79 are in the same class as 52. (In the terminology introduced in the Pre-lab section: "−15 and 66 are congruent to 3 mod 9" and "7 and 79 are congruent to 52 mod 9".) Here are the computations with 3 replaced by −15 and 66 and 52 replaced by 7 and 79:

```
[ > modp(-15+7, 9);
[ > modp(-15+79, 9);
[ > modp(66+7, 9);
[ > modp(66+79, 9);
[ >
```

Here's some fancier code that will check a bunch of choices from each congruence class at the same time:

```
[ > seq(seq(modp((3+9*i)+(52+9*j), 9), i=-5..5), j= -5..5);
[ >
```

This looks promising. For every selection we have made, the sum lands in the same congruence class. These examples illustrate the principle that addition of congruence classes is "well defined". Multiplication of congruence classes behaves in a similar manner. The recipe for multiplication is similar to that for addition: to multiply two congruence classes, we select any elements a and b from each of the classes and multiply them together. The product of the congruence classes is then defined to be the congruence class containing the product $a\,b$. We modify the above examples to illustrate multiplication of congruence classes.

```
[ > modp(3*52, 9);
[ > modp(-15*7, 9);
[ > modp(-15*79, 9);
[ > modp(66*7, 9);
[ > modp(66*79, 9);

[ > seq(seq(modp((3+9*i)*(52+9*j), 9), i= -5..5),j= -5..5);
[ >
```

Try some other examples, and when you feel comfortable with what is going on, move on to the research question and formalize your observations.

■ Research Question 2

State and prove a theorem which shows that arithmetic of congruence classes modulo n is well defined.

We now know that the arithmetic of congruence classes is well defined. You may be thinking "So what? What's in it for me?" A reasonable question. A significant part of this course involves studying the arithmetic of congruences, and the ability to use any member of a congruence class to perform computations can frequently be a big help. Here's an example that isn't too exciting, but does illustrate the point. Suppose that you wish to find the value of n % 23894857635998476, where

$$n = 23894857635998475^{4578}.$$

(Remember, we said this wouldn't be exciting. We'll get to the cool examples soon enough.) One way to proceed would be to do the exponentiation, divide by n, and find the remainder. This will work in principle, but is not at all practical (n has over 74,000 digits.) A much faster way to go is to observe that

$$23894857635998475 \equiv -1 \pmod{23894857635998476},$$

so that

$$23894857635998475^{4578} \equiv (-1)^{4578} \pmod{23894857635998476}.$$

It's easy to see that $(-1)^{4578} = 1$, and hence

$$23894857635998475^{4578} \equiv 1 \pmod{23894857635998476}.$$

Wasn't that much easier?

We close this section with a comment on the implications of what we have just seen. In moving from the set of integers to congruences modulo n, we go from an infinite set to a set with just n elements, the n

congruence classes. Concretely, we can think of the n congruence classes as being represented by the possible remainders for division by n: $0, 1, \ldots, n-1$. We can do addition, subtraction, and multiplication with this set. Since arithmetic for these operations is well defined for congruences, we can be somewhat lax about where we reduce to remainders modulo n. Moreover, if we are judicious in choosing when to reduce modulo n, we can do some interesting things as we will see in upcoming chapters.

■ $\displaystyle\sum_{j=1}^{n} j \ \% \ n$

■ Warming Up

The first experiment will help test your feel for congruences. The following command will simply list the first 30 consecutive integers (try it):

```
[ > seq(j, j=1..30);
[ >
```

We can see the congruence classes of these numbers, say modulo 6, with the command below. **But first**, try to predict what you think the result will be before executing the command.

```
[ > seq(modp(j, 6),   j= 1..30);
[ >
```

Did you predict it correctly? Try changing the 6 to other positive integers and see if you can predict the results. Once you have this down cold, you're ready for Exercise 1 below.

■ *Exercise 1*

Identify the pattern when counting modulo n, and explain why this pattern occurs.

■ Taking Sums

Recall that the Pre-lab section ended with a question about the remainder upon division by n of the sum $1 + 2 + 3 + \ldots + n$. To get a feel for what we might expect, we begin by trying a few simple examples. (Unless the answer to a question is pretty obvious, it is almost always a good idea to try some easy examples.) First, sum(j, j = 1..n) is the Maple command to compute the sum of the first n integers. Try it:

```
[ > sum(j, j= 1..15);
[ >
```

We find the remainder upon division by 15 with the modp command.

```
[ > modp(sum(j, j= 1..15), 15);
[ >
```

The command seesum defined below automates this process.

```
[ > seesum := n -> modp(sum(j,j=1..n), n):
[ >
```

Here it is in action:

```
[ > seesum(15);
[ >
```

Try it for several different values of *n*. Can you find the pattern? You may find it helpful to see several values at once. For this, we can use Maple's seq command. You can try this by executing the next command.

```
[ > seq(seesum(n), n=1..3);
[ >
```

This shows us the first 3 values of *n*. Is the pattern clear? If not, or if you want to confirm your guess, change the 3 to a larger value and execute it again. Once you think you know what's going on, you're ready to tackle Research Question 3.

▧ Research Question 3

Let *n* be a positive integer. Find a simple formula for

$$(1 + 2 + 3 + ... + n) \bmod n.$$

▧ Additive Orders

A basic question in the study of congruences is the following:

Given an integer *a* and a positive integer *n*, which integers *m* satisfy $m\,a \equiv 0 \pmod{n}$?

The *additive order* of *a* modulo *n* is defined to be the smallest positive integer *m* that satisfies the congruence equation $m\,a \equiv 0 \pmod{n}$. In order to get a feel for the above question, what's the first thing that we do? Repeat three times: "Try some examples." Here's what we get if we compute $m\,a\ \%\ n$, with $n = 10$, $a = 6$, and values of *m* between 0 and 20:

```
> n := 10:
  a := 6:
  for m from 0 to 20 do
   printf('%3d * %d %% %d = %d\n', m, a, n, m*a mod n);
  od;
[ >
```

As we can see, for these values of *m* we have $m\,a \equiv 0 \pmod{n}$ for $m = 0, 5, 10, 15$, and 20. (Thus the additive order of 6 modulo 10 is equal to 5.) It is also clear that there is a bunch of extra information in the above output that we don't need. The command findkillers defined below provides "just the facts, ma'am." It takes specific values of *a* and *n* as input, computes $m\,a\ \%\ n$ with lots of integers *m*, and prints out those values of *m* such that $m\,a \equiv 0 \pmod{n}$.

```
> findkillers := proc(a, n)
     local m;
     for m from -2*n to 8*n do
       if modp(a*m, n) = 0 then printf('%1d ', m)
       fi;
     od;
  end:
```

Here is `findkillers` in action using the values of *a* and *n* from above:

```
[ > findkillers(6, 10);
```

Here it is for $a = 5$ and $n = 14$:

```
[ > findkillers(5, 14);
[ >
```

■ Research Question 4

If we keep *n* fixed and replace *a* by another integer *b* congruent to *a* (mod *n*), how will the output from `findkillers(a,n)` be related to the output of `findkillers(b,n)`?

■ Research Question 5

Find the form of all values of *m* that satisfy $m\,a \equiv 0 \pmod{n}$.

Mathematica Lab: Congruences

■ Congruence Classes

We begin this section by reviewing the three different ways of thinking about congruence classes that were discussed in the Prelab section.

■ By Counting

Portions of the congruence classes modulo n can be viewed using the command **congclasses[n]**, which is defined as follows:

```
congclasses[n_] :=
  Nicetable[Table[i + n j, {i, 0, n-1}, {j, -3, 10}]]
```

♥ Mathematica Note

The function **Nicetable** is not a standard Mathematica command. It is defined in the initialization cells of this notebook. It is designed to produce more compact output for some large tables, such as the one produced by **congclasses**.

Here's what we get when $n = 7$:

```
congclasses[7]
```

Keep in mind that the output only shows part of each congruence class. Try changing the 7 to some other positive integers n to see the congruence classes modulo n.

■ By Differences Within Rows

Referring to the above description of congruence classes, we see that each number is exactly n more than the number to its left (where n is the number of rows). By extension, if we pick any number r and look c columns to the right, we will have to add n to r a total of c times. Thus, c columns directly to the right of r is the number $r + c n$. Therefore it follows that two numbers are in the same row if they differ by a multiple of n. Using the congruence notation introduced in the Prelab section, our observation may be expressed as follows:

$$a \equiv b \pmod{n} \text{ if and only if } n \mid (b - a).$$

This characterization of congruence is extremely useful in proofs, since it brings everything known about divisibility into the picture and frequently reduces congruence problems to simple algebra.

■ By Remainders

In this section, we consider a third way to think about congruence classes. Recall that the Division Algorithm states that if any integer a is divided by a positive integer n, then the remainder r is always between 0 and $n - 1$. (Recall also our notation for the remainder: $r = a \% n$.) Mathematica has a built-in command for computing remainders: **Mod[a,n]** computes $a \% n$. For example, **Mod[27,4]** gives the remainder when 27 is divided by 4:

```
Mod[27, 4]
```

Can you predict the result of each of the following before executing it?

```
Mod[4, 27]
```

```
Mod[12, 12]
```

```
Mod[-1, 6]
```

```
Mod[-1, 176]
```

The command **modcongclasses[n]** produces output similar to **congclasses[n]**, but with each entry replaced by its remainder when divided by n. Here is the definition for **modcongclasses**:

```
modcongclasses[n_] := Nicetable[Table[Mod[i + n j, n],
                {i, 0, n - 1}, {j, -2, 10}]]
```

Here's **congclasses[7]**:

```
congclasses[7]
```

And here's **modcongclasses[7]**:

```
modcongclasses[7]
```

Quite a striking pattern, isn't it? Try changing 7 to some other positive integers.

■ Research Question 1

> Form a conjecture that explains the output of **modcongclasses[n]**.

■ Summing Up

Above we have considered three ways of looking at congruence modulo n. Each is useful in its own way. The first description is somewhat visual and gives a good intuitive feel for congruence classes. The description in terms of differences frequently works the best in proofs. The characterization in terms of remainders makes it easy to compute modulo n since there are only n integers (the remainders $0, 1, \ldots, n-1$) to keep track of. (There will be more on computing modulo n later in the lab.)

■ A Tip

Mathematicians often think about and work with a concept in one way, but write their proofs in a different way. When reading proofs in a book, all one typically sees is the proof written in the most efficient manner, while the author may prefer to think about the problem in other terms. When there are several ways of describing the same thing, do not feel limited to work with only one of them. Learn how to use them all, so you can move back and forth between them depending on the situation.

■ Well-Defined Arithmetic

As mentioned in the Prelab section, it is possible to do arithmetic with congruence classes. To add two congruence classes modulo n, we just select any element a from the first class and any element b from the second class, and then compute $a + b$ as we would for normal integers. The sum of the two congruence classes is then defined to be equal to the congruence class containing the "usual" sum $a + b$. For example, here are the congruence classes modulo 9:

```
congclasses[9]
```

Suppose that we wish to add the congruence class containing 3 to the congruence class containing 52. According to the recipe described above, we just select an element from each class, add them together, and see which class the resulting sum is in. Clearly 3 and 52 are elements of our classes, so let's try them first:

```
3 + 52
```

As we can see, the sum is in the congruence class containing 55, which corresponds to the second row above. To simplify discussions involving congruence classes, it is helpful to specify the congruence classes by identifying them with the unique integer from the set $\{0, 1, 2, \ldots, n-1\}$ contained in a given class. Of course, from your work above you know that the correct integer is equal to $a \% n$, where a is *any* element of the congruence class. The command **Mod** can be used as shown to determine the congruence class in this way:

```
Mod[3 + 52, 9]
```

Things are fine so far, but suppose that we selected elements other than 3 and 52 from their respective congruence classes? After all, the recipe for addition says that we can use *any* elements from each class to do the addition. Will this work? Let's try some examples. From our table above, we can see that -15 and 66 are in the same class as 3 and that 7 and 79 are in the same class as 52. (In the terminology introduced in the Prelab section: "-15 and 66 are congruent to 3 mod 9" and "7 and 79 are congruent to 52 mod 9".) Here are the computations with 3 replaced by -15 and 66 and 52 replaced by 7 and 79:

```
Mod[-15 + 7, 9]
Mod[-15 + 79, 9]
Mod[66 + 7, 9]
Mod[66 + 79, 9]
```

Here's some fancier code that will check a bunch of choices from each congruence class at the same time:

```
Table[Mod[(3 + 9 i) + (52 + 9 j), 9], {i, -5, 5}, {j, -5, 5}]
```

This looks promising. For every selection we have made, the sum lands in the same congruence class. These examples illustrate the principle that addition of congruence classes is well defined. Multiplication of congruence classes behaves in a similar manner. The recipe for multiplication is similar to that for addition: to multiply two congruence classes, we select any elements a and b from each of the classes and multiply them together. The product of the congruence classes is then defined to be the congruence class containing the product $a\,b$. We modify the above examples to illustrate multiplication of congruence classes.

```
Mod[3 * 52, 9]
```

```
Mod[-15 * 7, 9]
Mod[-15 * 79, 9]
Mod[66 * 7, 9]
Mod[66 * 79, 9]
```

```
Table[Mod[(3 + 9 i) * (52 + 9 j), 9], {i, -5, 5}, {j, -5, 5}]
```

Try some other examples, and when you feel comfortable with what is going on, move on to the research question and formalize your observations.

■ Research Question 2

> State and prove a theorem which shows that arithmetic of congruence classes modulo n is well defined.

We now know that the arithmetic of congruence classes is well defined. You may be thinking, "So what? What's in it for me?" A reasonable question. A significant part of this course involves studying the arithmetic of congruences, and the ability to use any member of a congruence class to perform computations can frequently be a big help. Here's an example that isn't too exciting, but does illustrate the point. Suppose that you wish to find the value of n % 23894857635998476, where

$$n = 23894857635998475^{4578}.$$

(Remember, we said this wouldn't be exciting. We'll get to the cool examples soon enough.) One way to proceed would be to do the exponentiation, divide by 23894857635998476, and find the remainder. This will work in principle, but is not at all practical (n has over 74,000 digits.) A much faster way to go is to observe that

$$23894857635998475 \equiv -1 \pmod{23894857635998476},$$

so that

$$23894857635998475^{4578} \equiv (-1)^{4578} \pmod{23894857635998476}.$$

It's easy to see that $(-1)^{4578} = 1$, and hence

$$23894857635998475^{4578} \equiv 1 \pmod{23894857635998476}.$$

Wasn't that much easier?

We close this section with a comment on the implications of what we have just seen. In moving from the set of integers to congruences modulo n, we go from an infinite set to a set with just n elements, the n congruence classes. Concretely, we can think of the n congruence classes as being represented by the possible remainders for division by n: 0, 1, ..., $n - 1$. We can do addition, subtraction, and multiplication with this set. Since arithmetic for these operations is well defined for congruences, we can be somewhat lax about where we reduce to remainders modulo n. Moreover, if we are judicious in choosing when to reduce modulo n, we can do some interesting things as we will see in upcoming chapters.

■ $\sum_{j=1}^{n} j$ % n

■ Warming Up

The first experiment will help test your feel for congruences. The following command will simply list the first 30 consecutive integers (try it):

```
Table[j, {j, 30}]
```

We can see the congruence classes of these numbers, say modulo 6, with the command below. **But first**, try to predict what you think the result will be before executing the command.

```
Table[Mod[j, 6], {j, 30}]
```

Did you predict it correctly? Try changing the 6 to other positive integers and see if you can predict the results. Once you have this down cold, you're ready for Exercise 1 below.

■ Exercise 1

Identify the pattern when counting modulo n, and explain why this pattern occurs.

■ Taking Sums

We now consider the question of determining the remainder of the sum $1 + 2 + 3 + \cdots + n$ upon division by n. To get a feel for what we might expect, we begin by trying a few simple examples. (Unless the answer to a question is pretty obvious, it is almost always a good idea to try some easy examples.) First, **Sum[j,{j,n}]** is the Mathematica command to compute the sum of the first n integers. Try it:

```
Sum[j, {j, 15}]
```

We find the remainder upon division by 15 with the **Mod** command.

```
Mod[Sum[j, {j, 15}], 15]
```

The command **seesum** defined below automates this process.

```
seesum[n_] := Mod[Sum[j, {j, n}], n]
```

Here it is in action:

```
seesum[15]
```

Try it for several different values of n. Can you find the pattern?

You may find it helpful to see several values at once. For this, we can use Mathematica's **Table** command. You can try this by executing the next cell.

```
Table[seesum[n], {n, 3}]
```

This shows us the first 3 values of n. Is the pattern clear? If not, or if you want to confirm your guess, change the **3** to a larger value and execute it again. Once you think you know what's going on, you're ready to tackle Research Question 3.

■ Research Question 3

> Let n be a positive integer. Find a simple formula for
>
> $$(1 + 2 + 3 + \cdots + n) \; \% \; n.$$

■ Additive Orders

A basic question in the study of congruences is the following:

> Given an integer a and a positive integer n, which integers m satisfy $m\,a \equiv 0 \pmod n$?

The *additive order* of a modulo n is defined to be the smallest positive integer m that satisfies the congruence equation $m\,a \equiv 0 \pmod n$. In order to get a feel for the above question, what's the first thing that we do? Repeat three times: "Try some examples." Here's what we get if we compute $m\,a \% n$ with $n = 10$, $a = 6$, and values of m between 0 and 20:

```
n = 10;
a = 6;
Do[Print[m, " * ", a, " % ", n, " = ", Mod[m * a, n]], {m, 0, 20}]
```

As we can see, for these values of m we have $m\,a \equiv 0 \pmod n$ for $m = 0, 5, 10, 15,$ and 20. (Thus the additive order of 6 modulo 10 is equal to 5.) It is also clear that there is a bunch of extra information in the above output that we don't need. The command **findkillers** defined below provides "just the facts, ma'am". It takes specific values of *a* and *n* as input, computes *m a* % *n* with lots of integers *m*, and then makes a list of those values of *m* such that *m a* ≡ 0 (mod *n*).

```
findkillers[a_, n_] := Select[Range[ -2 n, 8 n], Mod[a #, n] == 0 &]
```

♡ Mathematica Note

Here's how **findkillers** works: The command **Range[-2n,8n]** makes a list of all the integers from **−2n** to **8n**. The **Select** command then takes each number in this list, plugs it in place of the **#** in the equation **Mod[a #, n] == 0**, and picks out those values that satisfy the equation. The **&** at the end of the equation tells Mathematica that this is a pure function. If you want to know more about these commands, check out the Mathematica manual or the on-line help.

Here is **findkillers** in action using the values of *a* and *n* from above:

```
findkillers[6, 10]
```

Here it is for *a* = 5 and *n* = 14:

```
findkillers[5, 14]
```

■ Research Question 4

> If we keep *n* fixed and replace *a* by another integer *b* congruent to *a* (mod *n*), how will the output from **findkillers[a,n]** be related to the output of **findkillers[b,n]**?

■ Research Question 5

> Find the form of all values of *m* that satisfy *m a* ≡ 0 (mod *n*).

Web Lab: Congruences

3.1 Congruence Classes

We begin this section by reviewing the three different ways of thinking about congruence classes that were discussed in the Prelab section.

3.1.1 By Counting

Portions of the congruence classes modulo n can be viewed using the applet below. For example, here's what we get when $n = 7$:

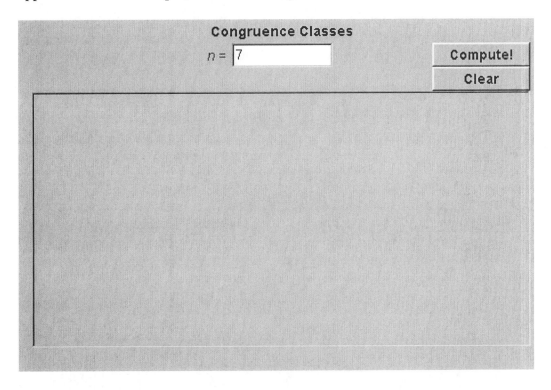

Keep in mind that the output only shows part of each congruence class. Try changing the 7 to some other positive integers n to see the congruence classes modulo n.

3.1.2 By Differences Within Rows

Referring to the above description of congruence classes, we see that each number is exactly n more than the number to its left (where n is the number of rows). By extension, if we pick any number r and look c columns to the right, we will have to add n to r a total of c times. Thus, c columns directly to the right of r is the number $r + cn$. Therefore it follows that two numbers are in the same row if they differ by a multiple of n. Using the congruence notation introduced in the Prelab section, our observation may be expressed as follows:

$$a \equiv b \pmod{n} \text{ if and only if } n \mid (b - a).$$

This characterization of congruence is extremely useful in proofs, since it brings everything known about divisibility into the picture and frequently reduces congruence problems to simple algebra.

3.1.3 By Remainders

In this section, we consider a third way to think about congruence classes. Recall that the Division Algorithm states that if any integer a is divided by a positive integer n, then the remainder r is always between 0 and $n - 1$. (Recall also our notation for the remainder: $r = a \% n$.) The on-line calculator can be used to compute remainders. To get the remainder when a is divided by n, we just type in "a % n". The "%" symbol is used by many computer languages to denote the remainder operation. For example, 27 % 4 gives the remainder when 27 is divided by 4:

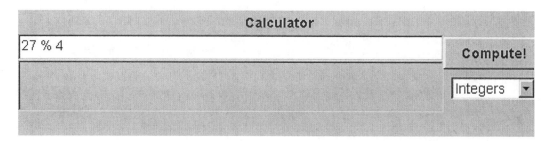

Can you predict the result of each of the following before executing it?

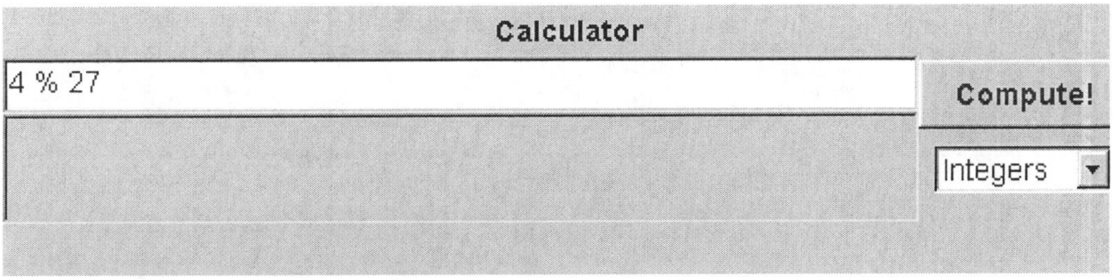

Calculator

| 4 % 27 | **Compute!** |
| | Integers ▾ |

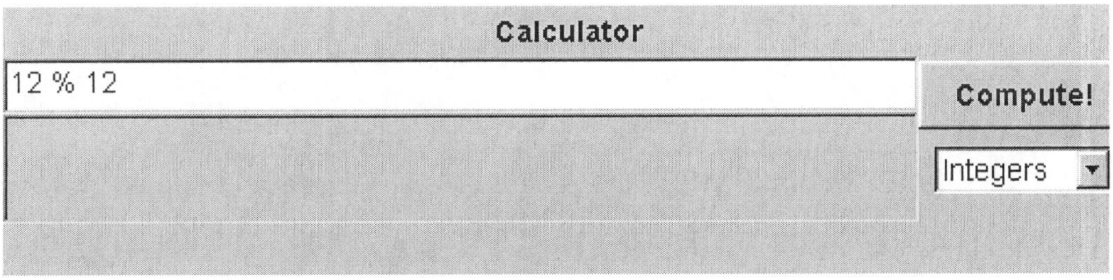

Calculator

| 12 % 12 | **Compute!** |
| | Integers ▾ |

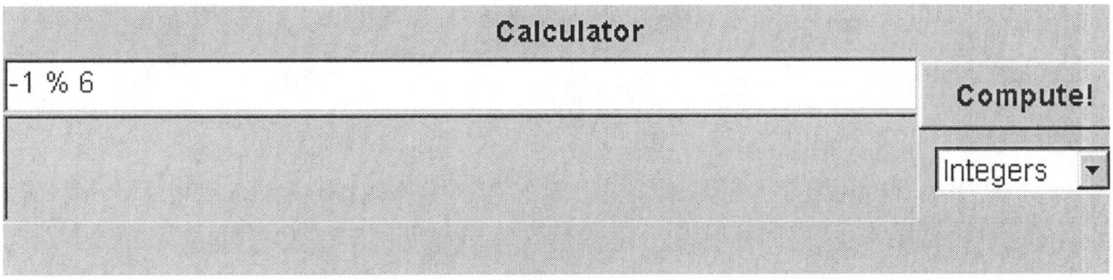

Calculator

| -1 % 6 | **Compute!** |
| | Integers ▾ |

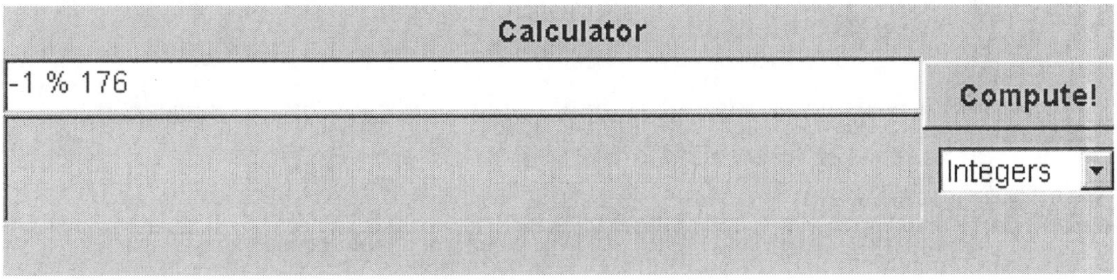

Calculator

| -1 % 176 | **Compute!** |
| | Integers ▾ |

The next applet produces output similar to "Congruence Classes" above. The difference is that each entry is replaced by its remainder when divided by n. Here's what we get when $n = 7$:

Quite a striking pattern, isn't it? Try changing 7 to some other positive integers.

Research Question 1

Form a conjecture that explains the output of "Congruence Classes modulo n."

3.1.4 Summing Up

Above we have considered three ways of looking at congruence modulo n. Each is useful in its own way. The first description is somewhat visual and gives a good intuitive feel for congruence classes. The description in terms of differences frequently works the best in proofs. The characterization in terms of remainders makes it easy to compute modulo n since there are only n integers (the remainders $0, 1, \ldots, n-1$) to keep track of. (There will be more on computing modulo n later in the lab.)

3.1.5 A Tip

Mathematicians often think about and work with a concept in one way, but write their proofs in a different way. When reading proofs in a book, all one typically sees is the proof written in the most efficient manner, while the author may prefer to think about the problem in other terms. When there are several ways of describing the same thing, do not feel limited to work with only one of them. Learn how to use them all, so you can move back and forth between them depending on the situation.

3.2 Well-Defined Arithmetic

As mentioned in the Prelab section, it is possible to do arithmetic with congruence classes. To add two congruence classes modulo n, we just select any element a from the first class and any element b from the second class, and then compute $a + b$ as we would for normal integers. The sum of the two congruence classes is then defined to be equal to the congruence class containing the "usual" sum $a + b$. For example, here are the congruence classes modulo 9:

Congruence Classes

$n =$ 9   Compute!

Clear

Suppose that we wish to add the congruence class containing 3 to the congruence class containing 52. According to the recipe described above, we just select an element from each class, add them together, and see which class the resulting sum is in. Clearly 3 and 52 are elements of our classes, so let's try them first:

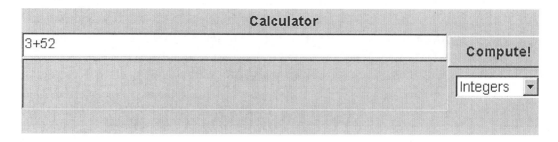

As we can see, the sum is in the congruence class containing 55, which corresponds to the second row above. To simplify discussions involving congruence classes, it is helpful to specify the congruence classes by identifying them with the unique integer from the set $\{0, 1, 2, \ldots, n-1\}$ contained in a given class. Of course, from your work in the previous section, you know that the correct integer is equal to $a \% n$, where a is *any* element of the congruence class. The on-line calculator can be used as shown to determine the congruence class in this way:

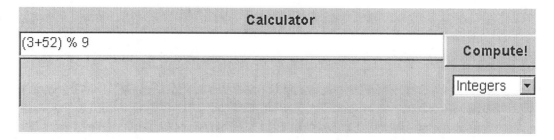

Things are fine so far, but suppose that we selected elements other than 3 and 52 from their respective congruence classes? After all, the recipe for addition says that we can use *any* elements from each class to do the addition. Will this work? Let's try some examples. From our table above, we can see that –15 and 66 are in the same class as 3 and that 7 and 79 are in the same class as 52. (In the terminology introduced in the Prelab section: "–15 and 66 are congruent to 3 mod 9" and "7 and 79 are congruent to 52 mod 9.") Here are the computations with 3 replaced by –15 and 66 and 52 replaced by 7 and 79:

```
                          Calculator
(-15+7) % 9                                        Compute!

                                                   Integers  ▾
```

```
                          Calculator
(-15+79) % 9                                       Compute!

                                                   Integers  ▾
```

```
                          Calculator
(66+7) % 9                                         Compute!

                                                   Integers  ▾
```

```
                          Calculator
(66+79) % 9                                        Compute!

                                                   Integers  ▾
```

Below is an applet that will check a bunch of choices from each congruence class at the same time:

```
        a = [3            ]      b = [52           ]   Compute!
        n = [9            ]      M = [6            ]     Clear
```

This looks promising. For every selection we have made, the sum lands in the same congruence class. These examples illustrate the principle that addition of congruence classes is well defined.

Multiplication of congruence classes behaves in a similar manner. The recipe for multiplication is similar to that for addition: to multiply two congruence classes, we select any elements a and b from each of the classes and multiply them together. The product of the congruence classes is then defined to be the congruence class containing the product ab. We modify the above examples to illustrate multiplication of congruence classes.

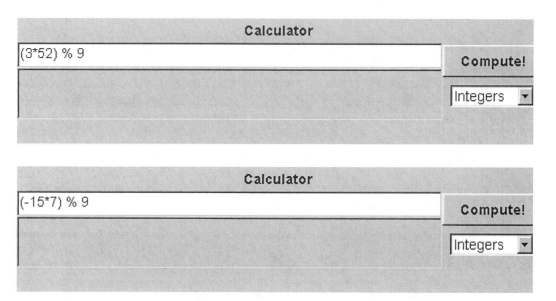

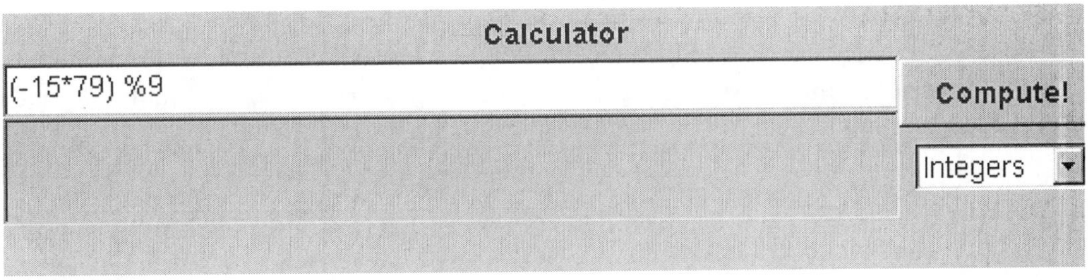

Calculator

(-15*79) %9

Compute!

Integers ▾

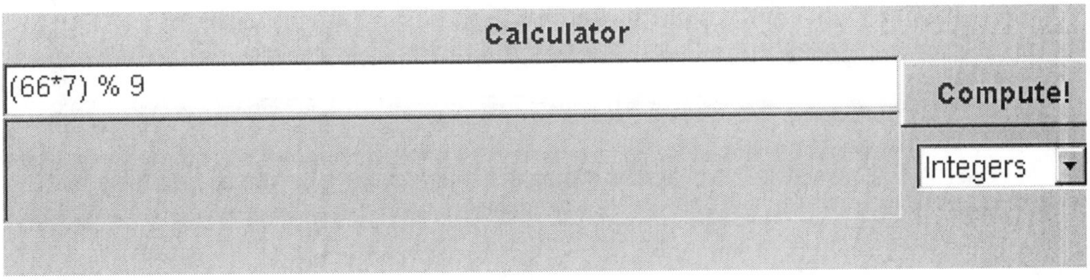

Calculator

(66*7) % 9

Compute!

Integers ▾

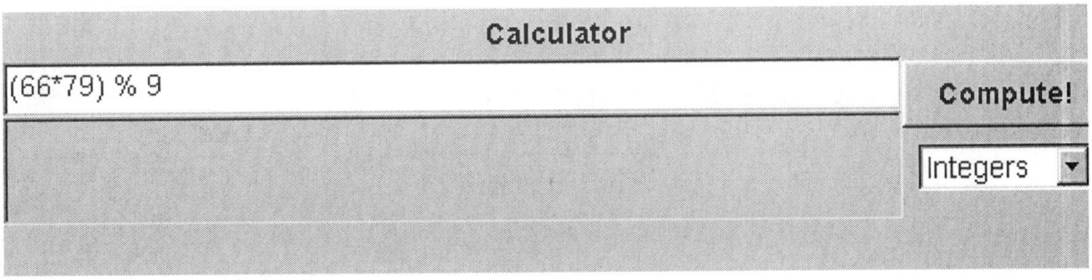

Calculator

(66*79) % 9

Compute!

Integers ▾

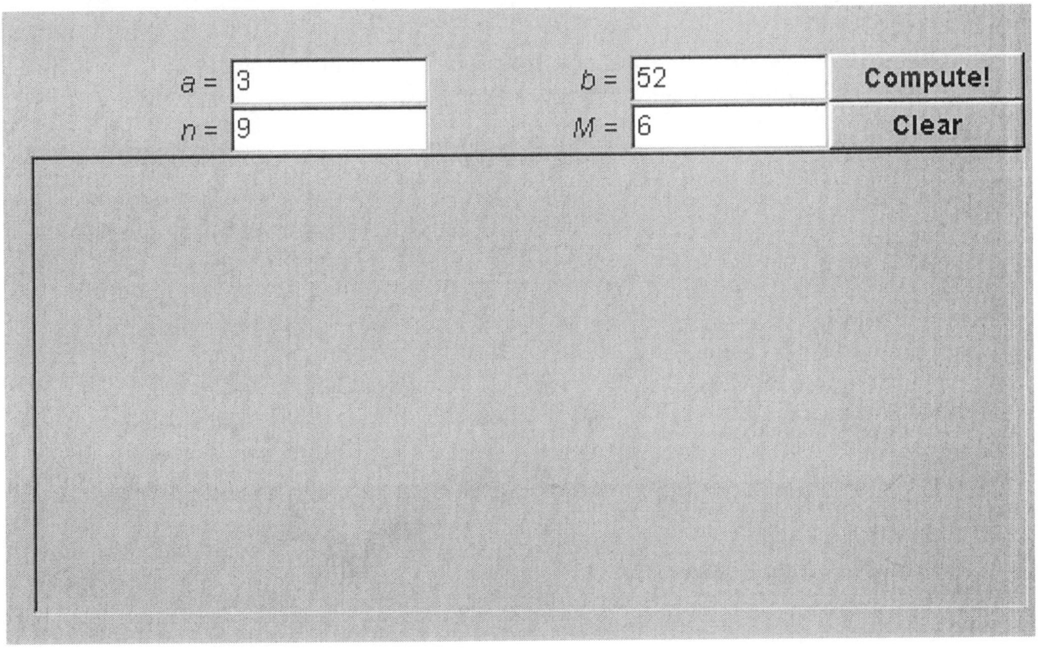

$a =$ 3 $b =$ 52   Compute!

$n =$ 9 $M =$ 6   Clear

Try some other examples, and when you feel comfortable with what is going on, move on to the research question and formalize your observations.

Research Question 2

State and prove a theorem which shows that arithmetic of congruence classes modulo n is well defined.

We now know that the arithmetic of congruence classes is well defined. You may be thinking, "So what? What's in it for me?" A reasonable question. A significant part of this course involves studying the arithmetic of congruences, and the ability to use any member of a congruence class to perform computations can frequently be a big help. Here's an example that isn't too exciting, but does illustrate the point. Suppose that you wish to find the value of n % 23894857635998476, where

$$n = 23894857635998475^{4578}.$$

(Remember, we said this wouldn't be exciting. We'll get to the cool examples soon enough.) One way to proceed would be to do the exponentiation, divide by 23894857635998476, and find the remainder. This will work in principle, but is not at all practical (n has over 74,000 digits.) A much faster way to go is to observe that

$$23894857635998475 \equiv -1 \pmod{23894857635998476},$$

so that

$$23894857635998475^{4578} \equiv (-1)^{4578} \pmod{23894857635998476},$$

It's easy to see that $(-1)^{4578} = 1$, and hence

$$23894857635998475^{4578} \equiv 1 \pmod{23894857635998476},$$

Wasn't that much easier?

We close this section with a comment on the implications of what we have just seen. In moving from the set of integers to congruences modulo n, we go from an infinite set to a set with just n elements, the n congruence classes. Concretely, we can think of the n congruence classes as being represented by the possible remainders for division by n: $0, 1, \ldots, n - 1$. We can do addition, subtraction, and multiplication with this set. Since arithmetic for these operations is well defined for congruences, we can be somewhat lax about where we reduce to remainders modulo n. Moreover, if we are judicious in choosing when to reduce modulo n, we can do some interesting things as we will see in upcoming chapters.

$$3.3 \quad \sum_{j=1}^{n} j \,\% \, n$$

3.3.1 Warming Up

The first experiment will help test your feel for congruences. Below are the first 30 positive integers:

$$1, 2, 3, 4, 5, 6, 7, 8, 9, 10, 11, 12, 13, 14, 15, 16, 17, 18, 19, 20, 21, 22, 23,$$
$$24, 25, 26, 27, 28, 29, 30$$

We can see the congruence classes of these numbers, say modulo 6, with the applet below. **But first**, try to predict what you think the result will be before executing the command.

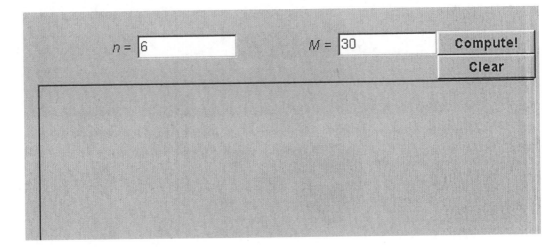

Did you predict it correctly? Try changing the 6 to other positive integers and see if you can predict the results. Once you have this down cold, you're ready for Exercise 1 below.

Exercise 1

Identify the pattern when counting modulo n, and explain why this pattern occurs.

3.3.2 Taking Sums

We now consider the question of determining the remainder of the sum $1 + 2 + 3 + \cdots + n$ upon division by n. To get a feel for what we might expect, we begin by trying a few simple examples. (Unless the answer to a question is pretty obvious, it is almost always a good idea to try some easy examples.) The applet below takes a positive integer n as input, and produces two pieces of output: the sum $1 + 2 + 3 + \cdots + n$, and the same sum reduced modulo n.

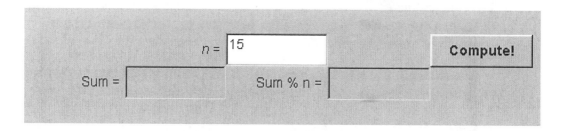

Try it for several different values of n. Can you find the pattern?

You may find it helpful to see several values at once. The applet below takes an integer n as input. The output is a list of the values of

$$1 + 2 + 3 + \cdots + M \pmod{M}$$

from $M = 1$ up to $M = n$. Here's what we get with $M = 3$:

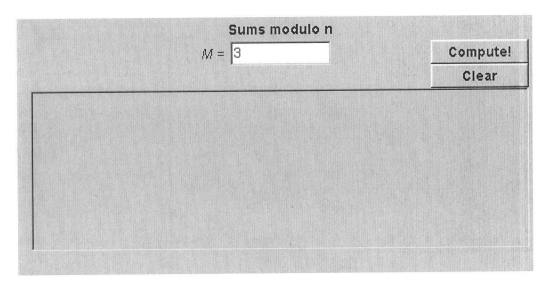

Is the pattern clear? If not, or if you want to confirm your guess, change the 3 to a larger value and execute the applet again. Once you think you know what's going on, you're ready to tackle Research Question 3.

Research Question 3

Let n be a positive integer. Find a simple formula for

$$(1 + 2 + 3 + \cdots + n) \% n.$$

3.4 Additive Orders

A basic question in the study of congruences is the following:

Given an integer a and a positive integer n, which integers m satisfy

$$ma \equiv 0 \pmod{n}?$$

The *additive order* of *a* modulo *n* is defined to be the smallest positive integer *m* that satisfies the congruence equation $ma \equiv 0 \pmod{n}$. In order to get a feel for the above question, what's the first thing that we do? Repeat three times: "Try some examples." Here's what we get if we compute *ma* % *n* with *n* = 10, *a* = 6, and values of *m* between 1 and 20:

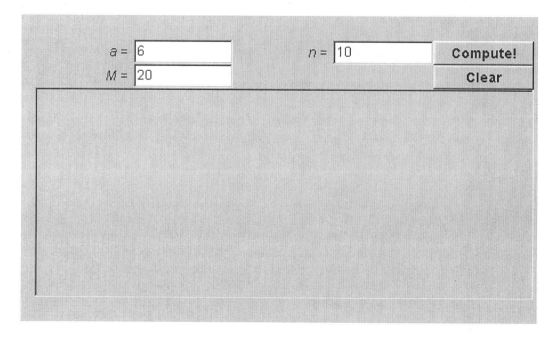

As we can see, for these values of *m* we have $ma \equiv 0 \pmod{n}$ for *m* = 5, 10, 15, and 20. (Thus the additive order of 6 modulo 10 is equal to 5.) It is also clear that there is a bunch of extra information in the above output that we don't need. The applet below provides "just the facts, ma'am." It takes specific values of *a* and *n* as input, computes *ma* % *n* with lots of integers *m*, and then makes a list of those values of *m* such that $ma \equiv 0 \pmod{n}$ Here it is in action using the values of *a* and *n* from above:

Here's what we get for $a = 5$ and $n = 14$:

Find Killers

$a =$ ⌷5 $n =$ ⌷14   Compute!

Clear

Research Question 4

If we keep n fixed and replace a by another integer b congruent to a (mod n), how will the output from the preceding applet change?

Research Question 5

Find the form of all values of m that satisfy $ma \equiv 0 \pmod{n}$.

Homework

1. Find all values a such that $0 \le a \le 33$ and $a \equiv 307 \pmod{17}$.

2. Find all values a such that $-20 \le a \le 50$ and $a \equiv 971 \pmod{23}$.

3. Find all values of n such that $237 \equiv 432 \pmod{n}$.

4. Find all values of n such that $-12 \equiv 22 \pmod{n}$.

5. If $a \equiv 7 \pmod{13}$, then what is $(5a - 3) \% 13$?

6. If $a \equiv -2 \pmod{17}$, then what is $(9 - 13a) \% 17$?

7. Show that every integer is congruent modulo 13 to a multiple of 5.

8. Prove or disprove: If $a \equiv b \pmod{n}$ and $b \equiv c \pmod{n}$, then $a \equiv c \pmod{n}$.

9. Prove or disprove: If $ac \equiv bc \pmod{n}$, then $a \equiv b \pmod{n}$.

10. Prove or disprove: If $a \equiv b \pmod{n}$, then $a^2 \equiv b^2 \pmod{n}$.

11. Prove or disprove: Suppose that $a^2 \equiv b^2 \pmod{p}$, where p is prime. Then either $a \equiv b \pmod{p}$ or $a \equiv -b \pmod{p}$.

12. Prove or disprove: Suppose that $a^2 \equiv b^2 \pmod{n}$, where n is composite. Then either $a \equiv b \pmod{n}$ or $a \equiv -b \pmod{n}$.

13. Prove or disprove: If $a \equiv b \pmod{d}$ and $d \mid n$, then $a \equiv b \pmod{n}$

14. Prove or disprove: If $a \equiv b \pmod{n}$ and $d \mid n$, then $a \equiv b \pmod{d}$

15. Show that if $a \equiv b \pmod{n}$, then $\gcd(a, n) = \gcd(b, n)$.

16. Suppose that $\gcd(a, 6) = 1$. What are the possible values of $a^2 \% 24$?

17. Suppose that $\gcd(a, 10) = 1$. What are the possible values of $a^2 \% 40$?

18. Let $d_2(k)$ denote the last two digits of 99^k for $k \ge 0$. Find a formula for $d_2(k)$.

19. What are the possible values of $(m^2 + n^2) \% 4$?

20. What are the possible values of $(r^3 + s^3 + t^3) \% 9$?

21. Suppose that $\gcd(m, n) = 1$, that the additive order of $a \pmod{m}$ equals u, and that the additive order of $a \pmod{n}$ equals v. What is the additive order of $a \pmod{mn}$?

22. Let $d_1(n)$ denote the last digit of n^2. What are the possible values of $d_1(n)$?

23. Suppose that p is a prime. What are the possible values of $p \% 6$?

24. For each value of a and n below, find the additive order of a (mod n).

    (a) $a = 51$, $n = 78$
    (b) $a = 41$, $n = 164$
    (c) $a = 60$, $n = 169$

25. For each value of d and n below, find all distinct a (mod n) such that a has additive order d modulo n.

    (a) $d = 6$, $n = 18$
    (b) $d = 8$, $n = 36$
    (c) $d = 9$, $n = 117$

26. (a) Find all possible additive orders for a (mod 52).
    (b) For each of the orders d in part (a), find one element a (mod 52) of order d.

27. (a) Find all possible additive orders for a (mod 78).
    (b) For each of the orders d in part (a), find one element a (mod 78) of order d.

4. Applications of Congruences

Prelab

Now that we are familiar with the notion of congruences mod n, we will consider some naturally occurring applications of congruences.

Periodic Sequences

In our first introduction to congruences, we considered the sequence generated by looking at the time of day every hour (on the hour). If we start looking at 1:00, the result would be the sequence

$$1, 2, 3, 4, 5, 6, 7, 8, 9, 10, 11, 12, 1, 2, 3, 4, 5, 6, 7, 8, 9, 10, 11, 12, \ldots .$$

Naturally, this sequence repeats. By recognizing that it repeats, we can easily extrapolate from the repeated pattern to determine values of the sequence. For example, to find the 10000th term of the sequence we note that $10000 = 12 \cdot 833 + 4$. Since the sequence repeats every 12 terms, the 10000th term will be the same as the 4th term. Thus the 10000th term of the sequence is 4.

Periodic sequences arise in many other contexts, and congruences give a good way of managing them. First, we fix some terminology. A sequence of numbers is typically written a_1, a_2, a_3, ..., and we will often abbreviate this by referring to simply the sequence a_i.

Definition A sequence a_i is *purely periodic* if there is an integer $P \neq 0$ such that $a_i = a_{i+P}$ for all i.

Naturally, if $a_i = a_{i+P}$ for all i, then

$$a_i = a_{i+P} = a_{(i+P)+P} = a_{i+2P}.$$

Repeating this process, we find that $a_i = a_{i+kP}$ for all i and all $k > 0$. As you might guess from this, the letter P in the definition above stands for *period*. Here is a formal definition:

Definition If a sequence a_i is purely periodic, then an integer P is a *period* for the sequence if $a_i = a_{i+P}$ for all i.

The presence of a period makes it easy to describe terms of a purely periodic sequence in terms of congruences.

189

1. **Suppose that P is a period for a purely periodic sequence a_i. Prove that if $j \equiv k \pmod{P}$, then $a_j = a_k$.**

We now consider the notion of a period more closely. Consider the sequence -1, 1, -1, 1, -1, 1, $\ldots$. A formula for producing this sequence is $a_i = (-1)^i$. Certainly, 2 is a period for this sequence, but it is not the only period according to our definition. The integers 4, 6, and 8 are other examples of periods. To show that 6 is a period, we would compute $a_{i+6} = (-1)^{i+6} = (-1)^i(-1)^6 = (-1)^i = a_i$. So, 6 is a period.

In this example there may be many periods but one of them is special, namely 2. One way to single it out is by the following definition.

Definition If a_i is a purely periodic sequence, the smallest positive period for a_i is called its *minimal period*.

There is a simple relationship between the minimal period of a sequence and the other periods. Part of this relationship is given in the following exercise:

2. **Suppose that a_i is a purely periodic sequence with minimal period m. Prove that if $P > 0$ and $m \mid P$, then P is a period for a_i.**

The converse to this statement is also true; verification of the converse appears as a homework exercise. Thus we have a complete characterization of all periods in terms of the minimal period:

Proposition *Suppose that a_i is a purely periodic sequence with minimal period m, and that $P > 0$. Then P is a period for a_i if and only if $m \mid P$.*

We have seen this relationship between periods and minimal periods before. If n is a positive integer and a is an integer, we can consider the sequence $(1 \cdot a) \% n$, $(2 \cdot a) \% n$, $(3 \cdot a) \% n$, $\ldots$. In other words, $a_i = (i \cdot a) \% n$. This sequence is purely periodic since $a_{i+n} = ((i+n) \cdot a) \% n$, and $i + n \equiv i \pmod{n}$. So, $a_{i+n} = a_i$, and the definition of purely periodic is satisfied with $P = n$. Moreover, this proves that n is a period for this sequence. Is it always the minimal period? The first step towards answering this question is to answer the related question stated below.

3. **Prove that the periods for the sequence $a_i = (i \cdot a) \% n$ are equal to the positive integers m such that $am \equiv 0 \pmod{n}$.**

As a consequence of this result, the minimal period for our sequence is the smallest positive integer m such that $am \equiv 0 \pmod{n}$. But this is exactly the *additive order* of a modulo n, as defined in Chapter 3. We found that all solutions to $am \equiv 0 \pmod{n}$ are given by the multiples of the additive order of a modulo n. So our previous result that the solutions to $am \equiv 0 \pmod{n}$ are the multiples of the additive order of a modulo n is a special case of the phenomenon that the periods of periodic sequence are the multiples of the minimal period.

There is one more variation on the idea of periodic sequence which we would like to consider. There are times when periodic sequences arise, but they are not purely periodic. For example, consider the decimal digits for the number 3/44.

$$3/44 = 0.06\overline{81} \approx 0.068181818181818181818181818181\ldots.$$

The digits eventually repeat with minimal period 2, but the repeating block does not start right away. To distinguish this situation, we call this sequence *ultimately* periodic instead of *purely* periodic. Here is the formal definition.

Definition A sequence a_i is *ultimately periodic* if there are integers N and P such that $a_{i+P} = a_i$ for all $i > N$.

Note the role of N in this definition; if we only look after the Nth term we will see a periodic sequence. The analogous definitions for *period* and *minimal period* apply here: they are just what you would expect but they only apply to the repeating part. In the decimal expansion for 3/44 above, we could take $N = 2$. The first two terms of this sequence, $a_1 = 0$ and $a_2 = 6$ are called the *preperiod* because they are the terms which occur before the repeating block begins.

Speedy Calculations

Using the arithmetic of congruences can greatly simplify some computations. Below are two examples.

Fast Powering If we want to find the remainder of a number when divided by 5, we can work with congruences modulo 5. Let's compute the remainder when 2^{1000} is divided by 5. We can start by considering the sequence $a_i = 2^i \% 5$:

$$2, 4, 3, 1, 2, 4, 3, 1, 2, 4, 3, 1, 2, 4, 3, 1, 2, 4, 3, 1, \ldots.$$

It looks like a periodic sequence. We are interested in the 1000th term, so we can extrapolate from the pattern. The sequence repeats with minimal period 4, so the 1000th term will be the same as the 4th term. The 4th term is 1, so the 1000th term is 1.

In the previous paragraph, we assumed that the sequence was periodic. Now we will deduce the same result without making any assumptions.[1] Obviously, $2^2 = 4$ and has remainder 4 when divided by 5. Now

$$2^4 = (2^2)^2 = 4^2 = 16 \equiv 1 \pmod{5}.$$

---

[1]Later in the course we will prove that this sequence is periodic and be able to reason as above in a completely rigorous way.

Therefore, $2^8 = (2^4)^2 \equiv 1^2 \pmod 5$. Here we use the fact that arithmetic modulo 5 is well defined, which allows us to substitute 1 for 2^4 because they are congruent mod 5. Since $1^2 \equiv 1 \pmod 5$, we can now finish easily:

$$2^{1000} = (2^4)^{250} \equiv 1^{250} \equiv 1 \pmod 5.$$

So, 2^{1000} has remainder 1 when divided by 5. Here we did the computation by hand, without having to do the much harder computation of first computing 2^{1000}, which has 302 digits!

4. **Use this method to compute the remainder when 3^{1000} is divided by 5. (If you first look for the period, it will make it easier to carry out the rigorous version of the computation.)**

5. **Use this method to compute the remainder when 3^{1001} is divided by 5.**

This approach can be useful for computing the remainders of large powers modulo n in some cases. For example, it is the basis for the standard divisibility tests discussed below.

Divisibility Tests In this section, we shall use congruences to deduce a well-known test for divisibility by 3. We start by considering the values of $10^n \pmod 3$. Clearly, since $10^0 = 1$, we have $10^0 \equiv 1 \pmod 3$. Furthermore, since $10 \equiv 1 \pmod 3$, it follows that for any positive integer n,

$$10^n \equiv 1^n \equiv 1 \pmod 3.$$

This observation is quite useful when we look at an arbitrary number in expanded notation:

$$
\begin{aligned}
6486073 &= 3 \cdot 10^0 + 7 \cdot 10^1 + 0 \cdot 10^2 + 6 \cdot 10^3 + 8 \cdot 10^4 + 4 \cdot 10^5 + 6 \cdot 10^6 \\
&\equiv 3 \cdot 1 \ + 7 \cdot 1 \ + 0 \cdot 1 \ + 6 \cdot 1 \ + 8 \cdot 1 \ + 4 \cdot 1 \ + 6 \cdot 1 \quad \pmod 3 \\
&\equiv 3 \ \ \ + 7 \ \ \ + 0 \ \ \ + 6 \ \ \ + 8 \ \ \ + 4 \ \ \ + 6 \quad \pmod 3
\end{aligned}
$$

Thus, 6486073 has the same remainder when divided by 3 as does the sum of its digits! In particular, this tells us that 6486073 is divisible by 3 if and only if the sum of its digits is divisible by 3.

So far we have determined that $6486073 \equiv 3+7+0+6+8+4+6 \equiv 34 \pmod 3$. Note that we can apply the test again to the result: $34 \equiv 3+4 \equiv 7 \pmod 3$. Clearly $7 \equiv 1 \pmod 3$, so that $6486073 \equiv 1 \pmod 3$, and hence 6486073 is not divisible by 3. Of course, there is nothing special about 6486073; the test can be applied to any integer.

6. **Use the test for divisibility by** 3 **to compute** $4856739844628202071234\%3$ **(showing the steps in using the test).**

The technique demonstrated in this section can be used to deduce tests for divisibility by other integers, as we will see in lab.

Check Digits

When information is transmitted or copied, there is always a chance of an error being made. In many cases, the information includes extra data to help detect such errors. When this information takes the form of a extra letter or digit, it is called a *check digit*. Check digits are used in many places including Vehicle Identification Numbers used to track automobiles, and the omnipresent bar codes found on most commercial merchandise. Here we will discuss two examples of check digits.

Computer Transmissions Computers continually transmit and receive data, to/from other computers, hard disks, floppy disks, cd-roms, etcetera. On a very basic level, one should think of this data as being a long sequence of 0s and 1s (known as *bits* in computer lingo). There is always a chance that information will be garbled along the way — a 0 comes through as a 1, or a bit is lost completely. If the receiving computer doesn't detect the error, it may misinterpret the remaining data, so that a single missing bit can be disastrous.

To minimize this problem, computers use congruences to help spot transmission errors. In one approach[2], the bits (the 0s and 1s) are grouped in batches of 7. (There are $2^7 = 128$ combinations of 7 bits — enough to distinguish and encode the letters and numbers on a standard keyboard.) For each group of 7 bits, one more is added called the *check bit* or *parity bit*. These 8 bits together form a *byte*. The first 7 bits can be anything and hold actual information. The 8th or check bit is typically set so that there are an even number of 1s in the byte. For example, if the first 7 bits are 0110101, then the 8th bit would be set to 0 to make 01101010 have an even number of 1s.[3] The receiving computer checks each byte for accuracy and has a good chance of detecting a transmission error.

Note that with this type of system, there is always a chance of bad data appearing good. For example, if 11010111 is sent and 10110111 is received, the byte appears to be correct even though two bits are wrong. However, if the likelihood of a single incorrect bit is small, then the likelihood of two incorrect bits is *really* small.

In some applications, it is important to discover even very unlikely errors. There are more elaborate schemes used in those cases. However, there is always a trade-off: the more redundancy included in the information, the easier it is to detect and fix errors, but the longer the transmission will take.

---

[2]The scheme described here is 7-*bit ASCII*. Often computers now use 8-*bit ASCII* where 8 bits are used to encode letters and a 9th bit is used as a check bit.

[3]In terms of congruences, the sum of the bits is congruent to 0 modulo 2.

ISBNs for Books Every book currently published is assigned a 10-digit ISBN (International Standard Book Number) by the publisher. With so many books being published, assigning each a unique number simplifies ordering and tracking of individual books. Naturally, when humans transcribe such a long number, there are opportunities for errors to be introduced. To help prevent such errors, the 10th digit is really a check digit.

One type of error which is typical with humans is to interchange two of the digits. The use of a check bit as described in the previous section would not be sufficient to catch this type of error, so a more sophisticated scheme is used.

The check digit in an ISBN is based on congruences mod 11. If $a_1, \ldots, a_9$ are the digits identifying the book, then the check digit a_{10} is set so that

$$a_{10} = \left(\sum_{i=1}^{9} i a_i \right) \% \, 11.$$

Since a_{10} is one of 0, 1, 2, ..., 10, there is a minor problem: there are 11 remainders but only 10 numeric digits. The solution is to use the letter X for the case $a_{10} = 10$.

7. **Select a book of your choice, and locate the ISBN. (The ISBN is usually located on the back of a book, or can be found near the bottom of the copyright page. The copyright page is usually on the back of the title page.) Give the title, author, ISBN, and verify that the check digit satisfies the relation above.**

Maple electronic notebook `04-apps.mws`

Mathematica electronic notebook `04-apps.nb`

Web electronic notebook Start with the web page `index.html`

Maple Lab: Applications of Congruences

■ Initializations (You must execute the initialization group.)

■ Parity Checking: ISBNs

Below we give a short Maple function to compute the check digit for an ISBN for a book. The following function takes the first nine digits from an ISBN and computes the tenth (check) digit.

```
> # The first line is a function which converts from ordinary
  remainders mod 11 to the use of X in place of a remainder of 10

  mymod := n-> if(n=10) then print('X') else n fi:

  # Compute the required sum, reduce it mod 11, then convert it to
  the 0, ... , 9, X code

  isbn := digitlist -> mymod(add(j*digitlist[j], j=1..9) mod 11):
```

Try it out:

```
> isbn([3,7,6,4,3,3,0,2,9]);
>
```

■ Exercise 1

Find two books of your choosing, and use the function isbn defined above to verify that the ISBN from each book has the correct check digit. Give the title, author, and ISBN for each book.

You can use this function to investigate the following two questions:

■ Research Question 1

(a) Does the check digit change if we change one digit of the input?

(b) Does the check digit change if we switch two digits of the input?

■ A Simple Game

You will have an opportunity to play the following game against the computer. We start with a pile of rocks. Two players take turns removing rocks from the pile. Each time, a player may remove 1, 2, or 3 rocks. The player who removes the last rock wins. The code which teaches the computer how to play is contained in the initialization groups. Be sure that you have executed it before playing the game.

In this version of the game, the computer always goes first. (It's only fair, since you are far smarter than the computer.) For example, to start a game with 10 rocks in the pile, execute the following command.

```
> start(10);
```

You take your turns with the take command:

```
[
```

```
[ > take(1);
[ >
```

Keep playing until the game is over. The computer's strategy is based on congruences. Can you discover what it is, and then beat the computer? Remember that you can vary the number of rocks at the beginning of a game with the start command.

■ Research Question 2

(a) For which starting values can you win, and for which starting values does the computer always win? What is the computer's winning strategy?

(b) Suppose that at each turn, a player may remove 1, 2, 3, 4, 5, or 6 rocks. What strategy should the computer use, and what strategy should you use to beat the computer?

■ Divisibility Tests

Recall how we developed the test for divisibility by 3 in the Prelab section. Suppose we wanted to calculate the remainder of 8456352 when divided by 3 quickly, and by hand. We think of 8456352 in expanded notation:

$$8456352 = 2 \cdot 10^0 + 5 \cdot 10 + 3 \cdot 10^2 + 6 \cdot 10^3 + 5 \cdot 10^4 + 4 \cdot 10^5 + 8 \cdot 10^6.$$

The divisibility test then arises by replacing each power of 10 with its value mod 3. We saw that

$$10^n \equiv 1 \ (\mathrm{mod} \ 3)$$

for all integers $n \geq 0$. Thus,

$$8456352 = 2 \cdot 10^0 + 5 \cdot 10 + 3 \cdot 10^2 + 6 \cdot 10^3 + 5 \cdot 10^4 + 4 \cdot 10^5 + 8 \cdot 10^6$$
$$\equiv 2 + 5 + 3 + 6 + 5 + 4 + 8 \ \ (\mathrm{mod} \ 3)$$
$$\equiv \ 33 \ \ (\mathrm{mod} \ 3).$$

Repeating the process, we discover that $33 \equiv 3 + 3 \equiv 6 \ (\mathrm{mod} \ 3)$, which is easily seen to be congruent to 0 $(\mathrm{mod} \ 3)$. Therefore, 8456352 is divisible by 3.

The key to this process is the values of the powers of 10 taken modulo 3. To deduce similar tests for other potential divisors, we can use the following function. It produces a list of the values of 10^j with j running from 0 to howmany.

```
[ > divtestmultipliers:=proc(m, howmany)
     local j;
     seq(10^j mod m, j=0..howmany);
   end:
```

Let's try it out with $m = 3$:

```
[ > divtestmultipliers(3, 50);
```

Can you connect the result of this calculation with the test for divisibility by 3?

■

■ Research Question 3

Devise a test for divisibility by 9.

Let's now construct a test for divisibility by 11. Using `divtestmultipliers`, we get:

```
[ > divtestmultipliers(11, 50);
[ >
```

Note that the sequence appears to be purely periodic. In fact, we can *prove* that the sequence is purely periodic by observing that $10^2 \equiv 1 \pmod{11}$, so that for any integer $k \geq 0$ we have

$$10^{(k+2)} \equiv 10^k \cdot 10^2 \pmod{11}$$

$$\equiv 10^k \cdot 1 \pmod{11}$$

$$\equiv 10^k \pmod{11}.$$

Since $10^{(k+2)} \equiv 10^k \pmod{11}$ for all $k \geq 0$, it follows that the sequence is purely periodic.

This result is not an accident; it will be true if we replace 11 with any prime p other than 2 or 5. So that you won't have any doubts about this assertion, it is left for you to prove in the next exercise.

■ *Exercise 2*

Prove that the sequence $10^j \bmod p$ ($j = 0, 1, 2, \ldots$) is purely periodic for any prime p except 2 and 5.

From our work above, we now *know* (as opposed to just suspect) that the values of $10^j \bmod 11$ alternate forever, beginning with 1, 10, 1, 10, ... On the basis of this, $n = d_k d_{k-1} \ldots d_1 d_0$, then

$$n = d_0 \cdot 10^0 + d_1 \cdot 10 + d_2 \cdot 10^2 + d_3 \cdot 10^3 + \cdots$$

$$\equiv d_0 \cdot 1 + d_1 \cdot 10 + d_2 \cdot 1 + d_3 \cdot 10 + \cdots \pmod{11}.$$

For example, the test applied to 64368 would be as follows:

$$64368 = 8 \cdot 10^0 + 6 \cdot 10 + 3 \cdot 10^2 + 4 \cdot 10^3 + 6 \cdot 10^4$$

$$\equiv 8 \cdot 1 + 6 \cdot 10 + 3 \cdot 1 + 4 \cdot 10 + 6 \cdot 1 \pmod{11}$$

$$\equiv 117 \pmod{11}$$

$$\equiv 7 \cdot 1 + 1 \cdot 10 + 1 \cdot 1 \pmod{11}$$

$$\equiv 18 \pmod{11},$$

which is obviously congruent to 7 (mod 11). (What happens if we try to continue the process? Is this a good idea?) Thus, we see that that $64368 \equiv 7 \pmod{11}$, and from this it follows that 64368 is not divisible by 11.

■ *Exercise 3*

The standard divisibility test for 11 uses multipliers 1 and −1 instead of 1 and 10. In other words, the test applied to 64368 would say that

$$64368 \equiv 8 \cdot 1 + 6 \cdot (-1) + 3 \cdot 1 + 4 \cdot (-1) + 6 \cdot 1 \pmod{11}$$

$$\equiv 7 \pmod{11},$$

which gets to the answer much more quickly. Explain why the two versions of the divisibility test for 11 are actually the same.

Research Question 4

Devise a test for divisibility by 37.

Research Question 5

The number 7 is infamous for not having a divisibility test. Show that this assertion is false by devising a test for divisibility by 7.

Exercise 4

Justify the standard tests for divisibility by $n = 2, 5$, and 10. In each of these three cases, the standard test states that a number is congruent to its units (last) digit modulo n.

Going Farther: Cryptography Basics

In this section, we lay the foundation for implementing a number of different types of secret codes. (One such code is described in this worksheet. Others will appear in later parts of the course.) In all cases, the basic situation is the same. There are two people named Alice and Bill who want to send messages to each other. Other people may intercept these messages, so Alice and Bill want to disguise the messages so that only they can read them. A method for disguising messages is called a *cryptosystem*. Many different cryptosystems have been used, but in this course we will concentrate only on those systems that have some connection with number theory.

To get started, we need a standard procedure for translating numbers into letters and vice versa. A standard conversion between letters and numbers is the *American Standard Code for Information Interchange*, also known as ASCII. Maple provides functions to convert back and forth between letters and their ASCII equivalents. ASCII reserves some numbers, such as the numbers between 0 and 31, for "nonprinting characters" like as Control characters, and we want to skip those. Here is our function to convert letters to numbers:

```
> texttonums := proc(a)
    local raw, j;
    raw := convert(a, 'bytes');
    [seq(raw[j]-32, j=1..nops(raw))]
  end:
>
```

Let's try it out. Note that the text has to be enclosed in "backquotes" for Maple to interpret it as a text string.

```
> texttonums('abc, ABC');
>
```

As you can see, we are using the convention that

$$a = 65, b = 66, ..., z = 90, \quad A = 33, B = 34, C = 35, ..., Z = 58.$$

A space is represented by the number 0, and our comma translated into the number 12. To convert a list of numbers back into text, we can use the analogous function:

```
> numstotext := proc(a)
    local raw, j;
```

```
      raw := [seq(a[j]+32, j=1..nops(a))];
      convert(raw, 'bytes')
    end:
```

We can use it on our output from up above:

```
[ > numstotext([65, 66, 67, 12, 0, 33, 34, 35]);
```

Now, let's put them together on some more meaningful text. We first convert from text to numbers:

```
[ > reid := texttonums('Both the Stanford and DEC uses of the ASCII
     control characters are in violation of the USA Standard Code, but
     no Federal marshal is likely to come running out and arrest people
     who type control-T to their computers.  Brian Reid (1978).');
```

Now, we convert the numbers back to text:

```
[ > numstotext(reid);
[ >
```

■ *Exercise 5*

What is the numerical equivalent of the punctuation mark "." (period)?

Now that we can represent text as numbers, we can move on to our first encryption/decription method.

■ Going Farther: Shift Ciphers

The simplest cipher we shall consider is called the *shift cipher*. When implementing the shift cipher, each letter of the message is shifted a fixed distance down the alphabet. Our "alphabet", including punctuation and both upper- and lowercase letters, is numbered from 0 to 94. Here it is:

```
[ > numstotext([seq(j,j=0..94)]);
[ >
```

If we let

P = a number representing a letter from the unencoded message

and

C = the corresponding number for encoded message,

then the shift cipher can be expressed mathematically as

$$C \equiv P + k \pmod{95}$$

where k is a fixed integer called the *key*. We work modulo 95 because we are using an "alphabet" with 95 "letters". This congruence is used to encode each letter in the original message. Shift ciphers are sometimes called Caesar ciphers, because Julius Caesar purportedly was the first to use one.

Let's look at an example. Suppose that our unencoded message (referred to as plaintext) is "Hello!", and we want to encipher this with a key $k = 4$. First, we take the plaintext and turn it into a number string.

```
[ > plaintext := `Hello!`;
[   t := texttonums(plaintext);
[ >
```

Next, we add 4 to each number in the string, and then reduce mod 95.

```
[ > c := [seq((t[j]+4) mod 95,j=1..nops(t))];
```

Finally, we turn this string back into the encoded text, called ciphertext.

```
[ > ciphertext := numstotext(c);
```

The command shift will do all of these steps automatically.

```
[ > shift := proc(plaintext, k)
[     local j, nums;
[     nums := texttonums(plaintext);
[     numstotext([seq((nums[j]+k) mod 95, j=1..nops(nums))]);
[   end:
```

For example, here is a shift cipher with $k = 7$.

```
[ > shift(` ABCDEFGHIJKLMNOPQRSTUVWXYZ`,7);
[ >
```

Decoding a shift cipher is easy if you know the key k. Since

$$C = (P + k) \% 95,$$

it is easy to see that

$$P = (C - k) \% 95.$$

In other words, a shift cipher with key k can be decoded with another shift cipher with key equal to $-k$. Here's an example.

```
[ > plaintext := `What's up, doc?`;
[   ciphertext := shift(plaintext,18);
[   shift(ciphertext,-18);
[ >
```

◼ ROT-13: Shift Ciphers and the Internet

A simpler version of the shift cipher called ROT-13 has been in use on the internet for many years. Specifically, it is used in Usenet, the collection of newsgroups where people discuss a multitude of topics. Consider, for example, the group rec.humor.funny. People submit jokes which are reviewed by a moderator, and then the funny ones are sent out over Usenet to the world. Sometimes the moderator wants to send out an "off-color" joke, but doesn't want to offend the more sensitive readers. Including a warning along with the joke does no good since a delicate reader may see the offending words before they read the warning.

The solution is that the naughty jokes are encoded by ROT-13. A plain text warning message is also included so that readers can decide if they want to "risk" decoding the joke. In ROT-13, the encryption only applies to letters of the alphabet; spaces and punctuation are left unchanged. Since there are 26 letters in the alphabet, the cipher works mod 26. The shift which is used is $k = 13$ (thus the name, because it ROTates letters 13 slots through the alphabet).

Most software used for reading Usenet newsgroups have the commands built into them for decoding such messages. The user just picks "Decode" from a menu or hits a special key. Rest assured that users do not have to understand shift ciphers to read dirty jokes on the internet.

Let's look at an example of ROT-13. Here is a Maple function to do the computations:

```
> rot13 := proc(inmessage)
    local innums, j;
    innums := texttonums(inmessage);
    for j from 1 to nops(innums) do
      if(innums[j]>64 and innums[j]<91) then
        innums[j] := ((innums[j]-65+13) mod 26)+65;
      else if(innums[j]>32 and innums[j]<59) then
        innums[j] := modp(innums[j]-33+13,26)+33;
      fi;
      fi;
    od;
    numstotext(innums);
    end:
>
```

Notice that we can use the same function for both encoding and decoding! The reason is that we are working modulo 26 in this case. To decode a shift by 13 we need to shift by -13. But, $-13 \equiv 13 \pmod{26}$. So, decoding is also a shift by 13. This is why 13 was chosen as the standard for this cipher in the first place.

To keep in the spirit of this application, here is a joke whose punch line has been encoded by ROT-13. The joke does not contain offensive words. However, we should warn the reader that understanding the joke requires a certain level of mathematical knowledge. Moreover, some readers may not find it funny.

Here is the joke:

Q: What did the mathematician say when epsilon went to zero?
N: Jryy, gurer tbrf gur arvtuobeubbq!

To decode the punch line, you need to run ROT-13 on this encrypted text:

```
> rot13('N: Jryy, gurer tbrf gur arvtuobeubbq!');
>
```

We now return to our standard shift ciphers, with a 95-letter alphabet.

Exercise 6

How many different shift ciphers are possible with a 95-letter alphabet?
How many different shift ciphers are possible with an n-letter alphabet?

Exercise 7

The following message has been encoded with a shift cipher with key 9.

```
> ciphertext := ']qn)xk rx~|)vj}qnvj}rlju)k{njt}q{x~pq)!x~um)kn)mn
```

```
   nuxyvnw})xo)jw)nj|#)!j#)}x)ojl}x{)uj{pn)y{rvn)w~vkn{|7))Kruu)Pj}n
   |`;
[ >
```

(a) Decode the message.

(b) The author of this message was trying to predict what important mathematical breakthroughs may occur in the near future. Explain why the statement is nonsensical.

Exercise 8

Below is a message that has been encoded with a shift cipher. We don't know what key *k* was used for the encoding, but we do know that the last character in the plain text message was a period. Use this information to decode the message.

```
> message :=
  'r5B5:3K9;:1EK@;K@41K3;B1>:91:@K5?K8571K35B5:3KC45?71EK-:0K/->?K@
  ;K@11:-31K.;E?YKXXK{YKuYKzR};A>71Y';
[ >
```

Hint: In Exercise 5, you determined the numerical equivalent of a period.

Exercise 9

Cryptanalysis is the art of deciphering an encoded message without knowing the key. Shift ciphers are easy to cryptanalyze because there are very few keys. Take, for example, the following message, which has been encoded with a shift cipher.

```
[ > ciphertext := 'i+An1/B';
```

Try all 95 different possible shift ciphers on this message. What is the plaintext?

Hint: The following code will make a list of all 95 different shifts.

```
> for j from 0 to 94 do
    printf('%2d %s\n', j, shift(ciphertext,j));
    od;
[ >
```

Exercise 10

Alice and Bill are sending messages to each other by using a shift cipher. They are very careful not to reveal their key to anyone, but they don't realize that shift ciphers are very unsecure.

Oscar has been observing their messages, and so far, he has intercepted the following coded texts.

```
> message1 :='Wxt&3T |vx?33[t*x3-#)3'(t&(xw3'()w-|"z3y#&3({x3t"t
  -'|'3(x'(3-x(R33U|  `;
> message2 :='Wxt&3U|  ?33a#(3-x(433j{t(3wt-3|'3((x3x'(R33T |vx';
> message3 :='Wxt&3T |vx?33\\3({|"~3((x3x'(3|'3#"3g{)&'wt-A33\\3+|
  3v{xv~3+|({3V{&|'A33U|  `;
> message4 :='Wxt&3U|
  ?33b"x3#({x&3({|"zA33j{t(3'3((x3wxy|"|(|#"3#3y3t3v#"(|")#)'3y)"v(
  |#"R33T |vx';
[ >
```

(a) Help Oscar decipher the messages. (Note that all of the messages start the same way.)

(b) Oscar wants to more than just eavesdrop; he wants to create some mischief by impersonating Bill and sending a message with false information to Alice. Devise such a message for Oscar, and encrypt it for him.

Mathematica Lab: Applications of Congruences

■ Parity Checking: ISBNs

Below we give a short Mathematica function to compute the check digit for an ISBN for a book. The following function takes the first nine digits from an ISBN and computes the tenth (check) digit.

```
(*  The first two lines set a function
    which converts from ordinary remainders modulo 11
    to the use of X in place of a remainder of 10 *)
mymod[n_] := n
mymod[10] := X

(*  Compute the required sum, reduce it modulo 11,
    then convert it to the 0, ..., 9, X code *)

isbn[digitlist_] := mymod[Mod[Sum[j digitlist[[j]], {j, 9}], 11]]
```

Try it out:

```
isbn[{3, 7, 6, 4, 3, 3, 0, 2, 9}]
```

■ Exercise 1

Find two books of your choosing, and use the function **isbn** defined above to verify that the ISBN from each book has the correct check digit. Give the title, author, and ISBN for each book.

You can use the function **isbn** to investigate the following two questions:

■ Research Question 1

> (a) Does the check digit change if we change one digit of the input?
> (b) Does the check digit change if we switch two digits of the input?

■ A Simple Game

You will have an opportunity to play the following game against the computer. We start with a pile of rocks. Two players take turns removing rocks from the pile. Each time, a player may remove 1, 2, or 3 rocks. The player who removes the last rock wins. The code which teaches the computer how to play is contained in the initialization cells. No peeking!

In this version of the game, the computer always goes first. (It's only fair, since you are far smarter than the computer.) For example, to start a game with 10 rocks in the pile, execute the following cell.

```
start[10]
```

You take your turns with the **take** command:

```
take[1]
```

Keep playing until the game is over. The computer's strategy is based on congruences. Can you discover what it is, and then beat the computer? Remember that you can vary the number of rocks at the beginning of a game with the **start** command.

■ **Research Question 2**

> (a) For which starting values can you win, and for which starting values does the computer always win? What is the computer's winning strategy?
>
> (b) Suppose that at each turn, a player may remove 1, 2, 3, 4, 5, or 6 rocks. What strategy should the computer use, and what strategy should you use to beat the computer?

■ Divisibility Tests

Recall how we developed the test for divisibility by 3 in the Prelab section. Suppose we want to calculate the remainder of 8456352 when divided by 3 quickly, and by hand. We think of 8456352 in expanded notation:

$$8456352 = 2 \cdot 10^0 + 5 \cdot 10^1 + 3 \cdot 10^2 + 6 \cdot 10^3 + 5 \cdot 10^4 + 4 \cdot 10^5 + 8 \cdot 10^6.$$

The divisibility test then arises by replacing each power of 10 with its remainder modulo 3. We saw that

$$10^n \equiv 1 \pmod 3$$

for all integers $n \geq 0$. Thus,

$$
\begin{aligned}
8456352 &= 2 \cdot 10^0 + 5 \cdot 10^1 + 3 \cdot 10^2 + 6 \cdot 10^3 + 5 \cdot 10^4 + 4 \cdot 10^5 + 8 \cdot 10^6 \\
&\equiv 2 + 5 + 3 + 6 + 5 + 4 + 8 \pmod 3 \\
&\equiv 33 \pmod 3.
\end{aligned}
$$

Repeating the process, we discover that $33 \equiv 3 + 3 \equiv 6 \pmod 3$, which is easily seen to be congruent to $0 \pmod 3$. Therefore, 8456352 is divisible by 3.

The key to this process is the values of the powers of 10 taken modulo 3. To deduce similar tests for other potential divisors, we can use the following function. It produces a list of the values of $10^j \% m$ with j running from 0 to **howmany**.

```
divtestmultipliers[m_, howmany_] :=
  Table[Mod[10^j, m], {j, 0, howmany}]
```

Let's try it out with $m = 3$:

```
divtestmultipliers[3, 50]
```

Can you connect the result of this calculation with the test for divisibility by 3?

■ **Research Question 3**

> Devise a test for divisibility by 9.

Let's now construct a test for divisibility by 11. Using **divtestmultipliers**, we get:

```
divtestmultipliers[11, 50]
```

Note that the sequence appears to be purely periodic. In fact, we can *prove* that the sequence is purely periodic by observing that $10^2 \equiv 1 \pmod{11}$, so that for any integer $k \geq 0$ we have

$$10^{k+2} \equiv 10^k \cdot 10^2 \pmod{11}$$
$$\equiv 10^k \cdot 1 \pmod{11}$$
$$\equiv 10^k \pmod{11}.$$

Since $10^{k+2} \equiv 10^k \pmod{11}$ for all $k \geq 0$, it follows that the sequence is purely periodic.

This result is not an accident; it will be true if we replace 11 with any prime p other than 2 or 5. So that you won't have any doubts about this assertion, it is left for you to prove in the next exercise.

■ **Exercise 2**

Prove that the sequence $10^j \% p$ ($j = 0, 1, 2, \ldots$) is purely periodic for any prime p except 2 and 5.

From our work above, we now *know* (as opposed to just suspect) that the values of $10^j \% 11$ alternate forever, beginning with $1, 10, 1, 10, \ldots$. On the basis of this, if $n = d_k\, d_{k-1} \ldots d_1\, d_0$, then

$$\begin{aligned} n \;=\; & d_0 \cdot 10^0 + d_1 \cdot 10^1 + d_2 \cdot 10^2 + d_3 \cdot 10^3 + \cdots \\ \equiv\; & d_0 \cdot 1 \;\;+\; d_1 \cdot 10 \;+\; d_2 \cdot 1 \;\;+\; d_3 \cdot 10 \;+\; \cdots \pmod{11}. \end{aligned}$$

For example, the test applied to 64368 would be as follows:

$$\begin{aligned} 64368 \;=\; & 8 \cdot 10^0 + 6 \cdot 10^1 + 3 \cdot 10^2 + 4 \cdot 10^3 + 6 \cdot 10^4 \\ \equiv\; & 8 \cdot 1 + 6 \cdot 10 + 3 \cdot 1 + 4 \cdot 10 + 6 \cdot 1 \pmod{11} \\ \equiv\; & 117 \pmod{11} \\ \equiv\; & 7 \cdot 1 + 1 \cdot 10 + 1 \cdot 1 \pmod{11} \\ \equiv\; & 18 \pmod{11}, \end{aligned}$$

which is obviously congruent to 7 (mod 11). Thus, we see that $64368 \equiv 7 \pmod{11}$, and from this it follows that 64368 is not divisible by 11.

■ **Exercise 3**

The standard divisibility test for 11 uses multipliers 1 and -1 instead of 1 and 10. In other words, the test applied to 64368 would say that

$$\begin{aligned} 64368 \;\equiv\; & 8 \cdot 1 + 6 \cdot (-1) + 3 \cdot 1 + 4 \cdot (-1) + 6 \cdot 1 \pmod{11} \\ \equiv\; & 7 \pmod{11}. \end{aligned}$$

Explain why the two versions of the divisibility test for 11 are actually the same.

■ **Research Question 4**

Devise a test for divisibility by 37.

■ **Research Question 5**

The number 7 is infamous for not having a divisibility test. Show that this assertion is false by devising a test for divisibility by 7.

■ **Exercise 4**

Justify the standard tests for divisibility by $n = 2$, 5, and 10. In each of these three cases, the standard test states that a number is congruent to its units (last) digit modulo n.

■ Going Farther: Cryptography Basics

In this section, we lay the foundation for implementing a number of different types of secret codes. (One such code is described in this notebook. Others will appear in later parts of the course.) In all cases, the basic situation is the same. There are two people named Alice and Bill who want to send messages to each other. Other people may intercept these messages, so Alice and Bill want to disguise the messages so that only they can read them. A method for disguising messages is called a *cryptosystem*. Many different cryptosystems have been used, but in this course we will concentrate only on those systems that have some connection with number theory.

To get started, we need a standard procedure for translating numbers into letters and vice versa. A standard conversion between letters and numbers is the *American Standard Code for Information Interchange*, also known as ASCII. Mathematica provides functions to convert back and forth between letters and their ASCII equivalents. ASCII reserves some numbers, such as the numbers between 0 and 31, for "nonprinting characters" such as Control characters, and we want to skip those. Here is our function to convert letters to numbers:

> `texttonums[text_] := ToCharacterCode[text] - 32`

Let's try it out. Note that the text has to be enclosed in "double quotes" for Mathematica to interpret it as a text string.

> `texttonums["abc, ABC"]`

As you can see, we are using the convention that

$$a = 65, b = 66, \ldots, z = 90, \quad A = 33, B = 34, C = 35, \ldots, Z = 58.$$

A space is represented by the number 0, and our comma translated into the number 12. To convert a list of numbers back into text, we can use the analogous function:

> `numstotext[nums_] := FromCharacterCode[nums + 32]`

We can use it on our output from up above:

> `numstotext[{65, 66, 67, 12, 0, 33, 34, 35}]`

Now, let's put them together on some more meaningful text. We first convert from text to numbers:

> ```
> reid =
>   texttonums["Both the Stanford and DEC uses of the ASCII control
>       characters are in violation of the USA Standard
>       Code, but no Federal marshal is likely to
>       come running out and arrest people who type
>             control-T to their computers. -- Brian Reid (1978)."]
> ```

Now, we convert the numbers back to text:

> `numstotext[reid]`

■ **Exercise 5**

What is the numerical equivalent of the punctuation mark "." (period)?

Now that we can represent text as numbers, we can move on to our first encryption/decription method.

■ Going Farther: Shift Ciphers

The simplest cipher we shall consider is called the *shift cipher*. When implementing the shift cipher, each letter of the message is shifted a fixed distance down the alphabet. Our "alphabet", including punctuation and both upper–and lowercase letters, is numbered from 0 to 94. Here it is:

```
numstotext[Range[0, 94]]
```

If we let

$$P = \text{a number representing a letter from the unencoded message}$$

and

$$C = \text{the corresponding number for encoded message,}$$

then the shift cipher can be expressed mathematically as

$$C = (P + k) \% 95,$$

where k is a fixed integer called the *key*. We work modulo 95 because we are using an "alphabet" with 95 "letters". This congruence is used to encode each letter in the original message. Shift ciphers are sometimes called Caesar ciphers, because Julius Caesar purportedly was the first to use one.

Let's look at an example. Suppose that our unencoded message (referred to as plaintext) is "Hello!", and we want to encipher this with a key $k = 4$. First, we take the plaintext and turn it into a number string.

```
plaintext = "Hello!";
t = texttonums[plaintext]
```

Next, we add 4 to each number in the string, and then reduce mod 95.

```
c = Mod[t + 4, 95]
```

Finally, we turn this string back into the encoded text, called ciphertext.

```
ciphertext = numstotext[c]
```

The command **shift** will do all of these steps automatically.

```
shift[plaintext_, k_] :=
    numstotext[Mod[texttonums[plaintext] + k, 95]];
```

For example, here is a shift cipher with $k = 7$.

```
shift[" ABCDEFGHIJKLMNOPQRSTUVWXYZ", 7]
```

Decoding a shift cipher is easy if you know the key k. Since

$$C = (P + k) \% 95,$$

it is easy to see that

$$P = (C - k) \% 95.$$

In other words, a shift cipher with key k can be decoded with another shift cipher with key equal to $-k$. Here's an example.

```
plaintext = "What's up, doc?";
ciphertext = shift[plaintext, 18]
shift[ciphertext, -18]
```

■ ROT–13: Shift Ciphers and the Internet

A simpler version of the shift cipher called ROT–13 has been in use on the internet for many years. Specifically, it is used in Usenet, the collection of newsgroups where people discuss a multitude of topics. Consider, for example, the group **rec.humor.funny**. People submit jokes which are reviewed by a moderator, and then the funny ones are sent out over Usenet to the world. Sometimes the moderator wants to send out an "off-color" joke, but doesn't want to offend the more sensitive readers. Including a warning along with the joke does no good since a delicate reader may see the offending words before they read the warning.

The solution is that the naughty jokes are encoded by ROT–13. A plain text warning message is also included so that readers can decide if they want to "risk" decoding the joke. In ROT–13, the encryption only applies to letters of the alphabet; spaces and punctuation are left unchanged. Since there are 26 letters in the alphabet, the cipher works modulo 26. The shift which is used is $k = 13$ (thus the name, because it ROTates letters 13 slots through the alphabet).

Most software used for reading Usenet newsgroups have the commands built into them for decoding such messages. The user just picks "Decode" from a menu or hits a special key. Rest assured that users do not have to understand shift ciphers to read dirty jokes on the internet.

Let's look at an example of ROT–13. Here is a Mathematica function to do the computations:

```
rot13[inmessage_] := Module[{innums},
   innums = texttonums[inmessage];
    Do[If[(innums[[j]] > 64 && innums[[j]] < 91),
     innums[[j]] = Mod[innums[[j]] - 65 + 13, 26] + 65];
    If[(innums[[j]] > 32 && innums[[j]] < 59),
     innums[[j]] = Mod[innums[[j]] - 33 + 13, 26] + 33],
    {j, 1, Length[innums]}];
   Return[numstotext[innums]]];
```

Notice that we can use the same function for both encoding and decoding! The reason is that we are working modulo 26 in this case. To decode a shift by 13, we need to shift by -13. But, $-13 \equiv 13 \pmod{26}$. So, decoding is also a shift by 13. This is why 13 was chosen as the standard for this cipher in the first place.

To keep in the spirit of this application, here is a joke whose punch line has been encoded by ROT–13. The joke does not contain offensive words. However, we should warn the reader that understanding the joke requires a certain level of mathematical knowledge. Moreover, some readers may not find it funny.

Here is the joke:

Q: What did the mathematician say when epsilon went to zero?
N: Jryy, gurer tbrf gur arvtuobeubbq!

To decode the punch line, you need to run ROT–13 on this encrypted text:

```
rot13["N: Jryy, gurer tbrf gur arvtuobeubbq!"]
```

We now return to our standard shift ciphers, with a 95–letter alphabet.

■ Exercise 6

How many different shift ciphers are possible with a 95 −letter alphabet?
How many different shift ciphers are possible with an *n* −letter alphabet?

■ Exercise 7

The following message has been encoded with a shift cipher with key 9.

```
ciphertext = "]qn)xk rx~|)vj}qnvj}rlju)
    k{njt}q{x~pq) !x~um)kn)mn nuxyvnw})xo)jw)nj|#) !j
    #) }x)ojl}x{()uj{pn)y{rvn)w~vkn{|7)66Kruu)Pj}n|";
```

(a) Decode the message.

(b) The author of this message was trying to predict what important mathematical breakthroughs may occur in the near future. Explain why the statement is nonsensical.

■ Exercise 8

Below is a message that has been encoded with a shift cipher. We don't know what key *k* was used for the encoding, but we do know that the last character in the plain text message was a period. Use this information to decode the message.

```
message = "r5B5:3K9;:1EK@;K@41K3;B1>:91:@K5?K8571K35B5:3KC45?
    71EK-:0K/->?K@;K11:-31K.;E?YKXXK{YKuYKzR};A>71Y";
```

Hint: In Exercise 5, you determined the numerical equivalent of a period.

■ Exercise 9

Cryptanalysis is the art of deciphering an encoded message without knowing the key. Shift ciphers are easy to cryptanalyze because there are very few keys. Take, for example, the following message, which has been encoded with a shift cipher.

```
ciphertext = "i+An1/B";
```

Try all 95 different possible shift ciphers on this message. What is the plaintext?

Hint: The following code will make a list of all 95 different shifts.

```
TableForm[Table[{j, shift[ciphertext, j]}, {j, 0, 94}],
    TableSpacing -> {0, 1}]
```

■ Exercise 10

Alice and Bill are sending messages to each other by using a shift cipher. They are very careful not to reveal their key to anyone, but they don't realize that shift ciphers are very unsecure.

Oscar has been observing their messages, and so far, he has intercepted the following coded texts:

```
message1 = "Wxt&3T |vx?33[t*x3-#)3'(t&
    (xw3'()w-|\"z3y#&3({x3t\"t -'|'3(x'(3-x(R33U|  ";
message2 = "Wxt&3U|  ?33a#(3-x(433j{t(3
    wt-3|'3({x3(x'(R33T |vx";
message3 = "Wxt&3T |vx?33\\3({|\"~3({x3(x'(3|'3#\"
    3g{)&'wt-A33\\3+|  3v{xv~3+|({3V{&|'A33U|  ";
message4 = "Wxt&3U|  ?33b\"x3#({x&3({|\"zA33j{t(3|'3
    ({x3wxy|\"|(|#\"3#y3t3v#\"(|\")#)'3y)\"v(|#\"R33T |vx";
```

(a) Help Oscar decipher the messages. (Note that all of the messages start the same way.)

(b) Oscar wants to do more than just eavesdrop; he wants to create some mischief by impersonating Bill and sending a message with false information to Alice. Devise such a message for Oscar, and encrypt it for him.

Web Lab: Applications of Congruences

4.1 Parity Checking: ISBNs

Below is a Java applet to compute the check digit for an ISBN for a book. The applet takes the first nine digits from an ISBN as input and computes the tenth (check) digit.

Note: ISBNs are usually printed with dashes between blocks of numbers. It is interesting that the dashes don't always occur in the same locations. The applet below doesn't care about the location of the dashes, or if you leave them out altogether.

ISBN Calculator

First 9 digits of ISBN #: `1-56592-149`

Check Digit: ` `

Compute!

Exercise 1

Find two books of your choosing, and use the ISBN applet to verify that the ISBN from each book has the correct check digit. Give the title, author, and ISBN for each book.

You can use this applet to investigate the following two questions:

Research Question 1

(a) Does the check digit change if we change one digit of the input?

(b) Does the check digit change if we switch two digits of the input?

212

4.2 A Simple Game

You will have an opportunity to play the following game against the computer. We start with a pile of rocks. Two players take turns removing rocks from the pile. Each time, a player may remove 1, 2, or 3 rocks. The player who removes the last rock wins.

In this version of the game, the computer always goes first. (It's only fair, since you are far smarter than the computer.) The rocks remaining in the pile are indicated by the solid black disks. The rocks taken by the computer are indicated by red circles, and the rocks taken by you are indicated by the blue circles.

To take 1, 2, or 3 rocks from the pile, just click on the corresponding number. The computer will then automatically make its move. Try it out!

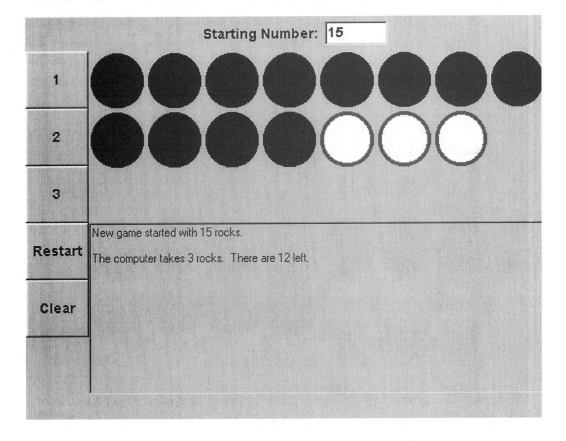

The computer's strategy is based on congruences. Can you discover what it is, and then beat the computer? You can vary the number of rocks at the beginning of a game by changing the "Starting Number" value and then clicking "Restart".

Research Question 2

(a) For which starting values can you win, and for which starting values does the computer always win? What is the computer's winning strategy?

(b) Suppose that at each turn, a player may remove 1, 2, 3, 4, 5, or 6 rocks. What strategy should the computer use, and what strategy should you use to beat the computer?

4.3 Divisibility Tests

Recall how we developed the test for divisibility by 3 in the Prelab section. Suppose we want to calculate the remainder of 8456352 when divided by 3 quickly, and by hand. We think of 8456352 in expanded notation:

$$8456352 = 2 \cdot 10^0 + 5 \cdot 10^1 + 3 \cdot 10^2 + 6 \cdot 10^3 + 5 \cdot 10^4 + 4 \cdot 10^5 + 8 \cdot 10^6.$$

The divisibility test then arises by replacing each power of 10 with its remainder modulo 3. We saw that

$$10^n \equiv 1 \pmod 3$$

for all integers $n \geq 0$. Thus,

$$\begin{aligned}
8456352 &= 2 \cdot 10^0 + 5 \cdot 10^1 + 3 \cdot 10^2 + 6 \cdot 10^3 + 5 \cdot 10^4 + 4 \cdot 10^5 \\
&\quad + 8 \cdot 10^6 \\
&\equiv 2 + 5 + 3 + 6 + 5 + 4 + 8 \pmod 3 \\
&\equiv 33 \pmod 3.
\end{aligned}$$

Repeating the process, we discover that $33 \equiv 3 + 3 \equiv 6 \pmod 3$, which is easily seen to be congruent to 0 (mod 3). Therefore, 8456352 is divisible by 3.

The key to this process is the values of the powers of 10 taken modulo 3. To deduce similar tests for other potential divisors, we can use the applet below. It produces a list of the values of $10^j \% n$ with j running from 0 to M. Let's try it out with $n = 3$:

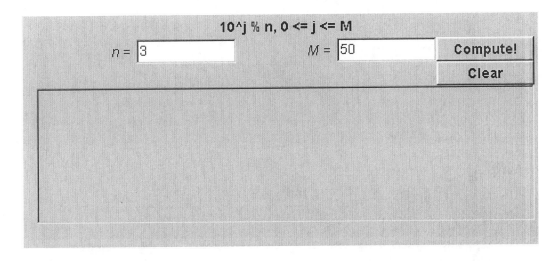

Can you connect the result of this calculation with the test for divisibility by 3?

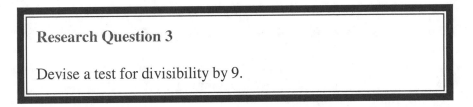

Research Question 3

Devise a test for divisibility by 9.

Let's now construct a test for divisibility by 11. Using our applet, we get:

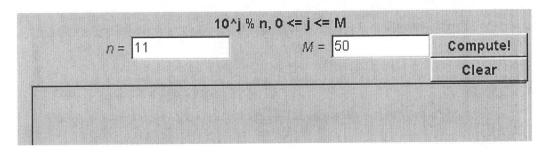

Note that the sequence appears to be purely periodic. In fact, we can *prove* that the sequence is purely periodic by observing that $10^2 \equiv 1 \pmod{11}$, so that for any integer $k \geq 0$ we have

$$
\begin{aligned}
10^{k+2} &\equiv 10^k \cdot 10^2 \pmod{11} \\
&\equiv 10^k \cdot 1 \pmod{11} \\
&\equiv 10^k \pmod{11}.
\end{aligned}
$$

Since $10^{k+2} \equiv 10^k \pmod{11}$ for all $k \geq 0$, it follows that the sequence is purely periodic.

This result is not an accident; it will be true if we replace 11 with any prime p other than 2 or 5. So that you won't have any doubts about this assertion, it is left for you to prove in the next exercise.

Exercise 2

Prove that the sequence $10^j \% p$ ($j = 0, 1, 2, \ldots$) is purely periodic for any prime p except 2 and 5.

From our work above, we now *know* (as opposed to just suspect) that the values of $10^j \% 11$ alternate forever, beginning with $1, 10, 1, 10, \ldots$. On the basis of this, if $n = d_k d_{k-1} \ldots d_1 d_0$, then

$$
\begin{aligned}
n &= d_0 \cdot 10^0 + d_1 \cdot 10^1 + d_2 \cdot 10^2 + d_3 \cdot 10^3 + \cdots \\
&\equiv d_0 \cdot 1 + d_1 \cdot 10 + d_2 \cdot 1 + d_3 \cdot 10 + \cdots \pmod{11}.
\end{aligned}
$$

For example, the test applied to 64368 would be as follows:

$$64368 = 8 \cdot 10^0 + 6 \cdot 10^1 + 3 \cdot 10^2 + 4 \cdot 10^3 + 6 \cdot 10^4$$
$$\equiv 8 \cdot 1 + 6 \cdot 10 + 3 \cdot 1 + 4 \cdot 10 + 6 \cdot 1 \quad (\text{mod } 11).$$
$$\equiv 117 \quad (\text{mod } 11).$$
$$\equiv 7 \cdot 1 + 1 \cdot 10 + 1 \cdot 1 \quad (\text{mod } 11).$$
$$\equiv 18 \quad (\text{mod } 11),$$

which is obviously congruent to 7 (mod 11). Thus, we see that $64368 \equiv 7$ (mod 11), and from this it follows that 64368 is not divisible by 11.

Exercise 3

The standard divisibility test for 11 uses multipliers 1 and −1 instead of 1 and 10. In other words, the test applied to 64368 would say that

$$64368 \equiv \begin{array}{l} 8 \cdot 1 + 6 \cdot (-1) + 3 \cdot 1 + 4 \cdot (-1) + 6 \cdot 1 \\ (\text{mod } 11) \end{array}$$
$$\equiv 7 \quad (\text{mod } 11).$$

Explain why the two versions of the divisibility test for 11 are actually the same.

Research Question 4

Devise a test for divisibility by 37.

Research Question 5

The number 7 is infamous for not having a divisibility test. Show that this assertion is false by devising a test for divisibility by 7.

Exercise 4

Justify the standard tests for divisibility by $n = 2$, 5, and 10. In each of these three cases, the standard test states that a number is congruent to its units (last) digit modulo n.

4.4 Going Farther: Cryptography Basics

In this section, we lay the foundation for implementing a number of different types of secret codes. (One such code is described in this chapter. Others will appear in later parts of the course.) In all cases, the basic situation is the same. There are two people named Alice and Bill who want to send messages to each other. Other people may intercept these messages, so Alice and Bill want to disguise the messages so that only they can read them. A method for disguising messages is called a *cryptosystem*. Many different cryptosystems have been used, but in this course we will concentrate only on those systems that have some connection with number theory.

To get started, we need a standard procedure for translating numbers into letters and vice versa. A standard conversion between letters and numbers is the *American Standard Code for Information Interchange*, also known as ASCII. Mathematica provides functions to convert back and forth between letters and their ASCII equivalents. ASCII reserves some numbers, such as the numbers between 0 and 31, for "nonprinting characters" such as Control characters, and we want to skip those.

The applet below will automatically perform conversions from text to numbers and from numbers to text. (It also has other functions that we will use in the next section.) To convert a text string to the numerical equivalents, just click on "Text->Nums":

Shift Cipher

ABCDEFGHIJKLMNOPQRSTUVWXYZ, abcdefghijklmnopqrstuvwxyz

Text->Nums | Nums->Text

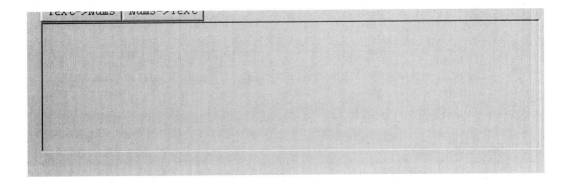

As you can see, we are using the convention that

$$A = 33, B = 34, C = 35, \ldots, Z = 58, a = 65, b = 66, \ldots, z = 90.$$

A space is represented by the number 0, and our comma translated into the number 12. To convert a list of numbers back into text, we just use "Nums->Text". Applied to our output from up above:

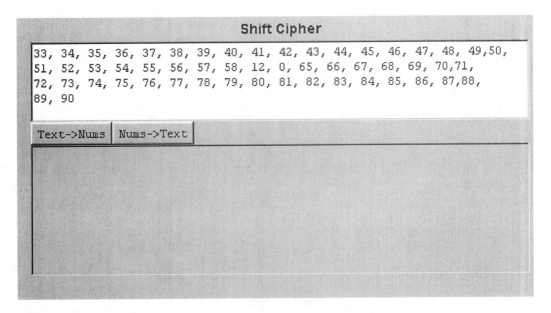

Now, let's put them together on some more meaningful text. We first convert from text to numbers:

Shift Cipher

Both the Stanford and DEC uses of the ASCII control characters are
in violation of the USA Standard Code, but no Federal marshal is likely
to come running out and arrest people who type control-T to their
computers. -- Brian Reid (1978).

| Text->Nums | Nums->Text |

Now, we convert the numbers back to text:

Shift Cipher

65, 82, 82, 69, 83, 84, 0, 80, 69, 79, 80, 76, 69, 0, 87, 72,79, 0,
84, 89, 80, 69, 0, 67, 79, 78, 84, 82, 79, 76, 13, 52, 0, 84,79, 0,
84, 72, 69, 73, 82, 0, 67, 79, 77, 80, 85, 84, 69, 82, 83,14, 0, 13,
13, 0, 34, 82, 73, 65, 78, 0, 50, 69, 73, 68, 0, 8, 17,25, 23, 24,
9, 14

| Text->Nums | Nums->Text |

Exercise 5

What is the numerical equivalent of the punctuation mark
"." (period)?

4.5 Going Farther: Shift Ciphers

The simplest cipher we shall consider is called the *shift cipher*. When implementing the shift cipher, each letter of the message is shifted a fixed distance down the alphabet. Our "alphabet," including punctuation and both upper- and lowercase letters, is numbered from 0 to 94. You can see it by clicking on "Nums->Text" below:

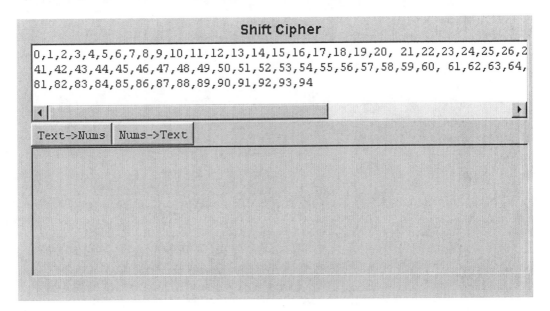

If we let

P = a number representing a letter from the unencoded message

and

C = the corresponding number for encoded message,

then the shift cipher can be expressed mathematically as

$$C = (P + k) \% 95,$$

where k is a fixed integer called the *key*. We work modulo 95 because we are using an "alphabet" with 95 "letters." This congruence is used to encode each letter in the original message. Shift ciphers are sometimes called Caesar ciphers, because Julius Caesar purportedly was the first to use one.

Let's look at an example. Suppose that our unencoded message (referred to as *plaintext*) is "Hello!", and we want to encipher this with a key $k = 4$. We start by turning the plaintext into numbers using "Text->Nums" in the "Shift Cipher" applet:

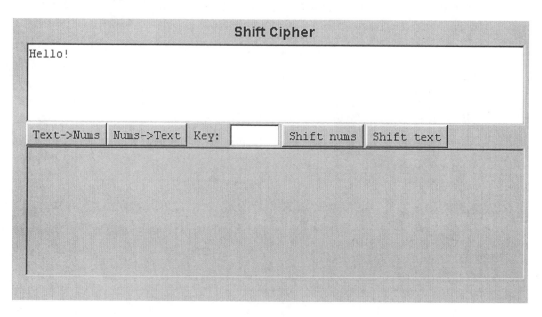

Next, we add 4 to each number in the string, and then reduce mod 95. This can be done by setting the value of "Key" to 4, and clicking on "Shift nums":

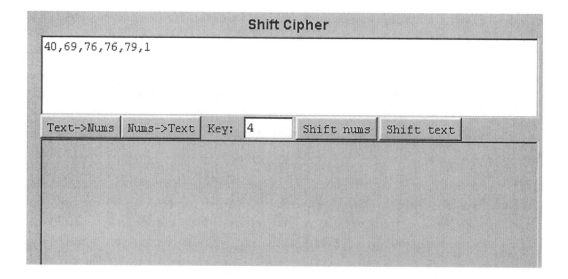

Finally, we turn this number string back into the encoded text, called *ciphertext*, by using "Nums->Text":

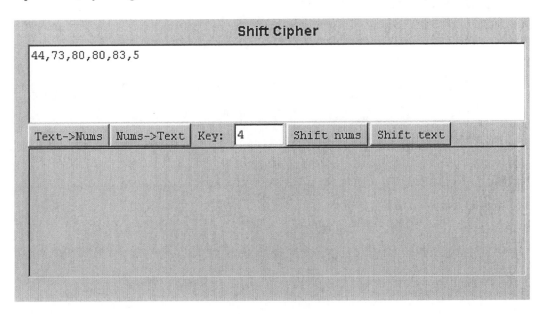

As you've probably already guessed, the applet will perform all of these steps automatically. You need only include the plaintext, the key, and click on "Text shift to Text":

Hello!

For example, here is a shift cipher with $k = 7$.

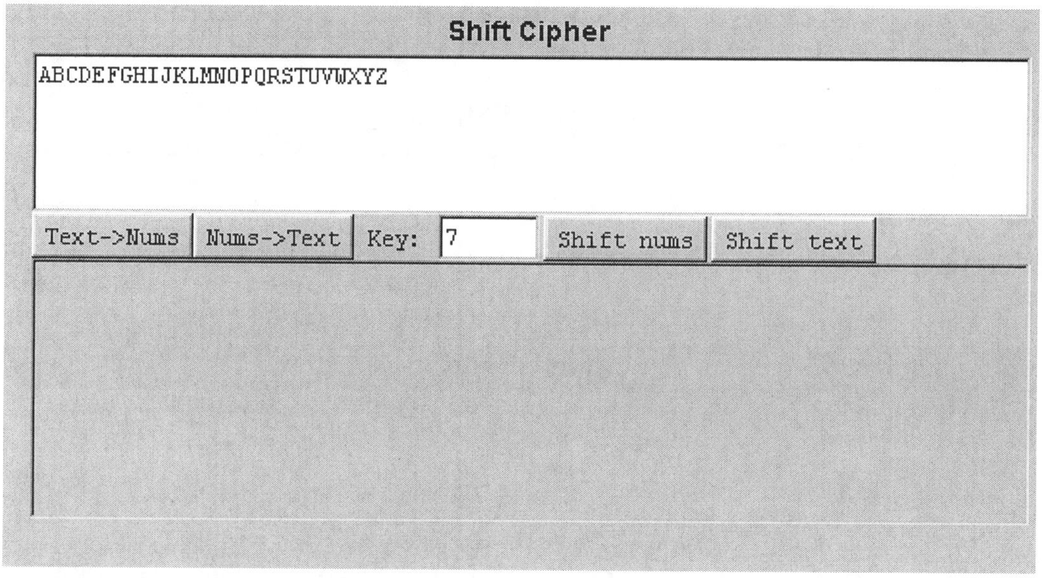

Shift Cipher

ABCDEFGHIJKLMNOPQRSTUVWXYZ

| Text->Nums | Nums->Text | Key: | 7 | Shift nums | Shift text |

Decoding a shift cipher is easy if you know the key k. Since

$$C = (P + k) \% 95,$$

it is easy to see that

$$P = (C - k) \% 95.$$

In other words, a shift cipher with key k can be decoded with another shift cipher with key equal to $-k$. Here's an example:

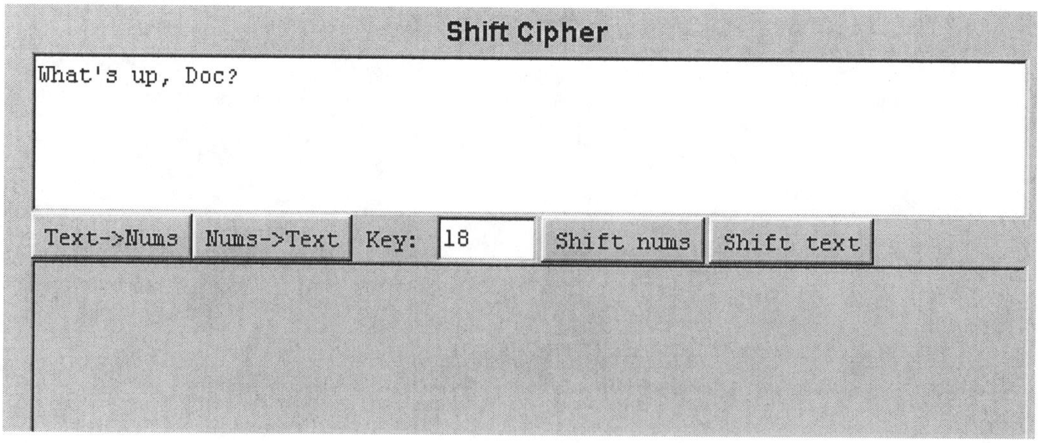

Shift Cipher

What's up, Doc?

| Text->Nums | Nums->Text | Key: | 18 | Shift nums | Shift text |

And here we decode by setting $k = -18$:

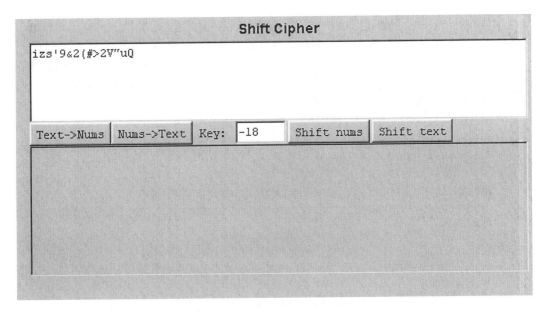

4.5.1 ROT-13: Shift Ciphers and the Internet

A simpler version of the shift cipher called ROT-13 has been in use on the internet for many years. Specifically, it is used in Usenet, the collection of newsgroups where people discuss a multitude of topics. Consider, for example, the group **rec.humor.funny**. People submit jokes which are reviewed by a moderator, and then the funny ones are sent out over Usenet to the world. Sometimes the moderator wants to send out an "off-color" joke, but doesn't want to offend the more sensitive readers. Including a warning along with the joke does no good since a delicate reader may see the offending words before they read the warning.

The solution is that the naughty jokes are encoded by ROT-13. A plain text warning message is also included so that readers can decide if they want to "risk" decoding the joke. In ROT-13, the encryption only applies to letters of the alphabet; spaces and punctuation are left unchanged. Since there are 26 letters in the alphabet, the cipher works modulo 26. The shift which is used is $k = 13$ (thus the name, because it ROTates letters 13 slots through the alphabet).

Most software used for reading Usenet newsgroups have the commands built into them for decoding such messages. The user just picks "Decode" from a menu or hits a special key. Rest assured that users do not have to understand shift ciphers to read dirty jokes on the internet.

One handy feature of ROT-13 is that we can use the same function for both encoding and decoding! The reason is that we are working modulo 26. To decode a shift by 13 we need to shift by -13. But, $-13 \equiv 13 \pmod{26}$. So, decoding is also a shift by 13. This is why 13 was chosen as the standard for this cipher in the first place.

Let's look at an example of ROT-13. To keep in the spirit of this discussion, below is a joke whose punch line has been encoded by ROT-13. The joke does not contain offensive words. However, we should warn the reader that understanding the joke requires a certain level of mathematical knowledge. Moreover, some readers may not find it funny. Here is the joke:

> Q: What did the mathematician say when epsilon went to zero?
> N: Jryy, gurer tbrf gur arvtuobeubbq!

To decode the punch line, you need to run ROT-13 on the encrypted text:

Shift Cipher

```
N: Jryy, gurer tbrf gur arvtuobeubbq!
```

```
Rot-13
```

4.5.2 Getting Some Exercise

We now return to our standard shift ciphers, with a 95-letter alphabet.

Exercise 6

> How many different shift ciphers are possible with
> a 95-letter alphabet?
>
> How many different shift ciphers are possible with
> an *n*-letter alphabet?

Exercise 7

The following message has been encoded with a shift cipher
with key 9.

ciphertext =]qn)xk rx~|)vj}qnvj}rlju)k{njt}q{x~pq)!x~um)kn)
mn \ nuxyvnw})xo)jw)nj|#)!j#)}x)ojl}x{)uj{pn)y{rvn)w~vkn
{|7)66Kruu)Pj}n|

> (a) Use the applet below to decode the message.
>
> (b) The author of this message was trying to predict
> what important mathematical breakthroughs may
> occur in the near future. Explain why the statement
> is nonsensical.

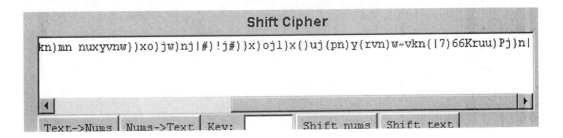

Shift Cipher

kn}mn nuxyvnw})xo)jw)nj|#)!j#)}x)ojl}x{)uj{pn)y{rvn)w~vkn{|7)66Kruu)Pj}n|

Text->Nums | Nums->Text | Key: | Shift nums | Shift text

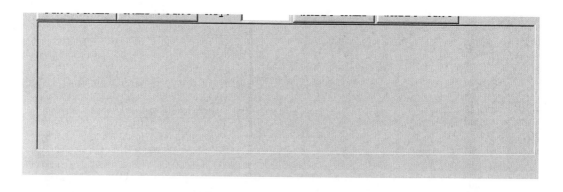

Exercise 8

Below is a message that has been encoded with a shift cipher. We don't know what key k was used for the encoding, but we do know that the last character in the plain text message was a period. Use this information to decode the message.

message = r5B5:3K9;:1EK@;K@41K3;B1>:91:@K5?
K8571K35B5:3KC45?71EK-:0K/->?K@;K@11:-31K.;E?
YKXXK{YKuYKzR};A>71Y

Hint: In Exercise 5, you determined the numerical equivalent of a period.

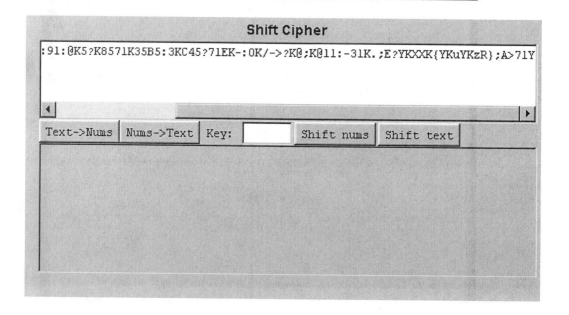

Shift Cipher

:91:@K5?K8571K35B5:3KC45?71EK-:0K/->?K@;K@11:-31K.;E?YKXXK{YKuYKzR};A>71Y

| Text->Nums | Nums->Text | Key: | Shift nums | Shift text |

Exercise 9

Cryptanalysis is the art of deciphering an encoded message without knowing the key. Shift ciphers are easy to cryptanalyze because there are very few keys. Take, for example, the following message, which has been encoded with a shift cipher.

$$\text{ciphertext} = \text{i+An1/B}$$

Try all 95 different possible shift ciphers on this message. What is the plaintext?

Hint: The applet below will make a list of all 95 different shifts.

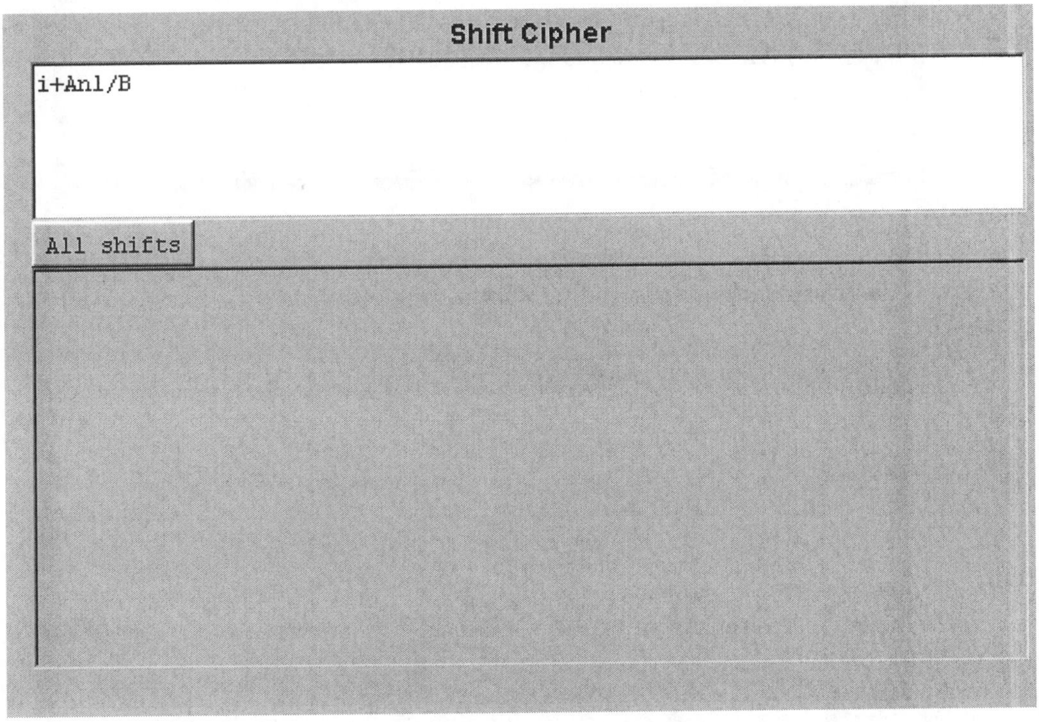

Shift Cipher

i+An1/B

All shifts

Exercise 10

Alice and Bill are sending messages to each other by using a shift cipher. They are very careful not to reveal their key to anyone, but they don't realize that shift ciphers are very unsecure.

Oscar has been observing their messages, and so far, he has intercepted the four coded texts. Each has been preloaded into an applet below.

> (a) Help Oscar decipher the messages. (Note that all of the messages start the same way.)

> (b) Oscar wants to do more than just eavesdrop; he wants to create some mischief by impersonating Bill and sending a message with false information to Alice. Devise such a message for Oscar, and encrypt it for him.

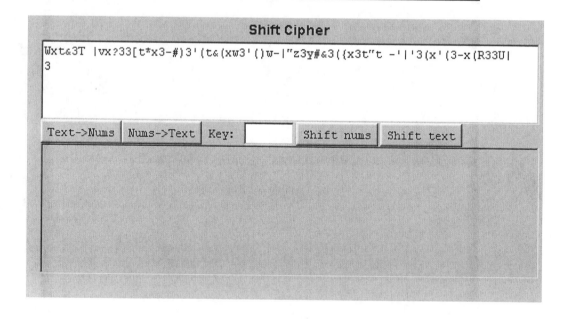

Shift Cipher

```
Wxt&3U|   ?33a#(3-x(433j{t(3wt-3|'3({x3(x'(R33T |vx
```

Text->Nums | Nums->Text | Key: [   ] | Shift nums | Shift text

Shift Cipher

```
Wxt&3T |vx?33\3({|"~3({x3(x'(3|'3#"3g{)&'wt-A33\3+|   3v{xv~3+|({3V{&|'A33
33
```

Text->Nums | Nums->Text | Key: [   ] | Shift nums | Shift text

Shift Cipher

```
Wxt&3U|   ?33b"x3#({x&3({|"zA33j{t(3|'3({x3wxy|"|(|#"3#y3t3v#"(|")#}#)'3y)"v
|vx
```

Text->Nums | Nums->Text | Key: [   ] | Shift nums | Shift text

Homework

1. The check digit for all ISBNs is one of the numbers 0, 1, 2, ... , 9, or the letter X. One of your fellow students comments "Gee, it sure is a pain to have to use that X all the time. Why don't they just compute the check sum modulo 10 instead of modulo 11, so that we can get rid of the X?" Would this plan work? Prove your answer.

2. Suppose that a publishing company hires a particularly poor typist who occasionally enters two digits of an ISBN incorrectly. (Note that these errors are not necessarily the result of interchanging the two digits.) Will the check digit always reflect the typist's error? Prove your answer.

3. Suppose $a_1 a_2 \ldots a_{10}$ is a complete ISBN number, where we treat the letter X as being equal to 10. Prove that the ISBN number has the correct check digit if and only if

$$\sum_{i=1}^{10} i a_i \equiv 0 \pmod{11}.$$

4. Suppose $a_1 a_2 \ldots a_{10}$ is a complete ISBN number, where we treat the letter X as being equal to 10. Prove that the ISBN number has the correct check digit if and only if

$$\sum_{i=1}^{10} (11 - i) a_i \equiv 0 \pmod{11}.$$

(Note, this sum is just like the one in the previous exercise, but the multipliers are used in reverse order.)

5. The following check digit scheme is used by major credit cards such as Visa, MasterCard, Discover, and American Express. Let $\mathbf{Z}_{10} = \{0, 1, \ldots, 9\}$ and define the function $f \colon \mathbf{Z}_{10} \to \mathbf{Z}_{10}$ given by

$$f(a) = \begin{cases} 2a & \text{if } a < 5, \\ 2a - 9 & \text{if } a \geq 5. \end{cases}$$

A credit card number $a_1 a_2 a_3 \ldots a_n$ has a valid check sum if[1]

$$a_n + f(a_{n-1}) + a_{n-2} + f(a_{n-3}) + \cdots \equiv 0 \pmod{10}.$$

The last term of the sum on the left side of the congruence is either a_1 or $f(a_1)$ depending on whether n is odd or even respectively.

---

[1]There are other restrictions for being a completely valid credit card number. For example, all Discover Card numbers begin with 6011.

(a) Prove that changing one digit of a valid credit card number invalidates the check sum.

(b) Does switching two adjacent distinct digits of a valid credit card number lead to an invalid number? Prove your answer.

6. The UPC bar code symbols on products use a check digit. A common system uses 12 digits where the 12th digit is the check digit. A 12 digit code $a_1 a_2 \ldots a_{12}$ passes the check sum test if

$$3a_1 + a_2 + 3a_3 + a_4 + \cdots + 3a_{11} + a_{12} \equiv 0 \pmod{10}.$$

In other words, a weighted sum of the digits is computed where the weighting factors alternate between 1 and 3.

(a) Will this system detect a change in one of the digits? Prove your answer.

(b) Will this system detect the transposition of two consecutive, but distinct, digits? Prove your answer.

(c) At the time of this writing, a site on the WWW explained this system but with weighting factors of 1 and 13 instead of 1 and 3 as above. Would this produce equivalent results? Prove your answer.

7. Suppose that a_i is a purely periodic sequence with period P, and that m is the minimal period for this sequence. Prove that $m \mid P$.

8. Suppose m is a positive integer. Complete the proof of Proposition 4.8 by showing that if the sequence $(10^0 \% m)$, $(10^1 \% m)$, $(10^2 \% m)$, ... is purely periodic, then $\gcd(m, 10) = 1$. (Hint: If the sequence is purely periodic, then there is a positive integer j so that 10^j has the same remainder when divided by m as does 10^0.)

9. Determine all primes p, with $p \neq 2$ or 5, such that the divisibility test for p requires exactly m multipliers for

(a) $m = 4$

(b) $m = 5$

(c) $m = 6$

10. We define a *rep-unit* of length n by $r(n) = (10^n - 1)/9$. It is the integer whose decimal expansion consists of the digit 1 repeated n times.

(a) Prove that $3 \mid r(n)$ if and only if $3 \mid n$.

(b) Prove that $11 \mid r(n)$ if and only if n is even.

(c) Prove that $41 \mid r(n)$ if and only if $5 \mid n$.

(d) If p is a prime with $p > 5$, prove that the length of the divisibility test for p is the smallest n such that $p \mid r(n)$.

(e) Suppose that m is a multiple of 5. Prove that $m \nmid r(n)$ for all $n \geq 1$.

11. Suppose that b is a positive integer. Then for any integer $n \geq 0$, there exist integers $a_0, a_1, a_2, \ldots, a_k$ such that

$$n = a_0 + a_1 b + a_2 b^2 + \cdots + a_k b^k, \tag{4.3}$$

and $0 \leq a_i \leq b - 1$ for each $i = 0, 1, 2, \ldots, k$. The right side of (4.3) is called the *base b expansion for n*, and we say that $a_0, a_1, a_2, \ldots, a_k$ are the *digits of n in base b*.

(a) Find the base 7 expansion for each of $n = 24$, $n = 59$, and $n = 117$.

(b) Find the base 13 expansion for each of $n = 24$, $n = 59$, and $n = 117$.

(c) Find an algorithm that uses congruences modulo b for determining the digits base b for an integer n.

(d) Let m be a fixed positive integer. Determine the positive values of b for which the following divisibility test holds:

  A positive integer n is divisible by m if and only if the sum of the digits for n in base b is divisible by m.

  We have already seen two examples of divisibility tests of this type. The first was for divisibility by 3, where we had $b = 10$ and $m = 3$, and the second was for divisibility by 9, where we had $b = 10$ and $m = 9$.

12. Suppose we start the rock game where the rocks are arranged to form a square where there are an odd number of rocks on each side of the square. In particular, this means that the starting number of rocks is a perfect square. When will the first player be able to win? Prove your answer.

13. Suppose we have a nonempty finite set S and a function $f \colon S \to S$. Fix a value $x_0 \in S$ and define a sequence by $x_{n+1} = f(x_n)$.

(a) Prove that the sequence x_i is ultimately periodic.

(b) Prove that the minimal period for the sequence is at most the number of elements in S.

(c) Prove that if f is bijective,[2] then the sequence x_i is purely periodic.

---

[2]In other words, f is one-to-one and onto.

14. Fix a positive integer n. Here we will prove that a rational number with denominator n has an ultimately periodic decimal expansion with minimal period of at most $n - 1$.

    Consider A/n with $A \in \mathbf{Z}$, and let $S = \{r \mid 0 \le r < n\}$.

    (a) Show that $A/n = q + r/n$ where $q \in \mathbf{Z}$ and $r \in S$.

    (b) Performing long division to compute the decimal expansion of r/n is an iterative process. To compute the jth digit to the right of the decimal point, d_j, for the expansion we use the remainder from the previous round, r_{j-1}. In the process we produce a new remainder $r_j \in S$. The process starts with $r_0 = r$. Show that d_j and r_j are the quotient and remainder when $10r_{j-1}$ is divided by n.

    (c) Prove that the sequence r_j is ultimately periodic with minimal period of length at most $n - 1$. (Hint: There are two cases to consider, and you may want to modify the set S for each case.)

    (d) Prove that the sequence of decimal digits d_j is ultimately periodic with minimal period of length at most $n - 1$.

    The next four exercises related to the Fibonacci numbers.[3] The *Fibonacci numbers* F_i are defined by $F_0 = 0$, $F_1 = 1$, and $F_i = F_{i-1} + F_{i-2}$ for $i \ge 2$.

15. Fix an integer $n > 1$. Consider the set S of ordered pairs (a, b) where $0 \le a < n$ and $0 \le b < n$.

    (a) Determine the order of S.

    (b) Let $f \colon S \to S$ be given by $f(a, b) = (b, (a + b) \% n)$, and define $x_0 = (0, 1)$. Define the sequence x_i by $x_i = f(x_{i-1})$. Prove that $x_i = (F_i \% n, F_{i+1} \% n)$.

    (c) Write out the first ten terms of the sequence x_i with $n = 5$.

    (d) Prove that the sequence $y_i = (F_i \% n)$ is a purely periodic sequence with minimal period of at most $n^2 - 1$.

16. Prove the following statements.

    (a) For positive integers n and i, $F_{n+i} \equiv F_{n+1}F_i \pmod{F_n}$. (Hint: The sequence is determined by two consecutive terms. What can we say about F_n and F_{n+1} when compared to F_0 and F_1 modulo F_n.)

    (b) For positive integers n, and k, $F_{nk} \equiv 0 \pmod{F_n}$.

    (c) If $m \mid n$, then $F_m \mid F_n$.

---

[3]This sequence of numbers are named for Leonardo Pisano, also known as Fibonacci. Fibonacci published books on arithmetic, geometry, and number theory in the early thirteenth century.

17. Let A be the matrix $\begin{pmatrix} 1 & 1 \\ 1 & 0 \end{pmatrix}$.

   (a) Prove that

   $$A^n \begin{pmatrix} F_1 \\ F_0 \end{pmatrix} = \begin{pmatrix} F_{n+1} \\ F_n \end{pmatrix}.$$

   (b) Diagonalize the matrix A. In other word, find an invertible matrix B such that $BAB^{-1} = D$ where D is a diagonal matrix.

   (c) Find a closed form formula for A^n. (Hint: First find a simple formula for D^n and use the fact that we can write A in terms of B and D.)

   (d) Use the previous parts to prove

   $$F_n = \frac{\left(\frac{1+\sqrt{5}}{2}\right)^n - \left(\frac{1-\sqrt{5}}{2}\right)^n}{\sqrt{5}}. \tag{4.4}$$

18. The preceeding exercise gives a derivation of formula (4.4). Use induction to give an alternative proof of this formula.

5. Solving Linear Congruences

Prelab

Being able to solve linear equations is such a useful technique in algebra that we will study the analogous question for congruences in this chapter. Specifically, we are interested in solving linear congruence equations of the form

$$ax \equiv b \pmod{n},$$

where a, b, and n are given.[1] In other words, given integers a and b and a positive integer n, we want to be able to completely answer the three following questions:

1. Are there any solutions?

2. If there are solutions, *what* are they?

3. If there are solutions, *how many* are there?

A good way to warm up is to answer the same questions in similar, more familiar settings.

Let's start with the equation $ax = b$, where a and b are rational numbers and we are looking for solutions $x \in \mathbf{Q}$. Then there are several cases:

- If $a \neq 0$, then there is exactly one solution $x \in \mathbf{Q}$, namely $x = b/a$.

- If $a = b = 0$, then there are infinitely many solutions. In particular, every rational number x is a solution to $ax = b$.

- If $a = 0$ and $b \neq 0$, then there are no solutions.

Notice that in each case, the statements answered all three questions above, but there are different answers in different cases. This is somewhat typical.

Now, let's treat the case when a and b are integers and we are looking for solutions to $ax = b$ with $x \in \mathbf{Z}$:

1. **Given two integers a and b, answer the three questions above for finding all integer solutions to $ax = b$.**

---

[1]In RQs4 and 5 of Chapter 3, we looked at the special case when $b = 0$. Recall that this amounts to finding what is known as the *additive order* of a.

Be careful to treat *all* cases.

We now return to our original congruence equation $ax \equiv b \pmod{n}$. There is one approach we can use to find solutions for this equation which is not available when working with rational numbers or integers. For fixed n, there are only finitely many congruence classes mod n, represented by the set $\{0, 1, 2, \ldots, (n-1)\}$. So one way to completely answer the questions posed above for a given congruence equation is to plug in each of $0, 1, 2, \ldots, (n-1)$ for x, and see which ones work. Use this approach to find all solutions to each of the following congruence equations:

2. **Which congruence classes x satisfy $2x \equiv 3 \pmod{5}$?**

3. **Which congruence classes x satisfy $5x \equiv 1 \pmod{7}$?**

4. **Which congruence classes x satisfy $2x \equiv 2 \pmod{6}$?**

5. **Which congruence classes x satisfy $2x \equiv 3 \pmod{6}$?**

Maple electronic notebook `05-lincong.mws`

Mathematica electronic notebook `05-lincong.nb`

Web electronic notebook Start with the web page `index.html`

Maple Lab: Solving Linear Congruences

◼ Techniques Used with Rationals and Integers

◼ Finding Rational Solutions for Linear Equations

If a and b are rational numbers and $a \neq 0$, the equation $a\,x = b$ has exactly one rational solution, namely b/a. We can find this by dividing by a, or equivalently, by multiplying both sides of the equation by $1/a$. So, solving the equation is easy if a has a *multiplicative inverse* (a number which when multiplied by a gives 1).

The remaining case is when $a = 0$. Here our equation takes the form $0\,x = b$. If $b \neq 0$, then there are no solutions to the equation, and if $b = 0$, then *every* rational number x is a solution to the equation.

◼ Finding Integer Solutions for Linear Equations

Determining integer solutions to $a\,x = b$, when a and b are integers, is similar to the rational case, but a bit more complicated. The degenerate cases are the same: if $a = b = 0$, then every integer x is a solution; if $a = 0$ and $b \neq 0$, then there are no integer solutions to $a\,x = b$.

Now suppose $a \neq 0$. It follows directly from the definition of divisibility that $a\,x = b$ has a solution if and only if $a \mid b$. Moreover, if $a \mid b$, then there is a unique solution, namely $x = b\,/\,a$.

◼ Multiplicative Inverses

Finding rational solutions to linear equations is easy because every nonzero rational number has a multiplicative inverse. Finding integer solutions to linear equations is a bit more complicated because the only integers which have integer multiplicative inverses are .

A pleasant surprise is that when working modulo n, one frequently finds that many elements have multiplicative inverses. In this context, the multiplicative inverse (or just "inverse" for short) of a modulo n, which is denoted by $a^{(-1)}$, satisfies

$$a \cdot a^{(-1)} \equiv 1 \pmod{n}.$$

For example, if $a = 29$, then $a^{(-1)} \equiv 35 \pmod{78}$. We'll get to how to compute the inverse later in the chapter. For now, we can verify the claim with the following computation:

```
[ > 29*35 mod 78;
[ >
```

One complication is that not all choices for a will have an inverse mod n. One way to determine if a has an inverse mod n is to simply multiply a by each of $0, 1, 2, \ldots, n - 1$, and then look to see if any of the products is congruent to 1 (mod n). For example, here's what we get for $a = 29$ and $n = 78$:

```
[ > seq(modp(29*j, 78), j=0..77);
[ >
```

As we can see, the output list contains a 1, which indicates that 29 has an inverse mod 78. Moreover, since the 1 appears in the 36th entry, this tells us that

251

$$29 \cdot 35 \equiv 1 \ (\text{mod } 78),$$

as shown above. On the other hand, here's what we get for $a = 32$ and $n = 78$:

```
[ > seq(modp(32*j, 78), j=0..77);
[ >
```

There's no 1 in the list, so 32 does not have an inverse mod 78. It's easy to have Maple automate this process. The function `hasinv` defined below uses the method illustrated above to determine if a has an inverse modulo n. The function `hasinvlist` produces a list of all values of a that have an inverse mod n.

```
[ > hasinv := (a,n) -> member(1, [seq(modp(a*j, n), j=0..n-1)]):

    hasinvlist := n -> select(hasinv, [$(0..n-1)], n):
```

For instance, here is a list of all values of a that have an inverse mod 10:

```
[ > hasinvlist(10);
```

You can also look at the complement of the output of `hasinvlist`: the congruence classes modulo n which do *not* have inverses modulo n. We call this function `noinvlist`:

```
[ > noinvlist := n-> remove(hasinv, [$(0..n-1)], n):
```

Try it out:

```
[ > noinvlist(10);
[ >
```

Try both functions for different values of n, and try to determine which values of a will appear for a given value of n.

■ Research Question 1

If $n > 0$, what values of a (between 0 and $n - 1$) will have an inverse mod n ?

Note: You may find it easier to think of this question in the following equivalent form: for which values of a does the congruence equation $a\,x \equiv 1 \ (\text{mod } n)$ have solutions?

As you can see, the problem of finding an inverse for a modulo n is really a special case of the general problem considered in the Prelab discussion, namely, to solve the linear congruence equation

$$a\,x \equiv b \ (\text{mod } n).$$

We take up this problem in the next section.

■ Linear Congruence Equations

In a previous chapter, you completely determined all solutions to the linear diophantine equation

$$a\,x + b\,y = c.$$

Our goal in this section will be to do the same for the linear congruence equation

$$a\,x \equiv b \ (\mathrm{mod}\ n).$$

Specifically, if a, b, and n are integers (with n positive), then we want to completely answer the following three questions for this congruence equation:

1. Are there any solutions?

2. If there are solutions, *what* are they?

3. If there are solutions, *how many* are there?

There are many approaches one might take to try to answer these questions, but it may be helpful to have a function for computing examples. Recall from your Prelab work that if a, b, and n are given, then the solutions can be determined by simply trying all possible congruence classes for x. The function congsolve defined below does just that.

```
> congsolve := proc(a, b, n)
    local j, solutions;
    solutions := NULL;
    for j from 0 to n-1 do
     if modp(a*j, n) = modp(b, n) then
      solutions := solutions,j;
     fi;
    od;
    {solutions};
   end:
```

For example, here are all of the solutions to $2\,x \equiv 5 \ (\mathrm{mod}\ 11)$:

```
> congsolve(2,5,11);
```

Here are all of the solutions to $6\,x \equiv 9 \ (\mathrm{mod}\ 15)$:

```
> congsolve(6,9,15);
```

And here are all of the all of the solutions to $10\,x \equiv 35 \ (\mathrm{mod}\ 42)$:

```
> congsolve(10,35,42);
>
```

Here we get the empty set because there are no solutions.

Try some other examples on your own. The following exercise gives you a chance to check the function congsolve against your solutions found on the Prelab assignment.

Exercise 1

Compute the answers to questions 2, 3, 4, and 5 from the Prelab exercises using congsolve, and then compare to your Prelab solutions.

■ Research Question 2

For the congruence equation $a\,x \equiv b \pmod{n}$, determine conditions on a, b, and n for there to be at least one solution.

■ Research Question 3

Suppose that a, b, and n satisfy the conditions you gave in response to Research Question 2. Describe a method for finding a solution to $a\,x \equiv b \pmod{n}$.

Hint: One possibility is trial and error; just plug each of $x = 0, 1, 2, \ldots, n-1$ into the congruence. This will work, but there is a better, more efficient answer.

■ Research Question 4

Suppose that a, b, and n satisfy the conditions you gave in response to Research Question 2, and that x_0 is a solution to $a\,x \equiv b \pmod{n}$. Find the form, in terms of x_0, of all solutions x modulo n.

■ Research Question 5

Suppose that a, b, and n satisfy the conditions you gave in response to Research Question 2. How many distinct solutions x modulo n are there to $a\,x \equiv b \pmod{n}$?

■ Hints

■ *General Advice*

The full solution to Research Questions 2 to 5 may not be obvious at first, but in the end there is an elegant answer. One of our standard course strategies is particularly important here: If the general case stumps you, then study special cases first.

Another suggestion: If you have trouble working with the congruence, then use the definition of congruence to convert it to an equation of integers. This will allow you to apply everything that you have learned (in earlier chapters and courses) to solve the problem.

■ *Multiplicative Inverses and Additive Orders*

You have already worked extensively with two special cases of the congruence equation $a\,x \equiv b \pmod{n}$. One was in connection with multiplicative inverses. Specifically, finding the multiplicative inverse of a modulo n is equivalent to solving the congruence equation

$$a\,x \equiv 1 \pmod{n}.$$

In an earlier chapter you found a formula for the additive order of a modulo n. This required you to find all solutions to the equation

$$a\,x \equiv 0 \pmod{n}.$$

These are two special cases of our general congruence equation. You may be able to make use of what you know about these special cases to learn more about the general equation.

■ *Vary the value of b*

In our general problem, there are three parameters: a, b, and n. Our hint here is to test specific values of *a* and *n*, and in each case compute all possible *b* so that the congruence $ax \equiv b \pmod{n}$ has a solution. This is somewhat natural since it amounts to holding *a* and *n* fixed, and plugging in all values for *x* and then watching what values come out for *b*. Maple functions to do this are defined below:

```
> seemults := proc(a, n)
    local j;
    printf('\n    x |');
    for j from 0 to n-1 do printf('%3d',j); od;
    printf('\n ------');
    for j from 0 to n-1 do printf('---'); od;
    printf('\n %2d x |', a);
    for j from 0 to n-1 do printf('%3d', modp(a*j, n)); od;
  end:

  multlist := proc(a,n)
    local j;
    {seq(a*j mod n, j=0..(n-1))};
  end:
```

The function `seemults(a, n)` makes a table of values of *x* and the corresponding value of *a x* % *n*. Here is an example:

```
[ > seemults(6,9);
```

The function `multlist(a, n)` produces a list of all values of *b* for which the congruence $ax \equiv b \pmod{n}$ has a solution. Try it:

```
[ > multlist(6,9);
[ >
```

Both functions are useful. Try them with different values and use them in good health.

■ Applications

A special case of what you have discovered in the course of investigating the nature of solutions to the congruence equation $ax \equiv b \pmod{p}$ is the following:

If *p* is prime and *a* is not a multiple of *p*, then there exists an integer *b* such that $ab \equiv 1 \pmod{p}$.

Now fix *p* as a prime number, and suppose that *a* is an integer with $0 < a < p$. Then there is an integer *b*, also between 0 and *p*, such that $ab \equiv 1 \pmod{p}$. Since $ab \equiv 1 \pmod{p}$ implies that $ba \equiv 1 \pmod{p}$, we can try to pair off the integers between 1 and $p - 1$ so that the product of each pair is 1. The function `showinvs` defined below takes a prime number as input, and displays the inverse for each value of *a* between 1 and $p - 1$:

```
> showinvs := proc(p)
    local j;
    for j to p-1 do
      printf('%3d has inverse %2d\n', j, congsolve(j,1,p)[1]);
    od;
  end:
```

Here it is in action with $p = 11$:

```
[ > showinvs(11);
[ >
```

You may find this command useful when working on Research Question 6.

◼ Factorials

Recall the definition of n factorial: $n! = 1 \cdot 2 \bullet \bullet \bullet n$. Maple will compute factorials using the usual notation. Try it:

```
[ > 17!;
[ >
```

In each of the Research Questions below, you are asked to investigate the behavior of factorials modulo n.

◼ Research Question 6

(a) Find a formula for $n! \% n$.
(b) Find a formula for $(n - 2)! \% n$.
(c) Find a formula for $(n - 1)! \% n$.

◼ Hint:

If you have a conjecture that you believe to be correct, but are having trouble finding a proof, it may be helpful to review the proof of the formula for the sum

$$(1 + 2 + 3 + \bullet\bullet\bullet + n) \% n.$$

◼ Going Farther: Affine Ciphers

In this section, we shall consider a generalization of the shift cipher called the *affine cipher*. Recall that to encode a message using a shift cipher, we convert our text to a number list, rotate each number by the key k, and then convert the encoded number list back to letters.

When working with any cipher, we need the following two basic functions to convert between text and numbers. Execute the commands below to define these functions.

```
[ > texttonums := proc(a)
     local raw, j;
     raw := convert(a, 'bytes');
     [seq(raw[j]-32, j=1..nops(raw))]
   end:

   numstotext := proc(a)
     local raw, j;
     raw := [seq(a[j]+32, j=1..nops(a))];
     convert(raw, 'bytes')
     end:
```

Here's a quick example of the shift cipher, using the key $k = 52$.

```
> plaintext := 'Math is my life!';
  p := texttonums(plaintext);
  c := map('a'->modp('a'+52, 95), p);
  ciphertext := numstotext(c);
>
```

In terms of congruences, the relationship between p and c is

$$c \equiv p + k \pmod{95}.$$

The congruence is modulo 95 because we have a 95-character alphabet (including upper- and lowercase letters, digits, and punctuation). Decoding a message requires us to solve for p in the above congruence. Using simple algebra, we see that

$$p \equiv c - k \pmod{95}.$$

Thus decoding looks a lot like encoding.

```
> ciphertext := ',=DT=6HTBNTHA>9:TGJA:s';
  c := texttonums(ciphertext);
  p := map('a'->modp('a'-52, 95), c);
  plaintext := numstotext(p);
>
```

The affine cipher works in a similar manner, except that the key consists of two numbers a and b. We encode a message by applying the formula

$$c \equiv a\,p + b \pmod{95}.$$

For example, if we take $a = 41$ and $b = 12$, then we encipher as follows.

```
> a := 41:  b:= 12:
  plaintext := 'It's next to your pocket protector.';
  p := texttonums(plaintext);
  c := map('letter'->modp(a*'letter'+b, 95), p);
  ciphertext := numstotext(c);
>
```

The command affine defined below will let us do everything in one shot.

```
> affine := proc(plaintext, a, b)
     local p, c, j;
     p := texttonums(plaintext);
     c := [seq((a*p[j]+b) mod 95, j=1..nops(p))];
     numstotext(c);
  end:
>
```

Here it is in action.

```
> plaintext:='What about my mechanical pencil?';
  affine(plaintext, 41, 12);
>
```

To decode a message encoded with an affine cipher, we must solve for p in the congruence

$$c \equiv a\,p + b \pmod{95}.$$

Subtracting b from both sides and then multiplying by $a^{(-1)} \pmod{95}$ yields

$$p \equiv a^{(-1)}(c - b) \equiv a^{(-1)}c - a^{(-1)}b \pmod{95}.$$

Note that this is just another affine cipher with key $a^{(-1)}$ and $-a^{(-1)}b$. Below is a message encoded with an affine cipher taking $a = 41$ and $b = 12$ as above.

```
[ > ciphertext := 'n,x57v,kmBZvQ,D3v5QSU';
```

To decode, we first need to compute $a^{(-1)} \pmod{95}$. As noted earlier, this is equivalent to solving the congruence equation $a\,x \equiv 1 \pmod{95}$, which can be accomplished using `congsolve`.

```
[ > congsolve(41,1,95);
```

We can now finish the decoding.

```
[ > affine(ciphertext, 51, -51*12);
```

As you know, for some values of a there will be a multiplicative inverse modulo 95, and for others there will not. Suppose that we had selected a value of a for which there is no inverse modulo 95? One such choice is $a \equiv 38 \pmod{95}$, as can be seen below.

```
[ > congsolve(38,1,95);
```

Check out what happens.

```
[ > plaintext := 'abcdefghijklmnopqrstuvwxyz';
    affine(plaintext, 38,12);
[ >
```

The problem is clear. Since the letters b, f, k, p (and more) are all sent to R when encoded, we have no way of knowing how to decode R. This ambiguity renders the encription system useless. After all, what fun is a secret code if the messages can't be decoded?

■ *Exercise 2*

The message below was encoded using an affine cipher with key $a = 59$ and $b = 34$. Decode the message.

```
[ > ciphertext :=
    '^'4Bv_.e;4BHoBR'4B'\\$HR41.v4BbvB4_.eUBRHBR'4Bv.1vBHoBR'4Bv_.e;4
    vBHoBR'4BHR'4;BREHBvbX4v%BI^'4B#|e;4|;HE';
[ >
```

■ Going Farther: Cryptanalysis

■ The Basics

We begin this section with a simple example of cryptanalysis, the practice of breaking codes.

Suppose that Alice and Bill have been exchanging messages using an affine cipher. Oscar suspects that they have been talking about him, and desperately wants to know the contents of their messages. As usual, we may assume that Oscar knows that an affine cipher is being used, and that the message is being encoded one letter at a time. Thus Oscar knows that all he has to do is determine the values of a and b that form the key, and then he can decode everything passed between Alice and Bill. Bill is confident that the code can't be broken, and decides to taunt Oscar by telling him "In our code, the 'a' is encoded to 'M' and the 'w' is encoded to 'b'."

Although he doesn't realize it, Bill has actually given Oscar all the information that is needed to break the code. To see why, recall that the encoding formula is

$$c \equiv a\,p + b \pmod{95}.$$

Here are the numerical equivalents for the four letters given by Bill.

```
> texttonums('aM');
  texttonums('wb');
>
```

Plugging into our encoding formula, we see that

$$45 \equiv 65\,a + b \pmod{95}$$

and

$$66 \equiv 87\,a + b \pmod{95}.$$

We now have two congruences in two unknowns a and b. Although these are congruence equations, we can use the same methods as we would for algebraic equations to find a solution. Subtracting the first congruence equation from the second eliminates the unknown b, and leaves us with the single congruence

$$21 \equiv 22\,a \pmod{95}.$$

We know all about this type of congruence equation. Let's use `congsolve` to find the solution.

```
> congsolve(22,21,95);
>
```

This tells us that $a \equiv 83 \pmod{95}$. We can then find b by back substitution (i.e., plugging back into either of our original congruence equations). Using the first congruence equation, we have

$$45 \equiv 65 \cdot 83 + b \pmod{95}.$$

After simplifying, we find that $b \equiv 65 \pmod{95}$.

▮ *Exercise 3*

The message below was sent to Bill from Alice using an affine cipher with the keys found above. Help Oscar decode the messsage.

```
> ciphertext:='EL((Fahz|33abXM'iaPapczo)aMao|ba5?JW'cd?MWXJaAcc4
  aLoa'X|a(LA?M?Jwaascbab|\
  a5Moa(|M?oaMAcz'aA|''|?a5?JW'c3J3'|{30aMo)a?|M((Japcc(ag35M?UU
```

```
  L   `:
  [ >
```

Exercise 4

Alice and Bill have changed their keys. However, Bill has left a scrap of paper lying around
which shows that "OK" is now encoded as "na". Use this information to help Oscar decode the
message below.

```
[ > ciphertext
  :='p&4P'qL+L&wwj'kL4\\L@x'L&45iyihsL5[@L+LPw[&k\\K@LB4\\kL@x'L
  5wwjJLLbwLhw[Lj\\w2L2x'i'L4@L4%:`:
  [ >
```

Frequency Analysis

Alice and Bill have learned more about cryptosystems, and now know that they should never let
anyone else know both the coded and unencoded versions of the same text. However, it is still
possible for Oscar to break their code using a basic method of cryptanalysis called frequency
analysis. This involves counting the number of occurrences of each character in the coded message, guessing at
the correspondence to the original message, and then using this information as we did earlier to find
the key.

For example, suppose that Alice has sent the following encoded message to Bill:

```
[ > ciphertext :=
  'i{,,^RORV{!R7VFR6<<fR<vR7VFRqF[<v!Rk,<<LR6FV{v!R7VFRVFp7FLR\"Fv7
  }`:
  [ >
```

In this message, the most common character is "R" and the second most common is "F". We can see
this by counting by hand, or by using the function `freq`, which will give the frequency of all
characters occuring in the message. Here is the definition of the function. You need not worry about
the details of its programming.

```
[ > freq:= proc(intext)
    local countlist, j, nums, pairlist;
    countlist:= [seq(0, j=1..95)];
    nums := texttonums(intext);
    for j from 1 to nops(nums) do
      countlist[nums[j]+1] := countlist[nums[j]+1]+1;
    od;
    pairlist:=NULL;
    for j from 1 to nops(countlist) do
     if(countlist[j]>0) then
       pairlist := pairlist, [j, countlist[j]];
     fi;
    od;
    pairlist := sort([pairlist], (a,b)->evalb(a[2]>b[2]));
    [seq([numstotext([pairlist[j][1]-1]),pairlist[j][2]],
  j=1..nops(pairlist))];
    end:
```

Now we use it to count the frequencies of each letter in the cipher text.

```
[ > freq(ciphertext);
  [ >
```

Thus "R" occurs 12 times, "F" occurs 8 times, and so on. In English, the most common characters are " " (a space), e, and t in that order. Thus we might guess that in the encoding process, " " goes to "R" and "e" goes to "F". Here are the numerical equivalents.

```
[ > texttonums(' R');
    texttonums('eF');
[ >
```

This leads us to the pair of congruences

$$50 \equiv 0\,a + b \pmod{95}$$

and

$$38 \equiv 69\,a + b \pmod{95}.$$

We see immediately from the first congruence equation that $b \equiv 50 \pmod{95}$. Plugging this into the second congruence gives us

$$83 \equiv 69\,a \pmod{95}.$$

Using `congsolve`, we get

```
[ > congsolve(69,83,95);
[ >
```

which tells us that $a \equiv 37 \pmod{95}$. We next determine (you may check that this is true) that $a^{(-1)} \equiv 18 \pmod{95}$, and then decode the message.

```
[ > affine(ciphertext,18, -18*50);
[ >
```

Sometimes frequency analysis requires more than one try to find the key. Here's an example where the characters occuring in the text do not have the same relative frequency as they generally do in English. Below is an encoded note from Bill to Alice responding to her earlier message.

```
[ > ciphertext := 'S^&tZO%~Zkw%J%3/UZ%U3Z%3Z/UZo%;ZDU-':
[ >
```

We'll follow the exact same procedure as above, starting with a frequency count for the encoded message.

```
[ > freq(ciphertext);
[ >
```

We see that "Z" occurs 7 times and "%" occurs 6 times. Thus we initially guess that in the encoding process, " " goes to "Z" and "e" goes to "%". Here are the numerical equivalents.

```
[ > texttonums(' Z');
    texttonums('e%');
[ >
```

This leads us to the pair of congruences

$$58 \equiv 0\, a + b \pmod{95}$$

and

$$5 \equiv 69\, a + b \pmod{95}.$$

We see immediately from the first congruence that $b \equiv 58 \pmod{95}$. Plugging this into the second congruence gives us

$$42 \equiv 69\, a + b \pmod{95}.$$

Using `congsolve`, we get

```
[ > congsolve(69,42,95);
[ >
```

Hence $a \equiv 13 \pmod{95}$, so that $a^{(-1)} \equiv 22 \pmod{95}$. Decoding gives us

```
[ > affine(ciphertext, 22, -22*58);
[ >
```

What happened? Well, as foreshadowed above, our first guess for the cipher text corresponding to " " and "e" was not correct. Let's try reversing things, and guess that " " goes to "%" and "e" goes to "Z". This gives us the pair of congruence equations

$$5 \equiv 0\, a + b \pmod{95}.$$

and

$$58 \equiv 69\, a + b \pmod{95}.$$

Solving (we leave the details to you) yields $a \equiv 82 \pmod{95}$ and $b \equiv 5 \pmod{95}$. Since $a^{(-1)} \equiv 73$ (mod 95), we decode as follows:

```
[ > affine(ciphertext, 73, -73*5);
[ >
```

Although Bill's response is a little bizarre, it is clearly English, and so we may conclude that the code has been broken.

▦ *Exercise 5*

Decode the following message:

```
[ > ciphertext := `
   SAAP.s/5cp.}k.$e|$Sq./e.$.Y$|;Akvv.Yk$pk|.dk5p9..
   R.pYk.^$RE.qSq.R/j.Yk$|.$}/jp.pYk.$qd$5wkq.5j;}k|.pYk/|R.wA$vv
   .}kS5_./eek|kq.5kXp.pk|;2`:
[ >
```

■ *Exercise 6*

Decode the following message:

```
> ciphertext := '+<g^aP)ta[<<
  Z()kaHcv)6pBa)gH)0pav?cH)daH)pKaE-ppja0>':
>
```

Mathematica Lab: Solving Linear Congruences

■ Techniques Used with Rationals and Integers

■ Finding Rational Solutions for Linear Equations

If a and b are rational numbers and $a \neq 0$, the equation $ax = b$ has exactly one rational solution, namely b/a. We can find this by dividing by a, or equivalently, by multiplying both sides of the equation by $1/a$. So, solving the equation is easy if a has a *multiplicative inverse* (a number which when multiplied by a gives 1).

The remaining case is when $a = 0$. Here our equation takes the form $0x = b$. If $b \neq 0$, then there are no solutions to the equation, and if $b = 0$, then *every* rational number x is a solution to the equation.

■ Finding Integer Solutions for Linear Equations

Determining integer solutions to $ax = b$, when a and b are integers, is similar to the rational case, but a bit more complicated. The degenerate cases are the same: if $a = b = 0$, then every integer x is a solution; if $a = 0$ and $b \neq 0$, then there are no integer solutions to $ax = b$.

Now suppose $a \neq 0$. It follows directly from the definition of divisibility that $ax = b$ has a solution if and only if $a \mid b$. Moreover, if $a \mid b$, then there is a unique solution, namely $x = b/a$.

■ Multiplicative Inverses

Finding rational solutions to linear equations is easy because every nonzero rational number has a multiplicative inverse. Finding integer solutions to linear equations is a bit more complicated because the only integers which have integer multiplicative inverses are ± 1.

A pleasant surprise is that when working modulo n, one frequently finds that many elements have multiplicative inverses. In this context, the multiplicative inverse (or just "inverse" for short) of a modulo n, which is denoted by a^{-1}, satisfies

$$a \cdot a^{-1} \equiv 1 \pmod{n}.$$

For example, if $a = 29$, then $a^{-1} \equiv 35 \pmod{78}$. We'll get to how to compute the inverse later in the chapter. For now, we can verify the claim with the following computation:

```
Mod[29 * 35, 78]
```

One complication is that not all choices for a will have an inverse mod n. One way to determine if a has an inverse mod n is to simply multiply a by each of $0, 1, 2, \ldots, n-1$, and then look to see if any of the products is congruent to $1 \pmod{n}$. For example, here's what we get for $a = 29$ and $n = 78$:

```
Mod[29 * Range[0, 77], 78]
```

264

∇ Mathematica Note

The command **Range[r,s]** produces a list containing the integers from **r** to **s**. Thus the command **Range[0,77]** produces the list of integers {0,1,2,...,77}, and the command **29*Range[0,77]** multiplies each element in the list by 29. Finally, the command **Mod[...,78]** takes the list **29*Range[0,77]** and gives the remainder upon division by 78.

As we can see, the output list contains a 1, which indicates that 29 has an inverse mod 78. Moreover, since the 1 appears in the 36th entry, this tells us that

$$29 \cdot 35 \equiv 1 \ (\mathrm{mod} \ 78),$$

as shown above. On the other hand, here's what we get for $a = 32$ and $n = 78$:

```
Mod[32 * Range[0, 77], 78]
```

There's no 1 in the list, so 32 does not have an inverse mod 78. It's easy to have Mathematica automate this process. The function **hasinv** defined below uses the method illustrated above to determine if a has an inverse modulo n. The function **hasinvlist** produces a list of all values of a that have an inverse mod n.

```
hasinv[a_, n_] := MemberQ[Mod[a Range[0, n - 1], n], 1]
hasinvlist[n_] := Select[Range[0, n - 1], hasinv[#, n] &]
```

For instance, here is a list of all values of a that have an inverse mod 10:

```
hasinvlist[10]
```

You can also look at the complement of the output of **hasinvlist**: the congruence classes modulo n which do *not* have inverses modulo n. We call this function **noinvlist**:

```
noinvlist[n_] :=
        Complement[Range[0, n - 1], hasinvlist[n]];
```

Try it out:

```
noinvlist[10]
```

Try both functions for different values of n, and try to determine which values of a will appear for a given value of n.

■ Research Question 1

If $n > 0$, what values of a (between 0 and $n - 1$) will have an inverse mod n ?

Note: You may find it easier to think of this question in the following equivalent form: for which values of a does the congruence equation $a\,x \equiv 1 \ (\mathrm{mod}\,n)$ have solutions?

As you can see, the problem of finding an inverse for a modulo n is really a special case of the general problem considered in the Prelab discussion, namely, to solve the linear congruence equation

$$a\,x \equiv b \ (\mathrm{mod}\,n).$$

We take up this problem in the next section.

■ Linear Congruence Equations

In a previous chapter, you completely determined all solutions to the linear diophantine equation

$$a\,x + b\,y = c.$$

Our goal in this section will be to do the same for the linear congruence equation

$$a\,x \equiv b \pmod{n}.$$

Specifically, if a, b, and n are integers (with n positive), then we want to completely answer the following three questions for this congruence equation:

1. Are there any solutions?
2. If there are solutions, *what* are they?
3. If there are solutions, *how many* are there?

There are many approaches one might take to try to answer these questions, but it may be helpful to have a function for computing examples. Recall from your Prelab work that if a, b, and n are given, then the solutions can be determined by simply trying all possible congruence classes for x. The function **congsolve** defined below does just that.

```
congsolve[a_, b_, n_] :=
  Select[Range[0, n - 1], Mod[a #, n] == Mod[b, n] &]
```

For example, here are all of the solutions to $2\,x \equiv 5 \pmod{11}$:

```
congsolve[2, 5, 11]
```

Here are all of the solutions to $6\,x \equiv 9 \pmod{15}$:

```
congsolve[6, 9, 15]
```

And here are all of the solutions to $10\,x \equiv 35 \pmod{42}$:

```
congsolve[10, 35, 42]
```

Here we get the empty set because there are no solutions.

Try some other examples on your own. The following exercise gives you a chance to check the function **congsolve** against your solutions found on the Prelab assignment.

■ Exercise 1

Compute the answers to questions 2, 3, 4, and 5 from the Prelab exercises using **congsolve**, and then compare to your Prelab solutions.

■ Research Question 2

> For the congruence equation $a\,x \equiv b \pmod{n}$, determine conditions on a, b, and n for there to be at least one solution.

■ Research Question 3

> Suppose that a, b, and n satisfy the conditions you gave in response to Research Question 2. Describe a method for finding a solution to $a\,x \equiv b \pmod{n}$.
>
> **Hint:** One possibility is trial and error; just plug each of $x = 0, 1, 2, \ldots, n-1$ into the congruence. This will work, but there is a better, more efficient answer.

■ Research Question 4

> Suppose that a, b, and n satisfy the conditions you gave in response to Research Question 2, and that x_0 is a solution to $a\,x \equiv b \pmod{n}$. Find the form, in terms of x_0, of all solutions x modulo n.

■ Research Question 5

> Suppose that a, b, and n satisfy the conditions you gave in response to Research Question 2. How many distinct solutions x modulo n are there to $a\,x \equiv b \pmod{n}$?

■ Hints

■ General Advice

The full solution to Research Questions 2 to 5 may not be obvious at first, but in the end there is an elegant answer. One of our standard course strategies is particularly important here: If the general case stumps you, then study special cases first.

Another suggestion: If you have trouble working with the congruence, then use the definition of congruence to convert it to an equation of integers. This will allow you to apply everything that you have learned in earlier chapters to solve the problem.

■ Multiplicative Inverses and Additive Orders

You have already worked extensively with two special cases of the congruence equation $a\,x \equiv b \pmod{n}$. One was in connection with multiplicative inverses. Specifically, finding the multiplicative inverse of a modulo n is equivalent to solving the congruence equation

$$a\,x \equiv 1 \pmod{n}.$$

In an earlier chapter you found a formula for the additive order of a modulo n. This required you to find all solutions to the congruence

$$a\,x \equiv 0 \pmod{n}.$$

These are two special cases of our general congruence equation. You may be able to make use of what you know about these special cases to learn more about the general equation.

■ Vary the value of *b*

In our general problem, there are three parameters: a, b, and n. Our hint here is to test specific values of a and n, and in each case compute all possible b so that the congruence $a x \equiv b$ (mod n) has a solution. This is somewhat natural since it amounts to holding a and n fixed, and plugging in all values for x and then watching what values come out for b. Mathematica functions to do this are defined below:

```
seemults[a_, n_] := DisplayForm[
  GridBox[Transpose[ Join[{{"", a}, {"x", "x"}, {"|", "|"}},
     Table[{j, Mod[a j, n]}, {j, 0, n - 1}]]],
   RowLines -> True, ColumnAlignments -> Right,
   ColumnSpacings -> Join[{0}, Table[1, {n + 1}]]]]]
multlist[a_, n_] := Union[Table[Mod[a j, n], {j, 0, n - 1}]]
```

The function **seemults[a, n]** makes a table of values of x and the corresponding value of $a x \% n$. Here is an example:

```
seemults[6, 9]
```

The function **multlist[a, n]** produces a list of all values of b for which the congruence $a x \equiv b$ (mod n) has a solution. Try it:

```
multlist[6, 9]
```

Both functions are useful. Try them with different values and use them in good health.

■ Applications

A special case of what you have discovered in the course of investigating the nature of solutions to the congruence equation $a x \equiv b$ (mod p) is the following:

> If p is prime and a is not a multiple of p, then there exists an integer b such that $a b \equiv 1$ (mod p).

Now fix p as a prime number, and suppose that a is an integer with $0 < a < p$. Then there is an integer b, also between 0 and p, such that $a b \equiv 1$ (mod p). Since $a b \equiv 1$ (mod p) implies that $b a \equiv 1$ (mod p), we can try to pair off the integers between 1 and $p - 1$ so that the product of each pair is 1. The function **showinvs** defined below takes a prime number as input, and displays the inverse for each value of a between 1 and $p - 1$:

```
showinvs[p_] := TableForm[Table[{a, "has inverse",
                    congsolve[a, 1, p][[1]]}, {a, p - 1}],
   TableSpacing -> {0, 1}]
```

Here it is in action with $p = 11$:

```
showinvs[11]
```

You may find this command useful when working on Research Question 6.

■ Factorials

Recall the definition of n factorial: $n! = 1 \cdot 2 \cdots n$. Mathematica will compute factorials using the usual notation. Try it:

```
17 !
```

In the Research Question below, you are asked to investigate the behavior of factorials modulo *n*.

■ Research Question 6

(a) Find a formula for *n*! % *n*.
(b) Find a formula for (*n* −2)! % *n*.
(c) Find a formula for (*n* −1)! % *n*.

■ Hint:

If you have a conjecture that you believe to be correct, but are having trouble finding a proof, it may be helpful to review the proof of the formula for the sum

$$(1 + 2 + 3 + \cdots + n) \% n.$$

■ Going Farther: Affine Ciphers

In this section, we shall consider a generalization of the shift cipher called the *affine cipher*. Recall that to encode a message using a shift cipher, we convert our text to a number list, rotate each number by the key *k*, and then convert the encoded number list back to letters.

When working with any cipher, we need the following two basic functions to convert between text and numbers. Execute the commands below to define these functions.

```
texttonums[text_] := ToCharacterCode[text] - 32
numstotext[nums_] := FromCharacterCode[nums + 32]
```

Here's a quick example of the shift cipher, using the key *k* = 52.

```
plaintext = "Math is my life!"
p = texttonums[plaintext]
c = Mod[p + 52, 95]
ciphertext = numstotext[c]
```

In terms of congruences, the relationship between *p* and *c* is

$$c \equiv p + k \pmod{95}.$$

The congruence is modulo 95 because we have a 95−character alphabet (including upper−and lowercase letters, digits, and punctuation). Decoding a message requires us to solve for *p* in the above congruence. Using simple algebra, we see that

$$p \equiv c - k \pmod{95}.$$

Thus decoding looks a lot like encoding.

```
ciphertext = ",=DT=6HTBNTHA>9:TGJA:s"
c = texttonums[ciphertext]
p = Mod[c - 52, 95]
plaintext = numstotext[p]
```

The affine cipher works in a similar manner, except that the key consists of two numbers *a* and *b*. We encode a message by applying the formula

$$c \equiv a\,p + b \pmod{95}.$$

For example, if we take $a = 41$ and $b = 12$, then we encipher as follows.

```
a = 41; b = 12;
plaintext = "It's next to your pocket protector."
p = texttonums[plaintext]
c = Mod[a*p + b, 95]
ciphertext = numstotext[c]
```

The command **affine** defined below will let us do everything in one shot.

```
affine[plaintext_, a_, b_] :=
        numstotext[Mod[a*texttonums[plaintext] +b, 95]]
```

Here it is in action.

```
plaintext = "What about my mechanical pencil?"
affine[plaintext, 41, 12]
```

To decode a message encoded with an affine cipher, we must solve for p in the congruence

$$c \equiv a\,p + b \pmod{95}.$$

Subtracting b from both sides and then multiplying by a^{-1} (mod 95) yields

$$p \equiv a^{-1}(c - b) \equiv a^{-1}\,c - a^{-1}\,b \pmod{95}.$$

Note that this is just another affine cipher with key a^{-1} and $-a^{-1}\,b$. Below is a message encoded with an affine cipher taking $a = 41$ and $b = 12$ as above.

```
ciphertext = "n,x57v,kmBZvQ,D3v5QSU";
```

To decode, we first need to compute a^{-1} (mod 95). As noted earlier, this is equivalent to solving the congruence equation $a\,x \equiv 1$ (mod 95), which can be accomplished using **congsolve**.

```
congsolve[41, 1, 95]
```

We can now finish the decoding.

```
affine[ciphertext, 51, -51*12]
```

As you know, for some values of a there will be a multiplicative inverse modulo 95, and for others there will not. Suppose that we had selected a value of a for which there is no inverse modulo 95? One such choice is $a \equiv 38$ (mod 95), as can be seen below.

```
congsolve[38, 1, 95]
```

Check out what happens.

```
plaintext = "abcdefghijklmnopqrstuvwxyz";
affine[plaintext, 38, 12]
```

The problem is clear. Since the letters **b**, **f**, **k**, **p** (and more) are all sent to **R** when encoded, we have no way of knowing how to decode **R**. This ambiguity renders the encription system useless. After all, what fun is a secret code if the messages can't be decoded?

■ **Exercise 2**

The message below was encoded using an affine cipher with key $a = 59$ and $b = 34$. Decode the message.

```
ciphertext = "^'4Bv_.e;4BHoBR'4B'\\$HR41.v4BbvB4_.eUBRHBR'4Bv.1
    vBHoBR'4Bv_.e;4vBHoBR'4BHR'4;BREHBvbX4v%BI^'4B#|e;4|;HE";
```

■ Going Farther: Cryptanalysis

■ The Basics

We begin this section with a simple example of *cryptanalysis*, the practice of breaking codes. Suppose that Alice and Bill have been exchanging messages using an affine cipher. Oscar suspects that they have been talking about him, and desperately wants to know the contents of their messages. As usual, we may assume that Oscar knows that an affine cipher is being used, and that the message is being encoded one letter at a time. Thus Oscar knows that all he has to do is determine the values of a and b that form the key, and then he can decode everything passed between Alice and Bill. Bill is confident that the code can't be broken, and decides to taunt Oscar by telling him "In our code, the 'a' is encoded to 'M' and the 'w' is encoded to 'b'."

Although he doesn't realize it, Bill has actually given Oscar all the information that is needed to break the code. (No one has ever accused Bill of being the sharpest tack in the box.) To see why, recall that the encoding formula is

$$c \equiv a\,p + b \pmod{95}.$$

Here are the numerical equivalents for the four letters given by Bill.

```
texttonums["aM"]
texttonums["wb"]
```

Plugging into our encoding formula, we see that

$$45 \equiv 65\,a + b \pmod{95}$$

and

$$66 \equiv 87\,a + b \pmod{95}.$$

We now have two congruences in two unknowns a and b. Although these are congruence equations, we can use the same methods as we would for algebraic equations to find a solution. Subtracting the first congruence equation from the second eliminates the unknown b, and leaves us with the single congruence

$$21 \equiv 22\,a \pmod{95}.$$

We know all about this type of congruence equation. Let's use **congsolve** to find the solution.

```
congsolve[22, 21, 95]
```

This tells us that $a \equiv 83 \pmod{95}$. We can then find b by back substitution (i.e., plugging back into either of our original congruence equations). Using the first congruence equation, we have

$$45 \equiv 65 \cdot 83 + b \pmod{95}.$$

After simplifying, we find that $b \equiv 65 \pmod{95}$.

■ Exercise 3

The message below was sent to Bill from Alice using an affine cipher with the keys found above. Help Oscar decode the messsage.

```
ciphertext =
 "EL((Fahz|33abXM'iaPapczo)aMao|ba5?JW'cd?MWXJaAcc4aLoa'
  X|a(LA?M?Jwaascbab|a5Moa(|M?oaMAcz'aA|''|?a5?JW'
  c3J3'|{30aMo)a?|M((Japcc(ag35M?UU";
```

■ Exercise 4

Alice and Bill have changed their keys. However, Bill has left a scrap of paper lying around which shows that "OK" is now encoded as "na". Use this information to help Oscar decode the message below.

```
ciphertext = "p&4P'qL+L&wwj'kL4\\L@x'L&45iyihsL5[@L+LPw[&
 k\\K@LB4\\kL@x'L5wwjJLLbwLhw[Lj\\w2L2x'i'L4@L4%:";
```

■ Frequency Analysis

Alice and Bill have learned more about cryptosystems, and now know that they should never let anyone else know both the coded and unencoded versions of the same text. However, it is still possible for Oscar to break their code using a basic method of cryptanalysis called *frequency analysis*. This involves counting the number of occurrences of each character in the coded message, guessing at the correspondence to the original message, and then using this information as we did earlier to find the key.

For example, suppose that Alice has sent the following encoded message to Bill:

```
ciphertext =
 "i{,,^RORV{!R7VFR6<<fR<
  vR7VFRqF[<v!Rk,<<LR6FV{v!R7VFRVFp7FLR\"Fv7}";
```

In this message, the most common character is "R" and the second most common is "F". We can see this by counting by hand, or by using the function **freq**, which will give the frequency of all characters occuring in the message.

```
freq[ciphertext]
```

Thus "R" occurs 12 times, "F" occurs 8 times, and so on. In English, the most common characters are " " (a space), e, and t in that order. Thus we might guess that in the encoding process, " " goes to "R" and "e" goes to "F". Here are the numerical equivalents.

```
texttonums[" R"]
texttonums["eF"]
```

This leads us to the pair of congruences

$$50 \equiv 0\,a \;+\; b \;(\mathrm{mod}\,95)$$

and

$$38 \equiv 69\,a \;+\; b \;(\mathrm{mod}\,95).$$

We see immediately from the first congruence equation that $b \equiv 50 \;(\mathrm{mod}\,95)$. Plugging this into the second congruence gives us

$$83 \equiv 69\,a \;\;(\mathrm{mod}\,95).$$

Using **congsolve**, we get

```
congsolve[69, 83, 95]
```

which tells us that $a \equiv 37 \pmod{95}$. We next determine (you may check that this is true) that $a^{-1} \equiv 18 \pmod{95}$, and then decode the message.

```
affine[ciphertext, 18, -18 * 50]
```

Sometimes frequency analysis requires more than one try to find the key. Here's an example where the characters occuring in the text do not have the same relative frequency as they generally do in English. Below is an encoded note from Bill to Alice responding to her earlier message.

```
ciphertext = "S^&tZO%~Zkw%J%3/UZ%U3Z%3Z/UZo%;ZDU-";
```

We'll follow the exact same procedure as above, starting with a frequency count for the encoded message.

```
freq[ciphertext]
```

We see that "Z" occurs 7 times and "%" occurs 6 times. Thus we initially guess that in the encoding process, " " goes to "Z" and "e" goes to "%". Here are the numerical equivalents.

```
texttonums[" Z"]
texttonums["e%"]
```

This leads us to the pair of congruences

$$58 \equiv 0\,a + b \pmod{95}$$

and

$$5 \equiv 69\,a + b \pmod{95}.$$

We see immediately from the first congruence that $b \equiv 58 \pmod{95}$. Plugging this into the second congruence gives us

$$42 \equiv 69\,a \pmod{95}.$$

Using **congsolve**, we get

```
congsolve[69, 42, 95]
```

Hence $a \equiv 13 \pmod{95}$, so that $a^{-1} \equiv 22 \pmod{95}$. Decoding gives us

```
affine[ciphertext, 22, -22 * 58]
```

What happened? Well, as foreshadowed above, our first guess for the cipher text corresponding to " " and "e" was not correct. Let's try reversing things, and guess that " " goes to "%" and "e" goes to "Z". This gives us the pair of congruence equations

$$5 \equiv 0\,a + b \pmod{95}$$

and

$$58 \equiv 69\,a + b \pmod{95}.$$

Solving (we leave the details to you) yields $a \equiv 82 \pmod{95}$ and $b \equiv 5 \pmod{95}$. Since $a^{-1} \equiv 73 \pmod{95}$, we decode as follows:

```
affine[ciphertext, 73, -73 * 5]
```

Although Bill's response is a little bizarre, it is clearly English, and so we may conclude that the code has been broken.

■ **Exercise 5**

Decode the following message:

```
ciphertext =
  " SAAP.s/5cp.}k.$e|$Sq./e.$.Y$|;Akvv.Yk$pk|.dk5p9.. R.
  pYk.^$RE.qSq.R/j.Yk$|.$}/jp.pYk.$qd$5wkq.5j;}k|.pYk/|
  R.wA$vv.}kS5_./eek|kq.5kXp.pk|;2";
```

■ **Exercise 6**

Decode the following message:

```
ciphertext =
  "*;f]'O(s'Z;;~Y'(j'Gbu(5oA'(fG(/o'u>bG(c'G(oJ'D,ooi'/=";
```

Web Lab: Solving Linear Congruences

5.1 Techniques Used with Rationals and Integers

5.1.1 Finding Rational Solutions for Linear Equations

If a and b are rational numbers and $a \neq 0$, the equation $ax = b$ has exactly one rational solution, namely b/a. We can find this by dividing by a, or equivalently, by multiplying both sides of the equation by $1/a$. So, solving the equation is easy if a has a *multiplicative inverse* (a number which when multiplied by a gives 1).

The remaining case is when $a = 0$. Here our equation takes the form $0x = b$. If $b \neq 0$, then there are no solutions to the equation, and if $b = 0$, then *every* rational number x is a solution to the equation.

5.1.2 Finding Integer Solutions for Linear Equations

Determining integer solutions to $ax = b$, when a and b are integers, is similar to the rational case, but a bit more complicated. The degenerate cases are the same: if $a = b = 0$, then every integer x is a solution; if $a = 0$ and $b \neq 0$, then there are no integer solutions to $ax = b$.

Now suppose $a \neq 0$. It follows directly from the definition of divisibility that $ax = b$ has a solution if and only if $a \mid b$. Moreover, if $a \mid b$, then there is a unique solution, namely $x = b/a$.

5.2 Multiplicative Inverses

Finding rational solutions to linear equations is easy because every nonzero rational number has a multiplicative inverse. Finding integer solutions to linear equations is a bit more complicated because the only integers which have integer multiplicative inverses are ± 1.

A pleasant surprise is that when working modulo n, one frequently finds that many elements have multiplicative inverses. In this context, the multiplicative inverse (or just "inverse" for short) of a modulo n, which is denoted by a^{-1}, satisfies

$$a \bullet a^{-1} \equiv 1 \pmod{n}.$$

For example, if $a = 29$, then $a^{-1} \equiv 35 \pmod{78}$. We'll get to how to compute the inverse later in the chapter. For now, we can verify the claim with the following computation:

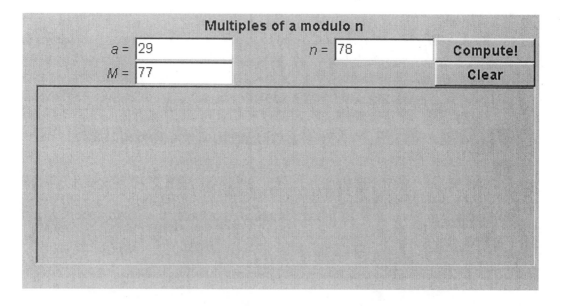

One complication is that not all choices for a will have an inverse mod n. One way to determine if a has an inverse mod n is to simply multiply a by each of $0, 1, 2, \ldots, n - 1$, and then look to see if any of the products is congruent to 1 (mod n). For example, here's what we get for $a = 29$ and $n = 78$:

As we can see, the output list contains a 1, which indicates that 29 has an inverse mod 78. Moreover, since the 1 appears in the 36th entry, this tells us that

$$29 \cdot 35 \equiv 1 \ (\text{mod } 78),$$

as above. On the other hand, here's what we get for $a = 32$ and $n = 78$:

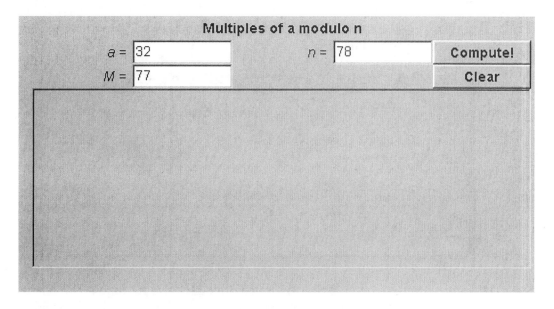

There's no 1 in the list, so 32 does not have an inverse mod 78.

In the first research question, we address the question of which integers a have a multiplicative inverse modulo n. Using the preceding applet to experiment would be tedious. To make matters easier, the applet below will take an integer n as input, and will produce two lists of numbers: those values of a between 0 and $n - 1$ that have inverses, and those values of a between 0 and $n - 1$ that do *not* have inverses. For example, here's what we get for $n = 10$:

Test out different values of n, and try to determine which values of a will appear for a given value of n.

> **Research Question 1**
>
> If $n > 0$, what values of a (between 0 and $n - 1$) will have an inverse mod n ?
>
> **Note:** You may find it easier to think of this question in the following equivalent form: for which values of a does the congruence equation $ax \equiv 1 \pmod{n}$ have solutions?

As you can see, the problem of finding an inverse for a modulo n is really a special case of the general problem considered in the Prelab discussion, namely, to solve the linear congruence equation

$$ax \equiv b \pmod{n}.$$

We take up this problem in the next section.

5.3 Linear Congruence Equations

In a previous chapter, you completely determined all solutions to the linear diophantine equation

$$ax + by = c.$$

Our goal in this section will be to do the same for the linear congruence equation

$$ax \equiv b \pmod{n}.$$

Specifically, if a, b, and n are integers (with n positive), then we want to completely answer the following three questions for this congruence equation:

1. Are there any solutions?
2. If there are solutions, *what* are they?
3. If there are solutions, *how many* are there?

There are many approaches one might take to try to answer these questions, but it may be helpful to have a function for computing examples. Recall from your Prelab work that if a, b, and n are given, then the solutions can be determined by simply trying all possible congruence classes for x. The applet below does just that. For example, here are all of the solutions to $2x \equiv 5 \pmod{11}$:

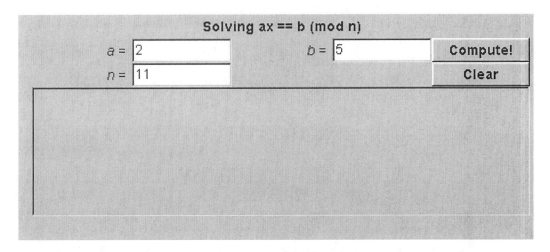

Here are all of the solutions to $6x \equiv 9 \pmod{15}$:

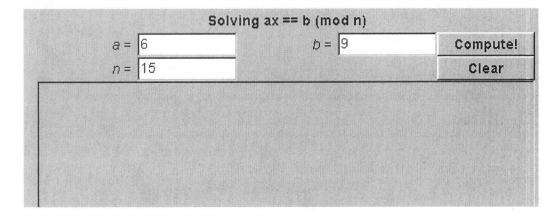

And here are all of the solutions to $10x \equiv 35 \pmod{42}$:

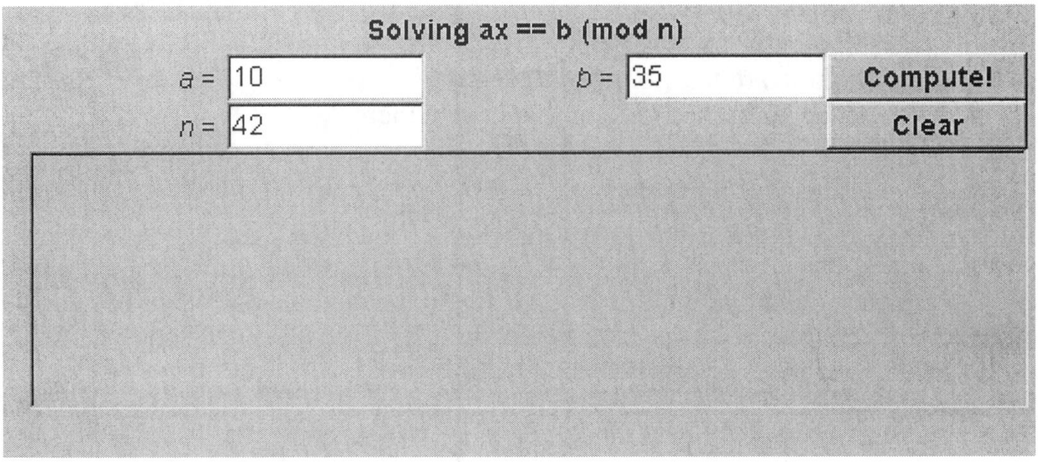

Solving ax == b (mod n)

a = 10   b = 35   Compute!

n = 42   Clear

Try some other examples on your own. The following exercise gives you a chance to check the function congsolve against your solutions found on the Prelab assignment.

Exercise 1

Compute the answers to questions 2, 3, 4, and 5 from the Prelab exercises using the above applet, and then compare to your Prelab solutions.

Research Question 2

For the congruence equation $ax \equiv b \pmod{n}$, determine conditions on a, b, and n for there to be at least one solution.

Research Question 3

Suppose that a, b, and n satisfy the conditions you gave in response to Research Question 2. Describe a method for finding a solution to $ax \equiv b \pmod{n}$.

Hint: One possibility is trial and error; just plug each of $x = 1$, $2, 3, \ldots, n - 1$ into the congruence. This will work, but there is a better, more efficient answer.

Research Question 4

Suppose that a, b, and n satisfy the conditions you gave in response to Research Question 2, and that x_0 is a solution to $ax \equiv b \pmod{n}$. Find the form, in terms of x_0, of all solutions x modulo n.

Research Question 5

Suppose that a, b, and n satisfy the conditions you gave in response to Research Question 2. How many distinct solutions x modulo n are there to $ax \equiv b \pmod{n}$?

Below are a collection of hints designed to help you as you work through the research questions from this section.

5.3.1 Hints

5.3.1.1 General Advice

The full solution to Research Questions 2 to 5 may not be obvious at first, but in the end there is an elegant answer. One of our standard course strategies is particularly important here: If the general case stumps you, then study special cases first.

Another suggestion: If you have trouble working with the congruence, then use the definition of congruence to convert it to an equation of integers. This will allow you to apply everything that you have learned in earlier chapters to solve the problem.

5.3.1.2 Multiplicative Inverses and Additive Orders

You have already worked extensively with two special cases of the congruence equation $ax \equiv b \pmod{n}$. One was in connection with multiplicative inverses. Specifically, finding the multiplicative inverse of a modulo n is equivalent to solving the congruence equation

$$ax \equiv 1 \pmod{n}.$$

In an earlier chapter you found a formula for the additive order of a modulo n. This required you to find all solutions to the congruence

$$ax \equiv 0 \pmod{n}.$$

These are two special cases of our general congruence equation. You may be able to make use of what you know about these special cases to learn more about the general equation.

5.3.1.3 Vary the value of b

In our general problem, there are three parameters: a, b, and n. Our hint here is to test specific values of a and n, and in each case compute all possible b so that the congruence $ax \equiv b \pmod{n}$ has a solution. This is somewhat natural since it amounts to holding a and n fixed, and plugging in all values for x and then watching what values come out for b. The applet below automates this process by taking each value of j satisfying $0 \leq j \leq M$, computing $aj \% n$, and printing out a list of the resulting values. Here is an example:

Multiples of a modulo n

$a =$ 6	$n =$ 9	Compute!
$M =$ 8		Clear

Try experimenting with different values of *a* and *n*.

5.4 Applications

A special case of what you have discovered in the course of investigating the nature of solutions to the congruence equation $ax \equiv b \pmod{n}$ is the following:

> If *p* is prime and *a* is not a multiple of *p*, then there exists an integer *b* such that $ab \equiv 1 \pmod{p}$.

Now fix *p* as a prime number, and suppose that *a* is an integer with $0 < a < p$. Then there is an integer *b*, also between 0 and *p*, such that $ab \equiv 1 \pmod{p}$. Since $ab \equiv 1 \pmod{p}$ implies that $ba \equiv 1 \pmod{p}$, we can try to pair off the integers between 1 and $p - 1$ so that the product of each pair is 1. The applet below takes a prime number as input, and displays the inverse for each value of *a* between 1 and $p - 1$. Here it is in action with $p = 11$:

Multiplicative inverses modulo p

$p =$ `11` | Compute! |
            | Clear |

You may find this applet useful when working on Research Question 6.

5.4.1(a) Factorials

Recall the definition of n factorial: $n! = 1 \cdot 2 \cdot 3 \cdots n$. The on-line calculator will compute factorials using the usual notation. Try it:

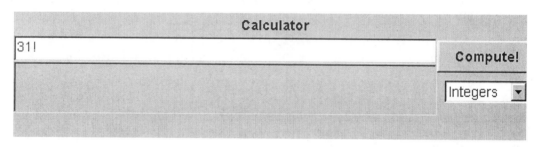

In Research Question 6, you are asked to investigate the behavior of factorials modulo n.

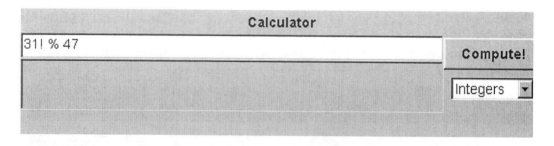

Research Question 6

(a) Find a formula for $n! \% n$.
(b) Find a formula for $(n - 2)! \% n$.
(c) Find a formula for $(n - 1)! \% n$.

5.4.1(b) Hint:

If you have a conjecture that you believe to be correct, but are having trouble finding a proof, it may be helpful to review the proof of the formula for the sum

$$(1 + 2 + 3 + \cdots + n) \,\%\, n.$$

5.5 Going Farther: Affine Ciphers

In this section, we shall consider a generalization of the shift cipher called the *affine cipher*. Recall that to encode a message using a shift cipher, we convert our text to a number list, rotate each number by the key k, and then convert the encoded number list back to letters.

Here's a quick example of the shift cipher, using the key $k = 52$. Click on "Shift text" to encode the message.

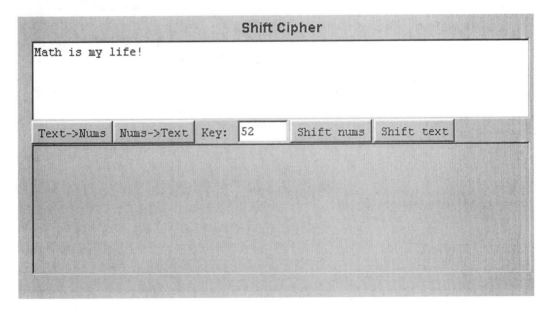

In terms of congruences, the relationship between p and c is

$$c \equiv p + k \,(\text{mod } 95).$$

The congruence is modulo 95 because we have a 95-character alphabet (including upper and lower case letters, digits, and punctuation). Decoding a message requires us to solve for p in the above congruence. Using simple algebra, we see that

$$p \equiv c - k \pmod{95}.$$

Thus decoding looks a lot like encoding.

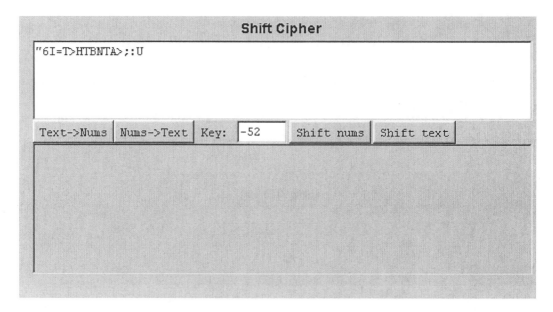

The affine cipher works in a similar manner, except that the key consists of two numbers a and b. We encode a message by applying the formula

$$c \equiv ap + b \pmod{95}.$$

For example, if we take $a = 41$ and $b = 12$, then we encipher by clicking on "Affine" below.

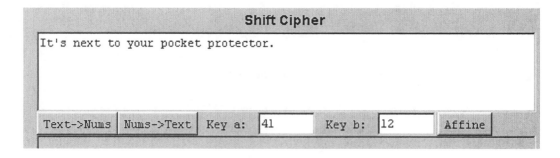

To decode a message encoded with an affine cipher, we must solve for p in the congruence

$$c \equiv ap + b \ (\text{mod } 95).$$

Subtracting b from both sides and then multiplying by a^{-1} (mod 95) yields

$$p \equiv a^{-1}(c - b) \equiv a^{-1}c - a^{-1}b \ (\text{mod } 95).$$

Note that this is just another affine cipher with key a^{-1} and $-a^{-1}b$.

Thus, to decode we first need to compute a^{-1} (mod 95). As noted earlier, this is equivalent to solving the congruence equation $ax \equiv 1$ (mod 95), which can be accomplished using the applet below. In this case, we have $a = 41$ and $b = 1$:

Solving ax == b (mod n)

$a =$ 41	$b =$ 1	**Compute!**
$n =$ 95		**Clear**

We can now decode the following message, which was encoded using $a = 41$ and $b = 12$. We decode using the affine applet with the decoding key that we have found.

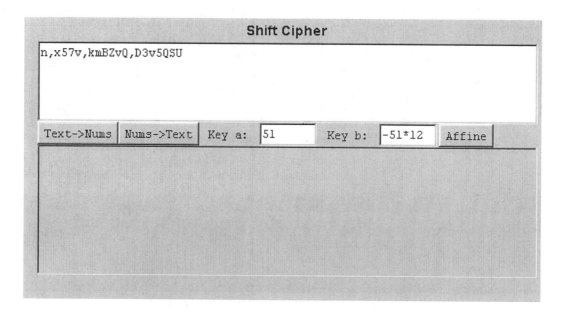

As you know, for some values of a there will be a multiplicative inverse modulo 95, and for others there will not. Suppose that we had selected a value of a for which there is no inverse modulo 95? One such choice is $a \equiv 38 \pmod{95}$, as can be seen below.

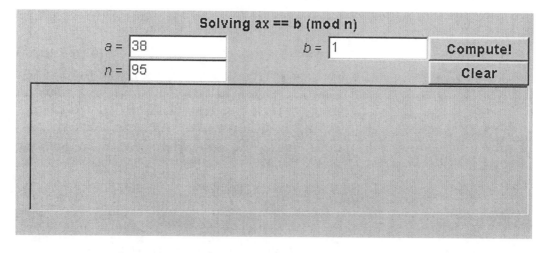

Check out what happens if we use this value of a.

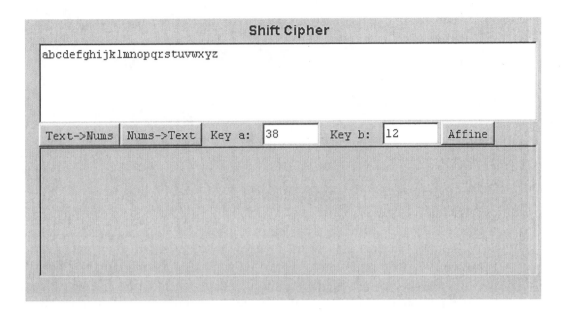

The problem is clear. Since the letters "b", "f", "k", "p" (and more) are all sent to "R" when encoded, we have no way of knowing how to decode "R". This ambiguity renders the encription system useless. After all, what fun is a secret code if the messages can't be decoded?

Exercise 2

The message recorded in the applet below was encoded using an affine cipher with key $a = 59$ and $b = 34$. Decode the message.

Shift Cipher

`vB4_.eUBRHBR'4Bv.lvBHoBR'4Bv_.e;4vBHoBR'4BHR'4;BREHBvbX4v%BI^'4B#|e;4|;HE`

| Text->Nums | Nums->Text | Key a: | | Key b: | | Affine |

5.6 Going Farther: Cryptanalysis

5.6.1 The Basics

We begin this section with a simple example of *cryptanalysis*, the practice of breaking codes. Suppose that Alice and Bill have been exchanging messages using an affine cipher. Oscar suspects that they have been talking about him, and desperately wants to know the contents of their messages. As usual, we may assume that Oscar knows that an affine cipher is being used, and that the message is being encoded one letter at a time. Thus Oscar knows that all he has to do is determine the values of a and b that form the key, and then he can decode everything passed between Alice and Bill. Bill is confident that the code can't be broken, and decides to taunt Oscar by telling him "In our code, the 'a' is encoded to 'M' and the 'w' is encoded to 'b'."

Although he doesn't realize it, Bill has actually given Oscar all the information that is needed to break the code. (No one has ever accused Bill of being the sharpest tack in the box.) To see why, recall that the encoding formula is

$$C \equiv aP + b \ (\mathrm{mod}\ 95).$$

Here are the numerical equivalents for the four letters given by Bill.

Shift Cipher

aMwb

Text->Nums

Plugging into our encoding formula, we see that

$$45 \equiv 65a + b \pmod{95}$$

and

$$66 \equiv 87a + b \pmod{95}.$$

We now have two congruences in two unknowns a and b. Although these are congruence equations, we can use the same methods as we would for algebraic equations to find a solution. Subtracting the first congruence equation from the second eliminates the unknown b, and leaves us with the single congruence

$$21 \equiv 22a \pmod{95}.$$

We know all about this type of congruence equation. Let's use our usual applet to find the solution.

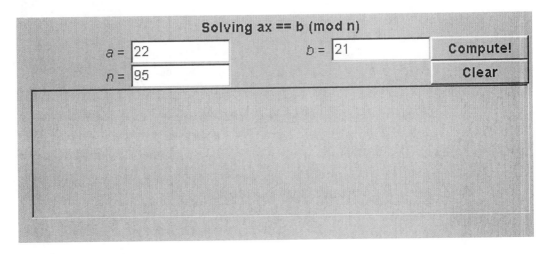

This tells us that $a \equiv 83 \pmod{95}$. We can then find b by back substitution (i.e., plugging back into either of our original congruence equations). Using the first congruence equation, we have

$$45 \equiv 65 \cdot 83 + b \pmod{95}.$$

After simplifying, we find that $b \equiv 65 \pmod{95}$.

Exercise 3

The message recorded in the applet below was sent to Bill from Alice using an affine cipher with the keys found above. Help Oscar decode the messsage.

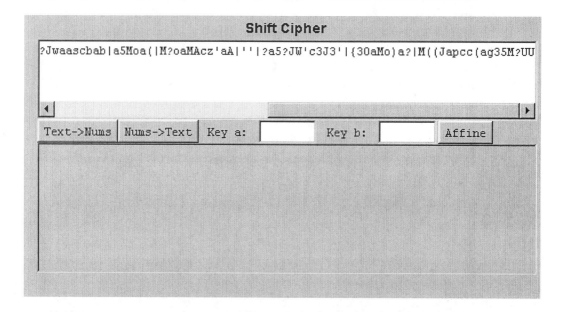

Exercise 4

Alice and Bill have changed their keys. However, Bill has left a scrap of paper lying around which shows that "OK" is now encoded as "na". Use this information to help Oscar decode the message below.

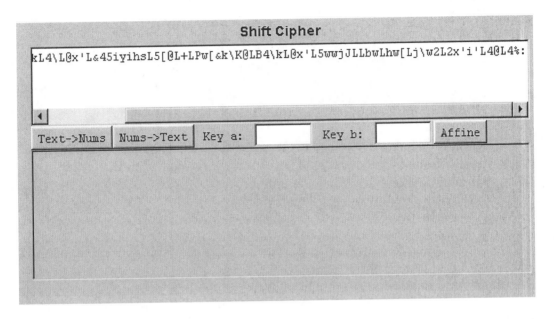

5.6.2 Frequency Analysis

Alice and Bill have learned more about cryptosystems, and now know that they should never let anyone else know both the coded and unencoded versions of the same text. However, it is still possible for Oscar to break their code using a basic method of cryptanalysis called *frequency analysis*. This involves counting the number of occurances of each character in the coded message, guessing at the correspondence to the original message, and then using this information as we did earlier to find the key.

For example, suppose that Alice has sent the encoded message below to Bill. In this message, the most common character is "R" and the second most common is "F". We can see this by counting by hand, or by using the applet, which will give the frequency of all characters occuring in the message.

Thus "R" occurs 12 times, "F" occurs 8 times, and so on. In English, the most common characters are " " (a space), "e", and "t" in that order. Thus we might guess that in the encoding process, " " goes to "R" and "e" goes to "F". Here are the numerical equivalents.

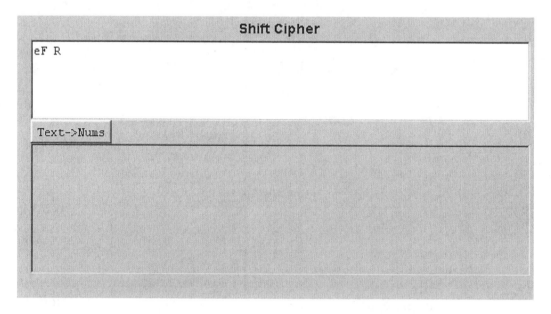

This leads us to the pair of congruences

$$38 \equiv 69 \cdot a + b \ (\mathrm{mod}\ 95)$$

and

$$50 \equiv 0 \cdot a + b \ (\mathrm{mod}\ 95).$$

We see immediately from the second congruence equation that $b \equiv 50$ (mod 95). Plugging this into the first congruence gives us

$$83 \equiv 69 \cdot a \ (\mathrm{mod}\ 95).$$

We can use our applet for solving congruences to handle this equation.

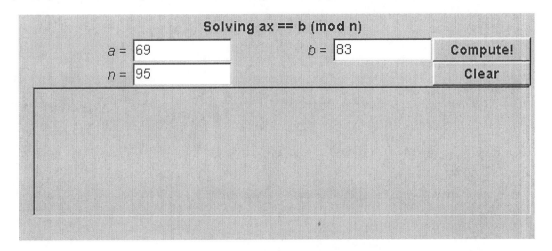

Thus we see that $a \equiv 37 \pmod{95}$. We next determine (you may check that this is true) that $a^{-1} \equiv 18 \pmod{95}$, and then decode the message.

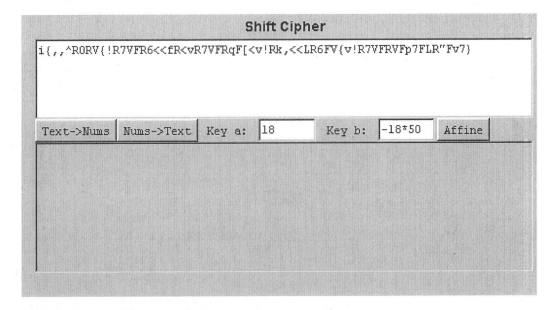

Sometimes frequency analysis requires more than one try to find the key. Here's an example where the characters occuring in the text do not have the same relative frequency as they generally do in English. Below is an encoded note from Bill to Alice responding to her earlier message. We'll follow the exact same procedure as above, starting with a frequency count for the encoded message.

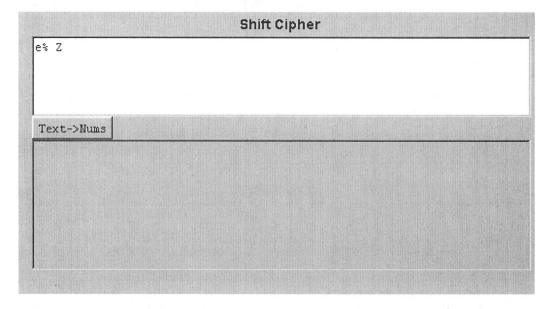

Shift Cipher

S^&tZ0%~Zkw%J%3/UZ%U3Z%3Z/UZo%;ZDU-

Frequencies

We see that "Z" occurs 7 times and "%" occurs 6 times. Thus we initially guess that in the encoding process, " " goes to "Z" and "e" goes to "%". Here are the numerical equivalents.

Shift Cipher

e% Z

Text->Nums

This leads us to the pair of congruences

$$58 \equiv 0 \cdot a + b \pmod{95}$$

and

$$5 \equiv 69 \cdot a + b \;(\mathrm{mod}\;95).$$

We see immediately from the first congruence that $b \equiv 58 \;(\mathrm{mod}\;95)$. Plugging this into the second congruence gives us

$$42 \equiv 69 \cdot a \;(\mathrm{mod}\;95).$$

Using our usual applet, we find that

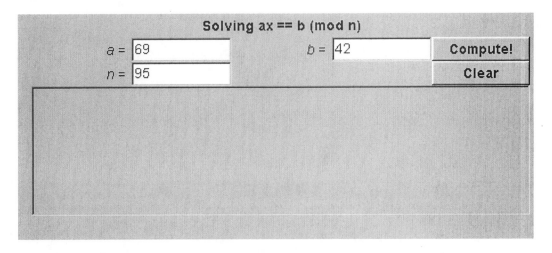

Hence $a \equiv 13 \;(\mathrm{mod}\;95)$, so that $a^{-1} \equiv 22 \;(\mathrm{mod}\;95)$. Decoding gives us

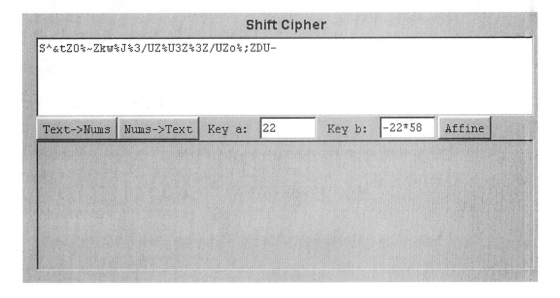

What happened? Well, as foreshadowed above, our first guess for the cipher text corresponding to " " and "e" was not correct. Let's try reversing things, and guess that " " goes to "%" and "e" goes to "Z". This gives us the pair of congruence equations

$$5 \equiv 0 \cdot a + b \pmod{95}$$

and

$$58 \equiv 69 \cdot a + b \pmod{95}.$$

Solving (we leave the details to you) yields $a \equiv 82 \pmod{95}$ and $b \equiv 5 \pmod{95}$ Since $a^{-1} \equiv 73 \pmod{95}$, we decode as follows:

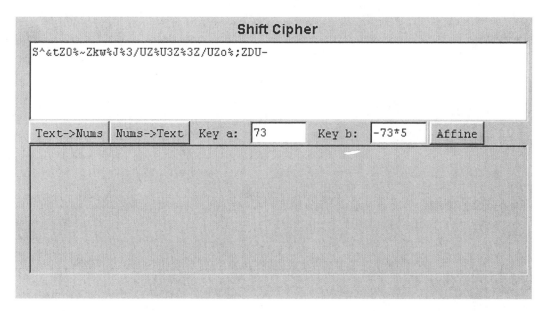

Although Bill's response is a little bizarre, it is clearly English, and so we may conclude that the code has been broken.

Exercise 5

Decode the following message:

Shift Cipher

.R/j.Yk$|.$}/jp.pYk.qd5wkq.5j;}k|.pYk/|R.wA$vv.}kS5_./eek|kq.5kXp.pk|;2

◄ ▶

| Text->Nums | Frequencies | Key a: | Key b: | Affine |

Exercise 6

Decode the following message:

Shift Cipher

*;f]`O(s`Z;;~Y'(j`Gbu(5oA`(fG(/o`u>bG(c`G(oJ`D,ooi`/=

| Text->Nums | Frequencies | Key a: | Key b: | Affine |

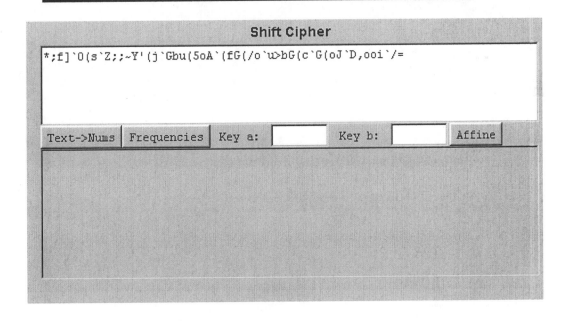

Homework

1. Suppose that $\gcd(a, n) = 1$ and that x_0 and y_0 are integers such that $ax_0 + ny_0 = 1$. Prove that x_0 is the multiplicative inverse of a modulo n.

2. Use the Euclidean Algorithm to find the multiplicative inverse for each of the following integers for the given modulus.

    (a) 7 (mod 19)   (b) 39 (mod 95)   (c) 91 (mod 191)

3. Suppose that p is an odd prime.

    (a) Show that $(p + 1)/2$ is an integer.

    (b) Show that $(p + 1)/2$ is the multiplicative inverse of 2 modulo p.

4. Suppose that $p \equiv 3 \pmod 4$.

    (a) Show that $(p + 1)/4$ is an integer.

    (b) Show that $(p + 1)/4$ is the multiplicative inverse of 4 modulo p.

5. Suppose that p is an odd prime.

    (a) Show that $(p + 1)^2/4$ is an integer.

    (b) Show that $(p + 1)^2/4$ is the multiplicative inverse of 4 modulo p.

6. Use the Euclidean Algorithm to find a solution for each of the following congruence equations.

    (a) $23x \equiv 1 \pmod{91}$

    (b) $117x \equiv 1 \pmod{241}$

    (c) $47x \equiv 1 \pmod{314}$

7. Use your work from the preceding exercise to find a solution for each of the following congruence equations.

    (a) $23x \equiv 7 \pmod{91}$

    (b) $117x \equiv 136 \pmod{241}$

    (c) $47x \equiv 310 \pmod{314}$

8. Use the Euclidean Algorithm to find a solution for each of the following congruence equations.

311

  (a) $14x \equiv 2 \pmod{100}$

  (b) $45x \equiv 5 \pmod{175}$

  (c) $36x \equiv 6 \pmod{78}$

9. Use your work from the preceding exercise to find a solution for each of the following congruence equations.

  (a) $14x \equiv 10 \pmod{100}$

  (b) $45x \equiv 95 \pmod{175}$

  (c) $36x \equiv 42 \pmod{78}$

In problems 10 to 14, solve the given congruences (i.e., find all distinct solutions, if any, for x modulo n).

10. Solve $15x \equiv 21 \pmod{108}$.

11. Solve $29x \equiv 8 \pmod{261}$.

12. Solve $42x \equiv 34 \pmod{163}$.

13. Solve $102x \equiv 37 \pmod{432}$.

14. Solve $121x \equiv 77 \pmod{253}$.

15. Suppose that x_0 is a solution to $ax \equiv b \pmod{n}$, and that $d = \gcd(a, n)$.

  (a) Prove that $d = \gcd(a, n, b)$.

  (b) Prove that x is a solution to $ax \equiv b \pmod{n}$ if and only if x is a solution to

$$\frac{a}{d}x \equiv \frac{b}{d} \quad \left(\bmod \frac{n}{d} \right).$$

(This exercise provides a slightly different procedure for solving a general linear congruence. If there are solutions to the original linear congruence, then "dividing through" by $\gcd(a, n)$ gives an equivalent congruence which is possibly simpler than the original one. In particular, since $\gcd(a/d, n/d) = 1$, the congruence in part (b) can be solved with a multiplicative inverse.)

16. What is the possible number of solutions for $8x \equiv b \pmod{20}$?

17. What is the possible number of solutions for $18x \equiv b \pmod{156}$?

18. What is the possible number of solutions for $ax \equiv b \pmod{18}$?

19. What is the possible number of solutions for $ax \equiv b \pmod{42}$?

20. For what values of $a \pmod{12}$ will $ax \equiv 6 \pmod{12}$ have exactly

    (a) 1 solution?   (b) 2 solutions?   (c) 3 solutions?   (d) 6 solutions?

21. For what values of $a \pmod{78}$ will $ax \equiv 26 \pmod{78}$ have exactly

    (a) 13 solutions?   (b) 26 solutions?

22. Solve the given system of linear congruences.

$$9x + 2y \equiv 3 \pmod{10}$$
$$4x + 5y \equiv 7 \pmod{10}$$

23. Solve the given system of linear congruences.

$$13x + 2y \equiv 1 \pmod{15}$$
$$10x + 9y \equiv 8 \pmod{15}$$

24. Bill says, "Let's improve our security by encoding each message with an affine cipher, and then running the encoded message through a second affine cipher." Alice responds, "That won't alter security; it's the same as just using a single affine cipher." Show that Alice is correct.

25. Suppose that messages are written in a 95-letter alphabet and then encoded using an affine cipher $ax + b \pmod{95}$.

    (a) If $a = 13$ and $b = 47$, how many letters will be unchanged by the encoding process?

    (b) If $a = 6$ and $b = 81$, how many letters will be unchanged by the encoding process?

    (c) In general, what conditions on a and b will ensure that all letters will be changed by the encoding process?

26. If we use an n-letter alphabet, an encoding scheme can be thought of a function $e \colon \mathbf{Z}_n \to \mathbf{Z}_n$ where $\mathbf{Z}_n = \{0, \ldots, n-1\}$. To decode the encoder e, one needs a function $d \colon \mathbf{Z}_n \to \mathbf{Z}_n$ such that $d(e(x)) = x$ (i.e., if we first encode x, then decode the result, we get back x).

    (a) Suppose a function $e \colon \mathbf{Z}_n \to \mathbf{Z}_n$ is fixed. Show that there exists a decoding function $d \colon \mathbf{Z}_n \to \mathbf{Z}_n$ if and only if the function e is one-to-one.

    (b) If e is an affine cipher, then $e(x) = (ax + b) \% n$ for some a and b in $\mathbf{Z}_n$. Show that e can be decoded if and only if $\gcd(a, n) = 1$.

27. The Hawaiian alphabet has 12 letters:[1] 5 vowels (a, e, i, o, u), and 7 consonants (h, k, l, m, n, p, w). How many distinct affine ciphers are possible for the Hawaiian alphabet?[2]

28. Arnold and Betty use affine ciphers with a 29-letter alphabet (they use the normal 26 letters, plus "space", "period", and "comma"). Charlie joins their group, but wants to add "exclamation point" to the group's alphabet to add a little zip to his messages, which would give them a 30-letter alphabet. Which would have more affine ciphers, a 29-letter alphabet or a 30-letter alphabet?[3]

29. The message below was encoded using an affine cipher. Decode the message.

```
ciphertext = "!:-i,-i,\:a,e6i\,1v>)eaX~"
```

(Hint: The original message ends with an exclaimation point.)

---

[1] The Hawaiian language was historically oral. The alphabet was written by nineteenth century missionaries.

[2] Naturally, we are only interested in affine ciphers which can be decoded.

[3] Naturally, we are only interested in affine ciphers which can be decoded.

6. Primes of Special Forms

Prelab

Prime numbers play such an important role in number theory that people have tried to find formulas for them for many years. The idea of having a formula which produces the primes in order is certainly appealing, but somewhat unrealistic.[1]

Instead, people have sought formulas which produced only prime numbers, or all of the prime numbers, or infinitely many of the prime numbers. To illustrate what we mean, let's analyze when primes can be given by the formula $f(n) = n^2 - 1$. If we compute the first few values of this function, we get: 0, 3, 8, 15, 24, 35. Of these, 3 is the only prime.

It should not be surprising that we did not find many primes of this form because $n^2 - 1 = (n-1)(n+1)$. If this is a prime number, the smaller factor has to be equal to 1. Thus, $n - 1 = 1$ which implies $n = 2$. The value $n = 2$ does give $f(2) = 3$, a prime number. So, by factoring our formula we are able to prove that the only prime which is one less than a perfect square is 3.

This attempt at a formula for producing primes was not very good (one may even say that it was spectacularly bad!). In the lab, you will investigate other formulas for producing primes which have been historically quite popular.

Here is an example which you should work out by hand, using the method employed above.

1. **For which integers n is $f(n) = n^2 - 10n + 24$ prime?**

Maple electronic notebook `06-primes.mws`

Mathematica electronic notebook `06-primes.nb`

Web electronic notebook Start with the web page `index.html`

---

[1]Mathematica has a function `Prime[n]` which produces the nth prime, and Maple has the analogous function `ithprime(n)`. From a computational point of view, it does exactly what we want. However, it cannot be manipulated in a proof the way an algebraic formula could.

Maple Lab: Primes of Special Forms

■ Introduction

In this chapter you will examine integers of specific forms. There are several questions which apply in all cases:

1. Under what conditions is an integer of this form always prime?

2. Under what conditions is an integer of this form always composite?

3. Are there infinitely many primes of this form?

Sometimes these questions are easy and sometimes they are quite difficult (e.g., currently unsolved after hundreds of years). Of the questions listed above, often question 2 is the easiest to answer. With that in mind, work on making good conjectures for all of the questions first. Then go back and see how much you can prove.

■ Mersenne Numbers

The mathematician Marin Mersenne (1588-1648) studied numbers of the form $2^j - 1$. In his honor, such numbers are known as *Mersenne numbers*, and any prime of the form $2^j - 1$ is called a *Mersenne prime*. Here's a test to see if the Mersenne number corresponding to $j = 6$ is a prime:

```
[ > isprime(2^6-1);
```

The function below makes it easy to test several Mersenne numbers at once:

```
> mersennetest := proc(bound)
    local j, num, torf;
    for j to bound do
     num := 2^j-1;
     if isprime(num) then torf := `true` else torf := `false` fi;
     printf(`%10d  %5s  %10d\n`,j, torf, num);
    od;
  end:
```

Here is what the output looks like for the first ten Mersenne numbers:

```
[ > mersennetest(10);
[ >
```

The first column shows the value of j, the second column indicates whether or not the jth Mersenne number is prime, and the third column shows the value of the jth Mersenne number. Use mersennetest to help you investigate Mersenne primes.

■ Research Question 1

Address the three questions given in the introduction of this notebook for Mersenne numbers.

316

An Application of Mersenne Primes

There is a connection between Mersenne primes and perfect numbers. A *perfect number* is an integer which is equal to the sum of its positive divisors less than itself. We can test if a number is perfect with the following function:

```
> with(numtheory):

  isperfect := n -> evalb(2*n = sigma(n)):
```

Do not worry if Maple issues a warning about there being a new definition of order; this is normal when loading the number theory library. The function `sigma` from Maple's number theory library computes the sum of all positive divisors of an integer.

It is easy to check by hand that 6 is a perfect number, but 7 is not. We can use this to check our function.

```
> isperfect(6);
> isperfect(7);
>
```

Maple Note

Here we opted for a compact version of this function. It uses the function `sigma(n)` which sums the positive divisors of *n*.

Our function `isperfect(n)` works by testing whether 2 times the integer *n* is equal to the sum of its positive divisors.

To find all perfect numbers in a range, use the following function:

```
> findperfect := bound -> select(isperfect, [$1..bound]):
```

So to find all perfect numbers up to 50:

```
> findperfect(50);
```

Try to find the connection between Mersenne primes and perfect numbers. Remember that you can factor integers with the command `ifactor`.

```
> ifactor(12);
>
```

Research Question 2

How are perfect numbers related to Mersenne numbers?

Fermat Numbers

The mathematician Pierre de Fermat (1601-1655) appears several times in this course. In search of a formula for primes, Fermat considered numbers of the form $2^j + 1$. Here is a function analogous to the one we used in the previous section:

```
> fermattest := proc(bound)
    local j, num, torf;
    for j to bound do
     num := 2^j+1;
     if isprime(num) then torf := `true` else torf := `false` fi;
     printf(`%10d  %5s  %10d\n`,j, torf, num);
    od;
  end:
```

Here it is in action for the first ten values of *j*:

```
> fermattest(10);
>
```

The format of the output is the same as for `mersennetest`.

◼ Research Question 3

Address the three questions given in the introduction of this worksheet for numbers of the form $2^j + 1$.

◼ Note on Terminology

We would like to refer to numbers of the form $2^j + 1$ as Fermat numbers, but this would not fit the standard usage. Unless *j* satisfies a condition (which you are supposed to discover), the number is definitely composite. When *j* satisfies this condition, $2^j + 1$ is called a *Fermat number*. In any case, a *Fermat prime* is a prime of the form $2^j + 1$.

◼ Arithmetic Progressions

An arithmetic progression is a sequence of the form $a k + b$ where *a* and *b* are fixed and *k* runs through integer values. The goal of this section is to discover whatever we can about primes in arithmetic progessions.

Let's start by looking at some terms of the arithmetic progression $3 k + 7$:

```
> seq(3*k+7, k= -10..10);
```

Can you guess what would happen if we considered the corresponding progression $-3 k + 7$?

```
> seq(-3*k+7, k= -10..10);
>
```

It was exactly the same, except for the ordering. It is not hard to see why this will happen in general, so we will restrict to positive values for *a*. Next observe that the integers in the progession $a k + b$ are the integers in the congruence class of *b* modulo *a*. For example, the numbers of the form $3 k + 7$ are the integers congruent to 7 modulo 3. We could get the same set by taking integers of the form $3 k + 1$ because $7 \equiv 1 \pmod 3$.

```
> seq(3*k+1, k= -10..10);
>
```

Keep in mind that we are seeing only part of an infinite sequence of numbers; adjust the bounds on the sequences to convince yourself that we are seeing the same set.

We can now safely restrict the values of *b* to be the possible remainders on division by *a*. Given an integer *a*, there are *a* congruence classes and each integer is in one of those congruence classes. Our questions about primes in arithmetic progressions can be thought of in terms of "how do the primes divide themselves among the different congruence classes modulo *a*?"

The next function looks at the first num primes and counts how many are in the different congruence classes modulo a.

```
> arith := proc(a, num)
    local p, j, counts, index;
    counts := array(1..a);
    for j to a do counts[j] := 0; od;
    for j to num do
     p := ithprime(j);
     index := modp(p, a);
     if(index = 0) then index := a; fi;
     counts[index] := counts[index] + 1;
    od;
    print(counts);
  end:
```

Let's try it on the congruence classes modulo 5 and the first 100 primes.

```
> arith(5,100);
>
```

The number of primes that are congruent to 1 is listed first, the number congruent to 2 is listed second, and so on. So, the computation tells us that 24 of the first 100 primes are congruent to 1 modulo 5. The last entry says that only one of the first 100 primes is congruent to 5 modulo 5.

■ Research Question 4

Try to answer the basic questions from the introduction for primes of the form *a k + b*.

Mathematica Lab: Primes of Special Forms

■ Introduction

In this chapter you will examine integers of specific forms. There are several questions which apply in all cases:

1. Under what conditions is an integer of this form always prime?

2. Under what conditions is an integer of this form always composite?

3. Are there infinitely many primes of this form?

Sometimes these questions are easy and sometimes they are quite difficult (e.g., currently unsolved after hundreds of years). Of the questions listed above, often question 2 is the easiest to answer. With that in mind, work on making good conjectures for all of the questions first. Then go back and see how much you can prove.

■ Mersenne Numbers

The mathematician Marin Mersenne (1588–1648) studied numbers of the form $2^j - 1$. In his honor, such numbers are known as *Mersenne numbers*, and any prime of the form $2^j - 1$ is called a *Mersenne prime*. Here's a test to see if the Mersenne number corresponding to $j = 6$ is a prime:

```
PrimeQ[2^6 - 1]
```

The function below makes it easy to test several Mersenne numbers at once:

```
mersennetest[bound_] :=
 TableForm[Table[{j, PrimeQ[2^j - 1], 2^j - 1}, {j, bound}]]
```

Here is what the output looks like for the first ten Mersenne numbers:

```
mersennetest[10]
```

The first column shows the value of j, the second column indicates whether or not the jth Mersenne number is prime, and the third column shows the value of the jth Mersenne number. Use **mersennetest** to help you investigate Mersenne primes.

■ Research Question 1

Address the three questions given in the introduction of this notebook for Mersenne numbers.

■ An Application of Mersenne Primes

There is a connection between Mersenne Primes and perfect numbers. A *perfect number* is an integer which is equal to the sum of its positive divisors less than itself. We can test if a number is perfect with the following function:

```
PerfectQ[n_] := 2*n == DivisorSigma[1, n]
```

It is easy to check by hand that 6 is a perfect number, but 7 is not. We can use this to check our function.

```
PerfectQ[6]
```

```
PerfectQ[7]
```

♡ Mathematica Note

Here we opted for a compact version of this function. It uses the function **DivisorSigma[j,n]** which takes all of the divisors of *n*, and adds their *j*th powers. Thus, **DivisorSigma[1,n]** simply gives the sum of the divisors of *n*.

Our function **PerfectQ[n]** works by testing whether 2 times the integer *n* is equal to the sum of its positive divisors.

To find all perfect numbers in a range, use the following function:

```
findperfect[bound_] := Select[Range[1, bound], PerfectQ[#] &]
```

So to find all perfect numbers up to 50:

```
findperfect[50]
```

Try to find the connection between Mersenne primes and perfect numbers. Remember that you can factor integers with the command **FactorInteger**.

```
FactorInteger[12]
```

■ Research Question 2

How are perfect numbers related to Mersenne numbers?

■ Fermat Numbers

The mathematician Pierre de Fermat (1601–1655) appears several times in this course. In search of a formula for primes, Fermat considered numbers of the form $2^j + 1$. Here is a function analogous to the one we used in the previous section:

```
fermattest[bound_] :=
  TableForm[Table[{j, PrimeQ[2^j + 1], 2^j + 1}, {j, bound}]]
```

Here it is in action for the first ten values of *j*:

```
fermattest[10]
```

The format of the output is the same as for **mersennetest**.

■ Research Question 3

Address the three questions given in the introduction of this notebook for numbers of the form $2^j + 1$.

■ Note on Terminology

We would like to refer to numbers of the form $2^j + 1$ as Fermat numbers, but this would not fit the standard usage. Unless j satisfies a condition (which you are supposed to discover), the number is definitely composite. When j satisfies this condition, $2^j + 1$ is called a Fermat number. In any case, a *Fermat prime* is a prime of the form $2^j + 1$.

■ Arithmetic Progressions

An arithmetic progression is a sequence of the form $ak + b$ where a and b are fixed and k runs through integer values. The goal of this section is to discover whatever we can about primes in arithmetic progressions.

Let's start by looking at some terms of the arithmetic progression $3k + 7$:

```
Table[3 k + 7, {k, -10, 10}]
```

Can you guess what would happen if we considered the corresponding progression $-3k + 7$?

```
Table[-3 k + 7, {k, -10, 10}]
```

It was exactly the same, except for the ordering. It is not hard to see why this will happen in general, so we will restrict to positive values for a. Next observe that the integers in the progession $ak + b$ are the integers in the congruence class of b modulo a. For example, the numbers of the form $3k + 7$ are the integers congruent to 7 modulo 3. We could get the same set by taking integers of the form $3k + 1$ because $7 \equiv 1 \pmod 3$.

```
Table[3 k + 1, {k, -10, 10}]
```

Keep in mind that we are seeing only part of an infinite sequence of numbers; adjust the bounds on the sequences to convince yourself that we are seeing the same set.

We can now safely restrict the values of b to be the possible remainders on division by a. Given an integer a, there are a congruence classes and each integer is in one of those congruence classes. Our questions about primes in arithmetic progressions can be thought of in terms of "how do the primes divide themselves among the different congruence classes modulo a?"

The next function looks at the first **num** primes and counts how many are in the different congruence classes modulo **a**.

```
arith[a_, num_] := Module[{p, j, counts, index},
     counts = Table[0, {a}];
     Do[p = Prime[j];
        index = Mod[p, a];
        If[index == 0, index = a];
        counts[[index]] = counts[[index]] + 1
        , {j, num}];
     counts]
```

Let's try it on the congruence classes modulo 5 and the first 100 primes.

```
arith[5, 100]
```

The number of primes that are congruent to 1 is listed first, the number congruent to 2 is listed second, and so on. So, the computation tells us that 24 of the first 100 primes are congruent to 1 modulo 5. The last entry says that only one of the first 100 primes is congruent to 5 modulo 5.

■ **Research Question 4**

Try to answer the basic questions from the introduction for primes of the form $a k + b$.

Web Lab: Primes of Special Forms

6.1 Introduction

In this chapter you will examine integers of specific forms. There are several questions which apply in all cases:

1. Under what conditions is an integer of this form always prime?
2. Under what conditions is an integer of this form always composite?
3. Are there infinitely many primes of this form?

Sometimes these questions are easy and sometimes they are quite difficult (e.g., currently unsolved after hundreds of years). Of the questions listed above, often question 2 is the easiest to answer. With that in mind, work on making good conjectures for all of the questions first. Then go back and see how much you can prove.

6.2 Mersenne Numbers

The mathematician Marin Mersenne (1588-1648) studied numbers of the form $2^j - 1$. In his honor, such numbers are known as *Mersenne numbers*, and any prime of the form $2^j - 1$ is called a *Mersenne prime*. The applet below takes a positive integer as input, and determines if the input is prime or composite. Here we test to see if the Mersenne number corresponding to $j = 6$ is a prime:

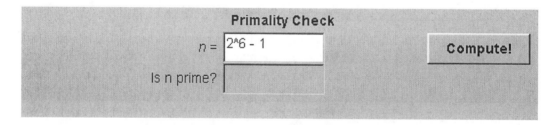

The applet below makes it easy to test several Mersenne numbers at once.

You simply enter a value M, and the output consists of three columns. The first column shows the value of j, the second column indicates whether or not the jth Mersenne number is prime, and the third column shows the value of the jth Mersenne number. The values of j run from 1 to M.

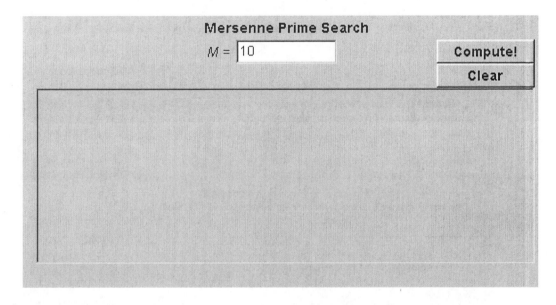

Use the above applet to help you investigate Mersenne primes.

Research Question 1

Address the three questions given in the introduction of this chapter for Mersenne numbers:

> 1. Under what conditions is an integer of this form always prime?
> 2. Under what conditions is an integer of this form always composite?
> 3. Are there infinitely many primes of this form?

6.2.1 An Application of Mersenne Primes

There is a connection between Mersenne primes and perfect numbers. A *perfect number* is an integer which is equal to the sum of its positive divisors less than itself. For example, if $n = 6$, then the positive divisors of n that are less than n are 1, 2, and 3. As

$$1 + 2 + 3 = 6,$$

we see that 6 is perfect. We can test if a number is perfect with the following applet:

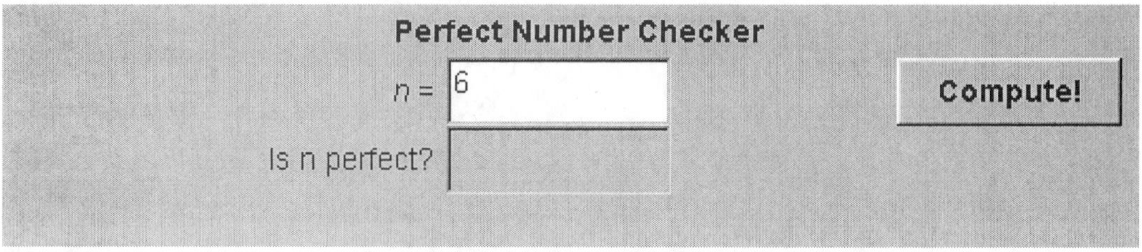

Perfect Number Checker

$n =$ 6   **Compute!**

Is n perfect?

On the other hand, 7 is not perfect. Verify this by hand, and then execute the applet.

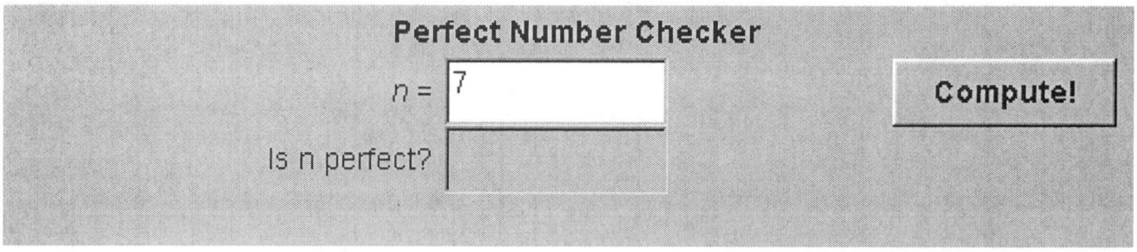

Perfect Number Checker

$n =$ 7   **Compute!**

Is n perfect?

The applet below does tests each number from 1 to M, and reports those that are perfect:

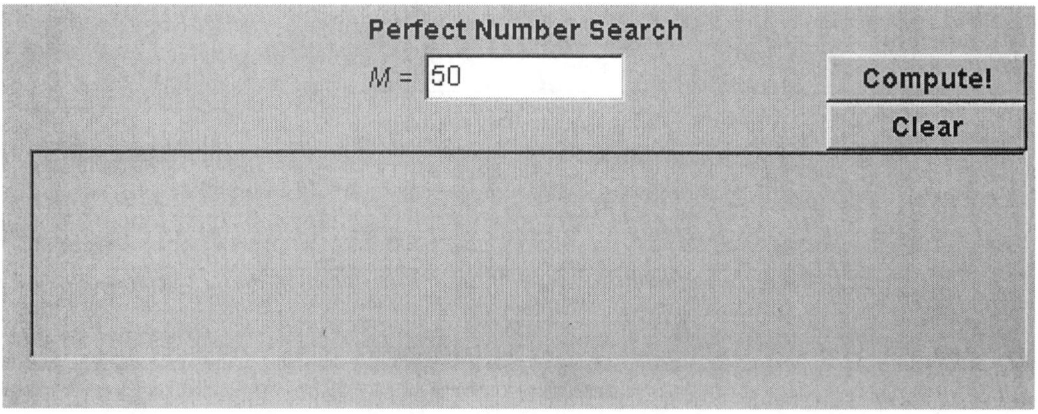

Perfect Number Search

$M =$ 50   **Compute!**

         Clear

In the next Research Question, you will be trying to find the connection between Mersenne primes and perfect numbers. In the course of your investigation, you may find it useful to factor integers. The applet below will do this automatically.

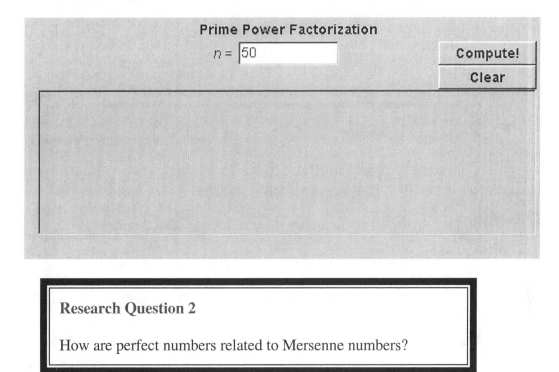

Research Question 2

How are perfect numbers related to Mersenne numbers?

6.3 Fermat Numbers

The mathematician Pierre de Fermat (1601-1655) appears several times in this course. In search of a formula for primes, Fermat considered numbers of the form $2^j + 1$. Below is an applet analogous to the one we used in the previous section when we studied Mersenne numbers. The format of the output is as before: the first column shows the value of j, the second column indicates whether $2^j + 1$ is prime, and the third column shows the value of the $2^j + 1$. The values of j run from 1 to M.

Fermat Prime Search

$M =$ 10   Compute!
                Clear

> **Research Question 3**
>
> Address the three questions given in the introduction of this chapter for numbers of the form $2^j + 1$:
>
> 1. Under what conditions is an integer of this form always prime?
> 2. Under what conditions is an integer of this form always composite?
> 3. Are there infinitely many primes of this form?

6.3.1 Note on Terminology

We would like to refer to numbers of the form $2^j + 1$ as Fermat numbers, but this would not fit the standard usage. Unless j satisfies a condition (which you are supposed to discover), the number is definitely composite. When j satisfies this condition, $2^j + 1$ is called a Fermat number. In any case, a *Fermat prime* is a prime of the form $2^j + 1$.

6.4 Arithmetic Progressions

An arithmetic progression is a sequence of the form $ak + b$ where a and b are fixed and k runs through integer values. The goal of this section is to discover whatever we can about primes in arithmetic progessions.

Let's start by looking at some terms of the arithmetic progression $3k + 7$:

```
┌──────────────────────────────────────────────────────────────┐
│                    Arithmetic Progression                      │
│   a = │3          │          b = │7    │       │  Compute!  │  │
│   L = │-10        │          M = │10   │       │   Clear    │  │
│  ┌──────────────────────────────────────────────────────┐    │
│  │                                                        │    │
│  │                                                        │    │
│  │                                                        │    │
│  │                                                        │    │
│  │                                                        │    │
│  └──────────────────────────────────────────────────────┘    │
└──────────────────────────────────────────────────────────────┘
```

Can you guess what would happen if we considered the corresponding progression $-3k + 7$?

```
┌──────────────────────────────────────────────────────────────┐
│                    Arithmetic Progression                      │
│   a = │-3         │          b = │7    │       │  Compute!  │  │
│   L = │-10        │          M = │10   │       │   Clear    │  │
│  ┌──────────────────────────────────────────────────────┐    │
│  │                                                        │    │
│  │                                                        │    │
│  │                                                        │    │
│  │                                                        │    │
│  │                                                        │    │
│  └──────────────────────────────────────────────────────┘    │
└──────────────────────────────────────────────────────────────┘
```

It was exactly the same, except for the ordering. It is not hard to see why this will happen in general, so we will restrict to positive values for a. Next observe that the integers in the progression $ak + b$ are the integers in the congruence class of b modulo a. For example, the numbers of the form $3k + 7$ are the integers congruent to 7 modulo 3. We could get the same set by taking integers of the form $3k + 1$ because $7 \equiv 1 \pmod 3$.

Arithmetic Progression

a = 3   *b* = 1   Compute!

L = -10   *M* = 10   Clear

Keep in mind that we are seeing only part of an infinite sequence of numbers; adjust the bounds on the sequences to convince yourself that we are seeing the same set.

We can now safely restrict the values of *b* to the possible remainders on division by *a*. Given an integer *a*, there are *a* congruence classes and each integer is in one of those congruence classes. Our questions about primes in arithmetic progressions can be thought of in terms of "how do the primes divide themselves among the different congruence classes modulo *a*?"

The applet below looks at the first *M* primes and counts how many are in each different congruence class modulo *a*. As a first example, let's try it on the congruence classes modulo 5 and the first 100 primes.

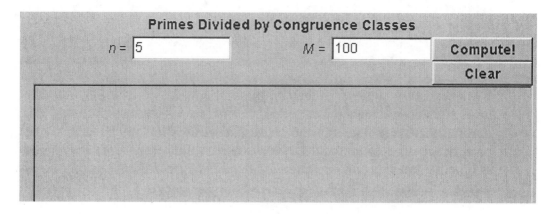

Primes Divided by Congruence Classes

n = 5   *M* = 100   Compute!

Clear

The number of primes that are congruent to 1 is listed first, the number congruent to 2 is listed second, and so on. So, the computation tells us that 24 of the first 100 primes are congruent to 1 modulo 5. The last entry says that only one of the first 100 primes is congruent to 5 modulo 5.

Research Question 4

Address the three questions given in the introduction of this chapter for numbers of the form $ak + b$:

1. Under what conditions is an integer of this form always prime?
2. Under what conditions is an integer of this form always composite?
3. Are there infinitely many primes of this form?

Homework

1. Prove that if a and k are integers, both greater than 1, such that $a^k - 1$ is prime, then $a = 2$.

2. Suppose that $f(x)$ is a polynomial with integer coefficients such that $f(n)$ is prime for every positive integer n. Prove that $f(x)$ is a constant polynomial.

3. Prove that if $(3^k + 1)/2$ is prime, then k is a power of 2.

4. Prove that if $(3^k - 1)/2$ is prime, then k is prime.

5. A rep-unit of length n, denoted $rep(n)$, is the number which has n digits all of which are 1. For example, $rep(4) = 1111$. Alternatively, one can use the formula $rep(n) = (10^n - 1)/9$. Prove that if $rep(n)$ is prime, then n is prime.

6. Carry out the heuristic shown in the chapter summary to predict whether there are finitely many prime rep-units or infinitely many prime rep-units. (Rep-units are defined in exercise 5.)

7. A prime p such that $2p + 1$ is also prime is known as a Germain prime.[1] Primes of the form $2p + 1$ are useful in cryptographic applications. It is not known whether there are infinitely many such primes. Carry out the heuristic shown in the chapter summary to predict whether there are infinitely many or finitely many Germain primes.

8. Prove that a power of a prime cannot be a perfect number.

9. Prove that there are no odd perfect numbers of the form pq where p and q are distinct primes.

10. Prove that the arithmetic progression $6k + 5$ contains infinitely many primes. (Hint: if $p_1, p_2, \ldots, p_n$ are primes of the form $6k + 5$, what can you say about prime divisors of $6p_1 p_2 \cdots p_n - 1$?)

11. Prove that if a and b are positive integers, then $\gcd(2^a - 1, 2^b - 1) = 2^{\gcd(a,b)} - 1$. (Hint: Use the Euclidean Algorithm for computing gcds with the identity $2^a - 1 = 2^{a-b}(2^b - 1) + 2^{a-b} - 1$.)

12. We know that a regular n-gon is constructible if and only if one can construct an angle with measure $360°/n$.

---

[1] These primes are named for Sophie Germain (1776–1831). She made progress on Fermat's Last Theorem for exponents p where $2p + 1$ is prime.

(a) Show that if m and n are relatively prime positive integers, then there are integer solutions (a, b) to

$$\frac{a}{m} + \frac{b}{n} = \frac{1}{mn}.$$

(b) Use part (a) to show that if one can construct a regular m-gon and a regular n-gon where $\gcd(m, n) = 1$, then one can construct a regular mn-gon. (Hint: Think about adding and subtracting angles.)

13. Recall that the Fibonacci numbers F_i are defined by $F_0 = 0$, $F_1 = 1$, and $F_i = F_{i-1} + F_{i-2}$ for $i \geq 2$. (See exercises 15–18 in Chapter 4.) Prove that if F_n is prime, then n is an odd prime or $n = 4$.

14. It is not known whether there are infinitely many prime Fibonacci numbers. Carry out the heuristic shown in the chapter summary to predict whether there are infinitely many or finitely many Fibonacci numbers which are prime. Be sure to account for exercise 13 above. You may also use the result of exercise 18 of Chapter 4.

7. The Chinese Remainder Theorem

Prelab

Problems related to the *Chinese Remainder Theorem* have been found in writings of Chinese mathematicians as far back as the first century. We will get to the exact statement of the Chinese Remainder Theorem (or "CRT" for short) in the lab portion of the chapter. For now, we'll try to get a feel for the types of situations addressed by the CRT through a couple of specific examples. Let's start with the following problem:

> Little Tommy has a small mound of pennies. His dad can tell that there is less than a quarter's worth, but cannot tell exactly how many there are. Tommy won't let his dad count the pennies (and it is driving him crazy). When Tommy arranges the pennies in piles of 4, there is one left over. When Tommy arranges the pennies in piles of 5, there are two left over. How many pennies does Tommy have?

One way to solve this problem is to hire some goons to take little Tommy's pennies, and then simply count them at our leisure. However, in this case we are interested in a more elegant solution. Here is another possibility: Let x be the number of pennies in Tommy's possession. Then the statements about piles of pennies translate to

$$x \equiv 1 \pmod 4 \qquad \text{and} \qquad x \equiv 2 \pmod 5.$$

We would like to determine the integers x which satisfy this pair of congruences. This is not difficult if we proceed algebraically and convert the congruences, one at a time, into equations. Let's see how this works.

From the congruence $x \equiv 1 \pmod 4$, we know that $x = 1 + 4k$ for some integer k. We now substitute this expression for x into the second congruence:

$$1 + 4k \equiv 2 \pmod 5 \implies 4k \equiv 1 \pmod 5$$

It's easy to determine that the solution to this congruence equation is given by $k \equiv 4 \pmod 5$. We could find this by trial and error, or apply our results from Chapter 5 on solving linear congruences. Now we convert this congruence into an equation:

$k = 4 + 5\ell$ for some integer ℓ. Substituting back, we find that solutions to our congruences are of the form

$$x = 1 + 4k$$
$$= 1 + 4(4 + 5\ell)$$
$$= 17 + 20\ell.$$

To answer the original question, we see that the only positive value of x of this form which is less than 25 (since Tommy has *less than a quarter's worth*) is 17. Thus Tommy has 17 pennies.

Notice that our two original congruences imply another congruence. The equation $x = 17 + 20\ell$ for some $\ell \in \mathbf{Z}$ is equivalent to the congruence $x \equiv 17 \pmod{20}$. In fact, the two original congruences are equivalent to this single congruence modulo 20. Going in the reverse direction is easier and left as an exercise.

1. **Show that if** $x \equiv 17 \pmod{20}$, **then** $x \equiv 1 \pmod 4$ **and** $x \equiv 2 \pmod 5$.

Use the method illustrated above to solve the following pairs of congruences. In each case, the solution should lead to a single congruence.

2. $x \equiv 0 \pmod 2$ **and** $x \equiv 1 \pmod 3$.

3. $x \equiv 1 \pmod 2$ **and** $x \equiv 1 \pmod 3$.

4. $x \equiv 1 \pmod 7$ **and** $x \equiv 5 \pmod{15}$.

As mentioned earlier, we will get to the statement of the Chinese Remainder Theorem in the lab. While working on the above exercises, you should think about the following questions, which will come up again later in the chapter:

- Under what conditions will a pair of congruences be solvable by this method?

- Can this method be used for proving the existence of solutions to a pair of congruences?

- How can this method be extended to three or more simultaneous congruences?

Maple electronic notebook `07-crt.mws`

Mathematica electronic notebook `07-crt.nb`

Web electronic notebook Start with the web page `index.html`

Maple Lab: Chinese Remainder Theorem

■ Solving Two Congruences

The Chinese Remainder Theorem is one of the oldest theorems in number theory. Your first job is to discover the right statement of this theorem. Most of the statement of the theorem is provided in the next section - you just need to fill in the missing part.

■ The Chinese Remainder Theorem for Two Congruences

Below is an incomplete statement of the Chinese Remainder Theorem for two congruences.

Chinese Remainder Theorem: If m_1 and m_2 are positive integers such that ????, then for any integers a_1 and a_2, the pair of congruences

$$x \equiv a_1 \pmod{m_1} \quad \text{and} \quad x \equiv a_2 \pmod{m_2}$$

has a unique solution x modulo $m_1 m_2$.

To complete Research Question 1, you need to figure out what condition is required in place of the ???? above to make the statement true. Notice that the ???? placement occurs after the introduction of m_1 and m_2 but before a_1 and a_2. This means that the missing condition should involve m_1 and m_2 but not a_1 or a_2. So to complete the statement of the theorem, you need to determine what condition on m_1 and m_2 allows one to find a unique solution to the congruences for all choices of a_1 and a_2.

■ Research Question 1
Complete the statement of the Chinese Remainder Theorem for two congruences.

■ Some Help

To find the correct statement of the theorem, you'll probably want to do some experimentation. To assist with this job, a function is provided that will find solutions to pairs of congruences. The function solvetwo(a1,a2,m1,m2) takes a_1, a_2, m_1, and m_2 as input, and finds all values of x (mod $m_1 m_2$) that satisfy the pair of congruences $x \equiv a_1$ (mod m_1) and $x \equiv a_2$ (mod m_2). Here is the definition:

```
> solvetwo := proc(a1,a2,m1,m2)
    local answer,j;
    answer := NULL;
    for j from 0 to m1*m2-1 do
     if modp(j-a1, m1)=0 and modp(j-a2, m2)=0 then
      answer := answer,j;
     fi;
    od;
    answer;
  end:
>
```

To see if the congruences $x \equiv 2$ (mod 4) and $x \equiv 1$ (mod 5) have any common solutions, we just execute the following:

347

```
[ > solvetwo(2,1,4,5);
```

As we can see from the output, in this case there is a unique solution given by $x \equiv 6 \pmod{20}$. Here's what we get for the pair of congruences $x \equiv 3 \pmod 6$ and $x \equiv 9 \pmod{10}$:

```
[ > solvetwo(3,9,6,10);
[ >
```

This time, the solution is not unique. You should compare the output from `solvetwo` to what you find algebraically (the method of the Prelab) in solving pairs of congruences. For example, try using `solvetwo` to find the solutions to the exercises given in the Prelab sheet.

One last suggestion: You may also want to try to generalize the algebraic method that you used in the Prelab when solving specific pairs of congruences. This can be helpful both in finding the correct statement of the theorem as well as the proof. In fact, this is one way in which mathematicians find the right statement for a theorem. They have an idea for a proof and see how generally the method can be applied. Whatever they get is the statement of the theorem.

■ A More General Theorem

In the previous section, you found the conditions required for the pair of congruences

$$x \equiv a_1 \pmod{m_1} \quad \text{and} \quad x \equiv a_2 \pmod{m_2}$$

to have a unique solution x modulo $m_1 m_2$. We now turn to the more general question of when this pair of congruences will have any solutions, unique or otherwise.

■ Research Question 2

For what values of a_1, a_2, m_1, and m_2 will the pair of congruences

$$x \equiv a_1 \pmod{m_1} \quad \text{and} \quad x \equiv a_2 \pmod{m_2}$$

have solutions modulo $m_1 m_2$? If there is a solution $x_0 \pmod{m_1 m_2}$, find the form of all other solutions $x \pmod{m_1 m_2}$ in terms of x_0.

■ Solving Lots of Congruences

Now that you have completed the first two Research Questions, you now know the exact nature of the solutions to the pair of congruences

$$x \equiv a_1 \pmod{m_1} \quad \text{and} \quad x \equiv a_2 \pmod{m_2}.$$

Recall that the cases where there is a *unique* solution modulo $m_1 m_2$ is covered by the Chinese Remainder Theorem for two congruences. Suppose instead you had three or more congruences. Is there a version of the Chinese Remainder Theorem in this case? There is indeed, and (surprise, surprise) it's your job to find it.

To help get you headed in the right direction, let's look at a specific example involving three specific congruences, such as

$$x \equiv 1 \quad (\text{mod } 5)$$
$$x \equiv 2 \quad (\text{mod } 6)$$
$$x \equiv 3 \quad (\text{mod } 7).$$

One way of proceeding is to begin by solving the first pair of congruences,

$$x \equiv 1 \quad (\text{mod } 5) \quad \text{and} \quad x \equiv 2 \quad (\text{mod } 6).$$

By applying the algebraic method of the Prelab, we can show that this pair of congruences is equivalent to the single congruence $x \equiv 26$ (mod 30). Let's check this with `solvetwo`:

```
[ > solvetwo(1,2,5,6);
```

OK, now we have reduced our problem to just two congruences:

$$x \equiv 26 \quad (\text{mod } 30) \quad \text{and} \quad x \equiv 3 \quad (\text{mod } 7).$$

We can employ our method for pairs again to reach the single congruence $x \equiv 206$ (mod 210). Again, let's check it with `solvetwo`:

```
[ > solvetwo(26,3,30,7);
[ >
```

So, we did it! We solved three congruences by twice applying what we know about solving a pair of congruences. Here's your chance to generalize this process.

■ Research Question 3

Give a statement of the Chinese Remainder Theorem (as in Research Question 1) for *n* congruences.

■ Getting Maple to Solve Lots of Congruences

The function `solvemany` defined below will solve a system of any number of congruences. It works in a manner similar to `solvetwo`, by systematically looking for integers that satisfy all of the congruences.

```
[ > solvemany := proc(alist, mlist)
    local j, k, M, good;
    M := 1;
    for j to nops(mlist) do
     M := M*mlist[j];
    od;
    for j from 0 to M-1 do
     good := true;
     k := 1;
     while((k <= nops(alist)) and good) do
      if(modp(j,mlist[k]) <> modp(alist[k],mlist[k])) then
       good := false;
      fi;
      k := k+1;
     od;
     if good then printf(` %d`, j) fi;
    od;
   end:
[
```

```
[ >
```

To solve the system of congruences

$$x \equiv 1 \pmod 5$$
$$x \equiv 2 \pmod 6$$
$$x \equiv 3 \pmod 7,$$

we enter:

```
[ > solvemany([1,2,3],[5,6,7]);
[ >
```

Notice that this agrees with what we got in the last section. The first argument of the function solvemany is a list of the a_i and the second argument is a list of the corresponding moduli m_i. The answer it returns is all solutions modulo $m_1 \, m_2 \cdots$ to the system of congruences. Try it with different numbers, but keep in mind that for large moduli the function may be slow in evaluating.

■ Explicit Formulas

Suppose we have a typical Chinese Remainder Theorem problem, where we wish to solve the pair of congruence equations

$$x \equiv a_1 \pmod{m_1} \quad \text{and} \quad x \equiv a_2 \pmod{m_2}$$

for x, as in Research Question 1. It would be nice to have a formula for x given in terms of a_1, a_2, m_1, and m_2. Such a formula might be useful both for computation and for theoretical purposes (for example, in the proof of Research Question 1).

One approach to finding a formula is to use what is known as the method of undetermined coefficients. The idea is to guess the general form of the formula leaving some coefficients unspecified, assume that the formula is correct, and see if you can deduce the correct values of the unknown coefficients. If you can solve for the coefficients, you would have a conjecture for the right formula. With the right formula in hand, it may not be too hard to prove that it is the correct formula.

In this case, we will guess that the solution has the form

$$x = c_1 \, m_1 + c_2 \, m_2,$$

where c_1 and c_2 are yet to be determined. If this formula for x is correct, what would we learn about c_1 and c_2 by plugging into the two congruence equations above? Substituting into the first equation, we would have

$$c_1 \, m_1 + c_2 \, m_2 \equiv a_1 \pmod{m_1},$$

which simplifies to just $c_2 \, m_2 \equiv a_1 \pmod{m_1}$. Can you solve this for c_2? Once you have done so, then use the second congruence equation in a similar manner to determine c_1 in terms of a_1, a_2, m_1, and m_2. You might want to compute some examples to test your formula using the functions solvetwo or chrem discussed above.

```
[ >
```

■ Research Question 4

With the assumptions of Research Question 1, find formulas for c_1 and c_2 so that

$$x = c_1 m_1 + c_2 m_2$$

is a solution to the pair of congruences in Research Question 1.

We now want to generalize the result of Research Question 4 to the case of several congruences. An important step in employing the method of undetermined coefficients is to guess the right general form of the answer. For Research Question 4, we suggested trying $x = c_1 m_1 + c_2 m_2$. For the general case, you will have to try to get the right formula.

Here is the setup. You are given n congruences of the form $x \equiv a_i \pmod{m_i}$ satisfying the hypotheses you conjectured in Research Question 3. You want a formula for the solution x in terms of the a_i and the m_i. In guessing the general form of the answer, think about what made the guess above work well for two congruences. One useful feature was that once the general form for x was substituted into one of the initial congruences, all but one term dropped out. That should have made it easier to solve for c_1 and c_2.

■ Research Question 5

With the assumptions of Research Question 3, find a formula for x so that x will be a solution to the system of congruences $x \equiv a_i \pmod{m_i}$.

Mathematica Lab: The Chinese Remainder Theorem

■ Solving Two Congruences

The Chinese Remainder Theorem is one of the oldest theorems in number theory. Your first job is to discover the right statement of this theorem. Most of the statement of the theorem is provided in the next section —you just need to fill in the missing part.

■ The Chinese Remainder Theorem for Two Congruences

Below is an incomplete statement of the Chinese Remainder Theorem for two congruences.

Chinese Remainder Theorem: If m_1 and m_2 are positive integers such that ????, then for any integers a_1 and a_2, the pair of congruences

$$x \equiv a_1 \pmod{m_1} \quad \text{and} \quad x \equiv a_2 \pmod{m_2}$$

has a unique solution x modulo $m_1 m_2$.

To complete Research Question 1, you need to figure out what condition is required in place of the ???? above to make the statement true. Notice that the ???? placement occurs after the introduction of m_1 and m_2 but before a_1 and a_2. This means that the missing condition should involve m_1 and m_2 but not a_1 or a_2. So to complete the statement of the theorem, you need to determine what condition on m_1 and m_2 allows one to find a unique solution to the congruences for all choices of a_1 and a_2.

■ Research Question 1

Complete the statement of the Chinese Remainder Theorem for two congruences.

■ Some Help

To find the correct statement of the theorem, you'll probably want to do some experimentation. To assist with this job, a function is provided that will find solutions to pairs of congruences. The function **solvetwo[a1,a2,m1,m2]** takes a_1, a_2, m_1, and m_2 as input, and finds all values of x (mod $m_1 m_2$) that satisfy the pair of congruences $x \equiv a_1$ (mod m_1) and $x \equiv a_2$ (mod m_2). Here is the definition:

```
solvetwo[a1_, a2_, m1_, m2_] :=
        Module[{jvalues},
        (* Pick out the values of j which work *)
        jvalues = Range[0, m1 * m2 - 1];
        jvalues = Select[jvalues, Mod[# - a1, m1] == 0 &];
        jvalues = Select[jvalues, Mod[# - a2, m2] == 0 &];
        jvalues]
```

To see if the congruences $x \equiv 2$ (mod 4) and $x \equiv 1$ (mod 5) have any common solutions, we just execute the following:

```
solvetwo[2, 1, 4, 5]
```

As we can see from the output, in this case there is a unique solution given by $x \equiv 6$ (mod 20). Here's what we get for the pair of congruences $x \equiv 3$ (mod 6) and $x \equiv 9$ (mod 10):

```
solvetwo[3, 9, 6, 10]
```

This time, the solution is not unique. You should compare the output from **solvetwo** to what you find algebraically (the method of the Prelab) in solving pairs of congruences. For example, try using **solvetwo** to find the solutions to the exercises given in the Prelab sheet.

One last suggestion: You may also want to try to generalize the algebraic method that you used in the Prelab when solving specific pairs of congruences. This can be helpful both in finding the correct statement of the theorem as well as the proof. In fact, this is one way in which mathematicians find the right statement for a theorem. They have an idea for a proof and see how generally the method can be applied. Whatever they get is the statement of the theorem.

■ A More General Theorem

In the previous section, you found the conditions required for the pair of congruences

$$x \equiv a_1 \pmod{m_1} \quad \text{and} \quad x \equiv a_2 \pmod{m_2}$$

to have a unique solution x modulo $m_1 m_2$. We now turn to the more general question of when this pair of congruences will have any solutions, unique or otherwise.

■ Research Question 2

For what values of a_1, a_2, m_1, and m_2 will the pair of congruences

$$x \equiv a_1 \pmod{m_1} \quad \text{and} \quad x \equiv a_2 \pmod{m_2}$$

have solutions modulo $m_1 m_2$? If there is a solution x_0 (mod $m_1 m_2$), find the form of all other solutions x (mod $m_1 m_2$) in terms of x_0.

■ Solving Lots of Congruences

Now that you have completed the first two Research Questions, you now know the exact nature of the solutions to the pair of congruences

$$x \equiv a_1 \pmod{m_1} \quad \text{and} \quad x \equiv a_2 \pmod{m_2}.$$

Recall that the cases where there is a *unique* solution modulo $m_1 m_2$ is covered by the Chinese Remainder Theorem for two congruences. Suppose instead you had three or more congruences. Is there a version of the Chinese Remainder Theorem in this case? There is indeed, and (surprise, surprise) it's your job to find it.

To help get you headed in the right direction, let's look at a specific example involving three specific congruences, such as

$$x \equiv 1 \pmod 5$$
$$x \equiv 2 \pmod 6$$
$$x \equiv 3 \pmod 7.$$

One way of proceeding is to begin by solving the first pair of congruences,

$$x \equiv 1 \pmod{5} \quad \text{and} \quad x \equiv 2 \pmod{6}.$$

By applying the algebraic method of the Prelab, we can show that this pair of congruences is equivalent to the single congruence $x \equiv 26 \pmod{30}$. Let's check this with **solvetwo**:

    solvetwo[1, 2, 5, 6]

OK, now we have reduced our problem to just two congruences:

$$x \equiv 26 \pmod{30} \quad \text{and} \quad x \equiv 3 \pmod{7}.$$

We can employ our method for pairs again to reach the single congruence $x \equiv 206 \pmod{210}$. Again, let's check it with **solvetwo**:

    solvetwo[26, 3, 30, 7]

So, we did it! We solved three congruences by twice applying what we know about solving a pair of congruences. Here's your chance to generalize this process.

■ Research Question 3

> Give a statement of the Chinese Remainder Theorem (as in Research Question 1) for *n* congruences.

■ Getting Mathematica to Solve Lots of Congruences

The function **solvemany** defined below will solve a system of any number of congruences. It works in a manner similar to **solvetwo**, by systematically looking for integers that satisfy all of the congruences.

```
solvemany[alist_, mlist_] :=
        Module[{jvalues},
        (* Pick out the values of j which work *)
        jvalues = Range[0, Apply[Times, mlist] - 1];
        Do[
        jvalues =
    Select[jvalues, Mod[# - alist[[i]], mlist[[i]]] == 0 &],
        {i, 1, Length[mlist]}];
        jvalues]
```

To solve the system of congruences

$$x \equiv 1 \pmod 5$$
$$x \equiv 2 \pmod 6$$
$$x \equiv 3 \pmod 7,$$

we enter:

    solvemany[{1, 2, 3}, {5, 6, 7}]

    {206}

Notice that the output agrees with what we got in the last section. The first argument of the function **solvemany** is a list of the a_i and the second argument is a list of the corresponding moduli m_i. The answer it returns is a list of all solutions modulo $m_1 m_2 \cdots$ to the system of congruences. Try it with different numbers, but keep in mind that for large moduli the function may be slow in evaluating.

■ Explicit Formulas

Suppose we have a typical Chinese Remainder Theorem problem, where we wish to solve the pair of congruence equations

$$x \equiv a_1 \pmod{m_1} \quad \text{and} \quad x \equiv a_2 \pmod{m_2}$$

for x, as in Research Question 1. It would be nice to have a formula for x given in terms of a_1, a_2, m_1, and m_2. Such a formula might be useful both for computation and for theoretical purposes (for example, in the proof of Research Question 1).

One approach to finding a formula is to use what is known as the method of undetermined coefficients. The idea is to guess the general form of the formula leaving some coefficients unspecified, assume that the formula is correct, and see if you can deduce the correct values of the unknown coefficients. If you can solve for the coefficients, you would have a conjecture for the right formula. With the right formula in hand, it may not be too hard to prove that it is the correct formula.

In this case, we will guess that the solution has the form

$$x = c_1 m_1 + c_2 m_2,$$

where c_1 and c_2, are yet to be determined. If this formula for x is correct, what would we learn about c_1 and c_2 by plugging into the two congruence equations above? Substituting into the first equation, we would have

$$c_1 m_1 + c_2 m_2 \equiv a_1 \pmod{m_1},$$

which simplifies to just $c_2 m_2 \equiv a_1 \pmod{m_1}$. Can you solve this for c_2? Once you have done so, then use the second congruence equation in a similar manner to determine c_1 in terms of a_1, a_2, m_1, and m_2. You might want to compute some examples to test your formula using the functions **solvetwo** or **solvemany** discussed above.

■ Research Question 4

> With the assumptions of Research Question 1, find formulas for c_1 and c_2 so that
>
> $$x = c_1 m_1 + c_2 m_2$$
>
> is a solution to the pair of congruences in Research Question 1.

We now want to generalize the result of Research Question 4 to the case of several congruences. An important step in employing the method of undetermined coefficients is to guess the right general form of the answer. For Research Question 4, we suggested trying $x = c_1 m_1 + c_2 m_2$. For the general case, you will have to try to get the right formula.

Here is the setup. You are given n congruences of the form $x \equiv a_i \pmod{m_i}$ satisfying the hypotheses you conjectured in Research Question 3. You want a formula for the solution x in terms of the a_i and the m_i. In guessing the general form of the answer, think about what made the guess above work well for two congruences. One useful feature was that once the general form for x was substituted into one of the initial congruences, all but one term dropped out. That should have made it easier to solve for c_1 and c_2.

■ **Research Question 5**

> With the assumptions of Research Question 3, find a formula for x so that x will be a solution to the system of congruences $x \equiv a_i \pmod{m_i}$.

Web Lab: The Chinese Remainder Theorem

7.1 Solving Two Congruences

The Chinese Remainder Theorem is one of the oldest theorems in number theory. Your first job is to discover the right statement of this theorem. Most of the statement of the theorem is provided in the next section - you just need to fill in the missing part.

7.1.1 The Chinese Remainder Theorem for Two Congruences

Below is an incomplete statement of the Chinese Remainder Theorem for two congruences.

> **Chinese Remainder Theorem**: If m_1 and m_2 are positive integers such that ????, then for any integers a_1 and a_2, the pair of congruences
>
> $$x \equiv a_1 \pmod{m_1} \quad \text{and} \quad x \equiv a_2 \pmod{m_2}$$
>
> has a unique solution x modulo $m_1 m_2$.

To complete Research Question 1, you need to figure out what condition is required in place of the ???? above to make the statement true. Notice that the ???? placement occurs after the introduction of m_1 and m_2 but before a_1 and a_2. This means that the missing condition should involve m_1 and m_2 but not a_1 or a_2. So to complete the statement of the theorem, you need to determine what condition on m_1 and m_2 allows one to find a unique solution to the congruences for all choices of a_1 and a_2.

7.1.2 Some Help

To find the correct statement of the theorem, you'll probably want to do some experimentation. To assist with this job, an applet is provided that will find solutions to pairs of congruences. The applet takes a_1, a_2, m_1, and m_2 as input, and finds all values of x (mod m_1m_2) that satisfy the pair of congruences $x \equiv a_1$ (mod m_1) and $x \equiv a_2$ (mod m_2). For example, to see if the congruences $x \equiv 2$ (mod 4) and $x \equiv 1$ (mod 5) have any common solutions, we just execute the following:

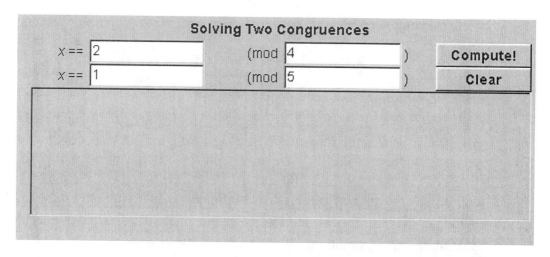

As we can see from the output, in this case there is a unique solution given by $x \equiv 6$ (mod 20). Here's what we get for the pair of congruences $x \equiv 3$ (mod 6) and $x \equiv 9$ (mod 10):

This time, the solution is not unique. You should compare the output from the applet to what you find algebraically (the method of the Prelab) in solving pairs of congruences. For example, try using applet to find the solutions to the exercises given in the Prelab sheet.

One last suggestion: You may also want to try to generalize the algebraic method that you used in the Prelab when solving specific pairs of congruences. This can be helpful both in finding the correct statement of the theorem as well as the proof. In fact, this is one way in which mathematicians find the right statement for a theorem. They have an idea for a proof and see how generally the method can be applied. Whatever they get is the statement of the theorem.

7.2 A More General Theorem

In the previous section, you found the conditions required for the pair of congruences

$$x \equiv a_1 \ (\text{mod } m_1) \quad \text{and} \quad x \equiv a_2 \ (\text{mod } m_2)$$

to have a unique solution x modulo $m_1 m_2$. We now turn to the more general question of when this pair of congruences will have any solutions, unique or otherwise.

Research Question 2

For what values of a_1, a_2, m_1, and m_2 will the pair of congruences

$$x \equiv a_1 \ (\text{mod } m_1) \quad \text{and} \quad x \equiv a_2 \ (\text{mod } m_2)$$

have a solution modulo m_1m_2? If there is a solution x_0 (mod m_1m_2), find the form of all other solutions x (mod m_1m_2) in terms of x_0.

Below is our applet for your use in investigating Research Question 2.

Solving Two Congruences

$x ==$ [   ]   (mod [   ])

$x ==$ [   ]   (mod [   ])

7.3   Solving Lots of Congruences

Now that you have completed the first two Research Questions, you know the exact nature of the solutions to the pair of congruences

$$x \equiv a_1 \ (\mathrm{mod}\ m_1) \quad \text{and} \quad x \equiv a_2 \ (\mathrm{mod}\ m_2).$$

Recall that the cases where there is a *unique* solution modulo m_1m_2 is covered by the Chinese Remainder Theorem for two congruences. Suppose instead you had three or more congruences. Is there a version of the Chinese Remainder Theorem in this case? There is indeed, and (surprise, surprise) it's your job to find it.

To help get you headed in the right direction, let's look at an example involving three specific congruences, such as

$$x \equiv 1 \ (\mathrm{mod}\ 5)$$
$$x \equiv 2 \ (\mathrm{mod}\ 6)$$
$$x \equiv 3 \ (\mathrm{mod}\ 7).$$

One way of proceeding is to begin by solving the first pair of congruences,

$$x \equiv 1 \ (\mathrm{mod}\ 5) \quad \text{and} \quad x \equiv 2 \ (\mathrm{mod}\ 6).$$

By applying the algebraic method of the Prelab, we can show that this pair of congruences is equivalent to the single congruence $x \equiv 26 \ (\mathrm{mod}\ 30)$. Let's check this with our applet:

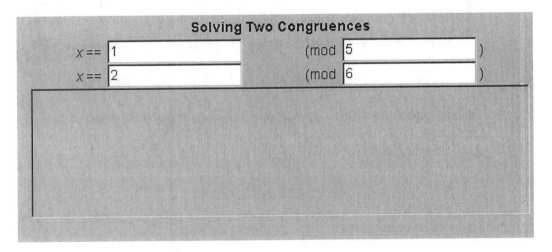

OK, now we have reduced our problem to just two congruences:

$$x \equiv 26 \ (\mathrm{mod}\ 30) \quad \text{and} \quad x \equiv 3 \ (\mathrm{mod}\ 7).$$

We can employ our method for pairs again to reach the single congruence $x \equiv 206 \ (\mathrm{mod}\ 210)$. Again, let's check this with the applet:

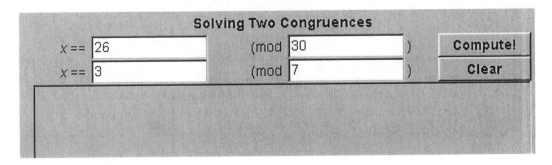

So, we did it! We solved three congruences by twice applying what we know about solving a pair of congruences. Here's your chance to generalize this process.

Research Question 3

Give a statement of the Chinese Remainder Theorem (as in Research Question 1) for *n* congruences.

7.3.1 Using Java to Solve Lots of Congruences

We now introduce a new and improved Java applet that can be used to solve more than two congruences simultaneously. For example, to solve the system of congruences

$$x \equiv 1 \pmod 5$$
$$x \equiv 2 \pmod 6$$
$$x \equiv 3 \pmod 7,$$

we enter

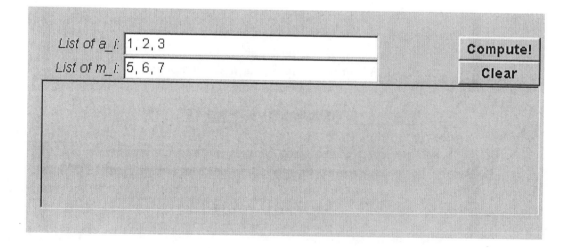

Notice that the output agrees with what we found above. Try it with different numbers.

7.4 Explicit Formulas

Suppose we have a typical Chinese Remainder Theorem problem, where we wish to solve the pair of congruence equations

$$x \equiv a_1 \ (\mathrm{mod}\ m_1) \quad \text{and} \quad x \equiv a_2 \ (\mathrm{mod}\ m_2).$$

for x, as in Research Question 1. It would be nice to have a formula for x given in terms of a_1, a_2, m_1, and m_2. Such a formula might be useful both for computation and for theoretical purposes. (For example, in the proof of Research Question 1.)

One approach to finding a formula is to use what is known as the method of undetermined coefficients. The idea is to guess the general form of the formula leaving some coefficients unspecified, assume that the formula is correct, and see if you can deduce the correct values of the unknown coefficients. If you can solve for the coefficients, you would have a conjecture for the right formula. With the right formula in hand, it may not be too hard to prove that it is the correct formula.

In this case, we will guess that the solution has the form

$$x = c_1 m_1 + c_2 m_2,$$

where c_1 and c_2 are yet to be determined. If this formula for x is correct, what would we learn about c_1 and c_2 by plugging into the two congruence equations above? Substituting into the first equation, we would have

$$c_1 m_1 + c_2 m_2 \equiv a_1 \ (\mathrm{mod}\ m_1),$$

which simplifies to just $c_2 m_2 \equiv a_1 \ (\mathrm{mod}\ m_1)$. Can you solve this for c_2? Once you have done so, then use the second congruence equation in a similar manner to determine c_1 in terms of a_1, a_2, m_1, and m_2. You might want to compute some examples to test your formula using the usual applet, which is included below the statement of Research Question 4.

Research Question 4

With the assumptions of Research Question 1, find formulas for c_1 and c_2 so that

$$x = c_1 m_1 + c_2 m_2,$$

is a solution to the pair of congruences in Research Question 1.

We now want to generalize the result of Research Question 4 to the case of several congruences. An important step in employing the method of undetermined coefficients is to guess the right general form of the answer. For Research Question 4, we suggested trying $x = c_1 m_1 + c_2 m_2$. For the general case, you will have to try to get the right formula.

Here is the setup. You are given n congruences of the form $x \equiv a_i \pmod{m_i}$ satisfying the hypotheses you conjectured in Research Question 3. You want a formula for the solution x in terms of the a_i and the m_i. In guessing the general form of the answer, think about what made the guess above work well for two congruences. One useful feature was that once the general form for x was substituted into one of the initial congruences, all but one term dropped out. That should have made it easier to solve for c_1 and c_2.

Research Question 5

With the assumptions of Research Question 3, find a formula for x so that x will be a solution to the system of congruences $x \equiv a_i \pmod{m_i}$.

List of a_i:

List of m_i:

Compute!

Clear

Homework

1. Find the smallest even integer m such that $3 \mid m$, $5 \mid (m+3)$, and $11 \mid (m+5)$.

2. Prove that if the set of integers $n_1, n_2, \ldots, n_k$ is pairwise relatively prime, then the set is relatively prime.

3. Bogie McGee, a professional golfer, has just received a shipment of new golf balls from one of his sponsors. The balls are just dumped into the shipping boxes, so that it's not clear how many there are. Bogie decides to use his knowledge of number theory to count the balls by placing them in rows of different lengths. When he places the balls in rows of 4 each, there are 3 remaining; when he places the balls in rows of 7 each, there are 4 remaining; when he places the balls in rows of 9 each, there is 1 remaining; and, when he places the balls in rows of 11 each, there are none remaining. In addition to this information, he also knows that each ball weighs at most 1.62 ounces avoirdupois (as required by the United States Golf Association),[1] and that the entire shipment (including the shipping crates) weighs approximately 420 pounds. Moreover, Bogie knows that golf balls are made to be very close in weight to the "legal limit" of 1.62 ounces. How many golf balls did he receive?

4. Describe all solutions to the pair of congruences

$$x \equiv 17 \pmod{23}$$
$$x \equiv 12 \pmod{93}.$$

5. Describe all solutions to the pair of congruences

$$x \equiv 81 \pmod{91}$$
$$x \equiv 37 \pmod{182}.$$

6. Describe all solutions to the pair of congruences

$$x \equiv 3 \pmod{12}$$
$$x \equiv 23 \pmod{140}.$$

7. Find three consecutive integers divisible by 2, 3, and 5, respectively.

8. Find four consecutive integers divisible by 2, 3, 5, and 7, respectively.

---

[1]The word *avoirdupois* means that, in this unit of measure, there are 16 ounces to a pound and 16 drams to an ounce. This eliminates any chance of confusion with other similarly named units of weight, such as troy ounces, in which there are 12 ounces to a pound and 480 grains to an ounce.

9. Find three consecutive even integers divisible by 3, 5, and 11, respectively.

10. Find a single congruence that is equivalent to the system of congruences

$$x \equiv r_1 \pmod{m}$$
$$x \equiv r_2 \pmod{m+1}.$$

11. Find three consecutive integers such that the first is divisible by 4, the second is divisible by 9, and the third is divisible by 25.

12. Prove that there are arbitrarily long sequences of integers which are all divisible by perfect squares greater than 1.

13. Prove that there are arbitrarily long sequences of integers which are all divisible by perfect cubes greater than 1.

14. Suppose that a positive integer n ends in 2 when expressed in base 4, ends in 1 when expressed in base 5, and ends in 7 when expressed in base 9. What is the smallest possible value for n? (You may give your answer in base 10.)

15. Suppose p is prime. Consider the pair of congruences

$$x \equiv a_1 \pmod{p^i} \qquad \text{and} \qquad x \equiv a_2 \pmod{p^j}$$

where $i \leq j$.

 (a) Prove that the pair of congruences have a common solution if and only if $a_1 \equiv a_2 \pmod{p^i}$.

 (b) Prove that if in part (a) we do have $a_1 \equiv a_2 \pmod{p^i}$, then the solutions to the initial pair of congruences are simply given by the congruence $x \equiv a_2 \pmod{p^j}$.

16. (a) Find a single congruence which is equivalent to the following system of congruences:

$$x \equiv 2 \pmod{2^3}, \quad x \equiv 3 \pmod{3^2}, \quad \text{and} \quad x \equiv 4 \pmod{5^4}.$$

 (b) For each of the following congruences, find an equivalent system of congruences where the moduli are prime powers. Note, the process in this part is the reverse of what you did in part (a).

   (i) $x \equiv 1956 \pmod{12150}$
   (ii) $x \equiv 12324 \pmod{12348}$
   (iii) $x \equiv 2181 \pmod{3675}$

(c) Find all solutions to the system of congruences in part (b). (Hint: Exercise 15 may help.)

17. Solve the following ancient problem from India. If eggs are removed from a basket 2, 3, 4, 5, or 6 at a time, then there will remain 1, 2, 3, 4, and 5 eggs respectively. If the eggs are removed 7 at a time, there are no eggs remaining in the basket. What is the least number of eggs which could be in the basket?

In the following problems, assume that m_1 and m_2 are positive integers which are relatively prime. Moreover, let $m_1', m_2' \in \mathbf{Z}$ such that $m_1 m_1' + m_2 m_2' = 1$. Finally, let

$$f(a) = (a \% m_1, a \% m_2)$$

and

$$g(a_1, a_2) = m_2 m_2' a_1 + m_1 m_1' a_2.$$

18. Show that if $a \equiv b \pmod{m_1 m_2}$, then $a \equiv b \pmod{m_1}$ and $a \equiv b \pmod{m_2}$. (This shows that the function f can be thought of as a well-defined function from $\mathbf{Z}_{m_1 m_2}$ to $\mathbf{Z}_{m_1} \times \mathbf{Z}_{m_2}$.)

19. Show that if $a_1 \equiv b_1 \pmod{m_1}$ and $a_2 \equiv b_2 \pmod{m_2}$, then $g(a_1, a_2) \equiv g(b_1, b_2) \pmod{m_1 m_2}$. (This shows that the function g is well-defined from $\mathbf{Z}_{m_1} \times \mathbf{Z}_{m_2}$ to $\mathbf{Z}_{m_1 m_2}$.)

20. Show that if $a \in \mathbf{Z}$, then

$$a \equiv g(f(a)) \pmod{m_1 m_2}.$$

(When combined with the next problem, this shows that f and g are inverse functions.)

21. Show that if $a_1, a_2 \in \mathbf{Z}$ and $(b_1, b_2) = f(g(a_1, a_2))$, then $a_1 \equiv b_1 \pmod{m_1}$ and $a_2 \equiv b_2 \pmod{m_2}$. (When combined with the previous problem, this shows that f and g are inverse functions.)

22. (a) Compute the function $g(a_1, a_2)$ which explicitly solves a pair of congruences $x \equiv a_1 \pmod{25}$ and $x \equiv a_2 \pmod{31}$. (The general form for g is given above problem 5; you just have to plug in values for m_1, m_2, m_1', and m_2'.)

    (b) Use your function from part (a) to solve the following pairs of congruences. In each case, find the least positive solution.

    (i) $x \equiv 19 \pmod{25}$ and $x \equiv 24 \pmod{31}$

    (ii) $x \equiv 23 \pmod{25}$ and $x \equiv 17 \pmod{31}$

(iii) $x \equiv 4 \pmod{25}$ and $x \equiv 4 \pmod{31}$

23. The idea of this problem is to investigate solutions to $x^2 \equiv 1 \pmod{pq}$ where p and q are distinct odd primes.

   (a) Show that if p is an odd prime, then there are exactly two solutions modulo p to $x^2 \equiv 1 \pmod{p}$.

   (b) Find all pairs $(a, b) \in \mathbf{Z}_p \times \mathbf{Z}_q$ such that $a^2 \equiv 1 \pmod{p}$ and $b^2 \equiv 1 \pmod{q}$.

   (c) Let $p = 17$ and $q = 23$. For each pair (a, b) from part (b), compute an integer modulo $17 \cdot 23$ such that $x \equiv a \pmod{17}$ and $x \equiv b \pmod{23}$.

   (d) Verify that each integer found in part (c) is a solution to

$$x^2 \equiv 1 \pmod{17 \cdot 23}.$$

8. Multiplicative Orders

Prelab

Our main interest in this chapter is to develop a deeper understanding of the powers of an integer reduced modulo n. To help you get a feel for this, we compute some powers by hand:

1. **Compute the values of $2^j \% 7$ for $0 \leq j \leq 8$.**

2. **Extrapolate from the list produced in question 1 to compute $2^{1234567} \% 7$.**

3. **Answer the two questions above for the powers of 2 modulo 12.**

In this chapter, we shall focus most of our attention on values of $a \pmod{n}$ such that $a^j \equiv 1 \pmod{n}$ for some positive integer j. Of particular importance is the concept of *multiplicative order*, which is defined as follows:

Definition Let a and n be integers, with n positive. Then the *multiplicative order of a modulo n* is the least integer $m > 0$ such that $a^m \equiv 1 \pmod{n}$, if such an integer exists. We denote m by $\mathrm{ord}_n(a)$, or more briefly by $\mathrm{ord}(a)$ when there is not likely to be confusion about the modulus n.

For the remainder of the book, when we use the term *order* without explicitly specifying whether we mean additive order or multiplicative order, we will mean multiplicative order.

We note that not all integers a will have an order. For instance, it is clear that $0^j \equiv 0 \pmod{n}$ for all positive integers j and n. Thus, $0^j \not\equiv 1 \pmod{n}$, so that 0 does not have an order modulo n.

4. **Determine (if it exists) the order of 2 modulo 7 and the order of 2 modulo 12. Explain your answers.**

5. **For each $a = 0$, 1, 2, 3, 4, 5, compute enough terms of the sequence $a^1 \% 6$, $a^2 \% 6$, ... to determine if a has an order modulo 6. For those a which do have an order, identify the order.**

One way to determine if a has an order modulo n is to examine the sequence $a^1, a^2, \ldots$ modulo n, as you did in the preceding exercise. However, there is an easier way to identify values of a that will have an order modulo n. Can you guess it based on the basis of your work above?

Maple electronic notebook `08-orders.mws`

Mathematica electronic notebook `08-orders.nb`

Web electronic notebook Start with the web page `index.html`

Maple Lab: Multiplicative Orders

■ Order, Order, Who Has an Order?

In the Prelab exercises, you started looking at powers of a modulo n (for fixed a and n). The first property these powers have is that they always repeat (eventually). The function seepows defined below allows us to see many powers of a modulo n at once to observe this.

```
> seepows:= proc(a, n, highestpower)
    local j;
     seq(modp(Power(a, j), n), j = 1..highestpower);
    end:
>
```

■ Maple Note

> The function Power is written with a capital P for a reason. It is a convention of Maple to use capitalized names of functions to indicate that you want to delay its execution. The net effect here is that Maple will know that it is working with the congruence modulo n before it computes the power. The result is somewhat faster since Maple will reduce intermediate results modulo n.

Let's see it in action. To compute the remainders of $3, 3^2, 3^3, \ldots, 3^{30}$ when divided by 180, execute the following cell.

```
> seepows(3, 180, 30);
>
```

Notice that the powers ultimately repeat. The repetition may not start with the very first term, as in the case shown. The terms before the repetition start are called the preperiod. In this case, the preperiod has a length of 1. After the preperiod, the sequence 9, 27, 81, 63 repeats from there on. We say that the sequence is periodic starting with the second term and that its minimal period is 4 (because the length of this repeating block is 4). Experiment with different values for a and n to see different combinations of periods and preperiods.

■ *Exercise 1*

> Consider the sequence $a^0, a, a^2, a^3, \ldots$ taken modulo n for a fixed integer n.
>
> (a) Show that there exist distinct integers i and j such that $a^i \equiv a^j \pmod{n}$.
>
> (b) Show that the sequence above is ultimately periodic. In other words, if we disregard the first few terms, there exists an integer P such that $a^k \equiv a^{(k+P)} \pmod{n}$ for all k sufficiently large.

The fact that powers of a fixed integer ultimately repeat modulo n makes the powers much more accessible. It makes it easy to compute very large powers of an integer mod n since one can extrapolate from the small powers.

We will now focus on powers of elements which have an order. As defined in the Prelab section, an integer a has order m if m is the least positive integer such that $a^m \equiv 1 \pmod{n}$. So, an integer a has an order if some power of a is congruent to 1 modulo n. Our first order of business is to determine which elements have an order. To assist you in your investigation, we provide the function `seeallpows` defined below. This function shows the powers modulo n for every congruence class a.

```
> seeallpows := proc(n, highestpower)     local a;
    for a from 0 to n-1 do
     print(seepows(a, n, highestpower));
    od;
   end:
[ >
```

For example, to see the first 15 powers for each of the congruence classes modulo 6, execute the next cell:

```
[ > seeallpows(6, 15);
[ >
```

Each row of the output corresponds to a different congruence class a. In a given row, we have a, a^2, a^3, a^4 $, \ldots, a^{15}$. You can easily see the value of a for a given row by looking at the first entry.

A congruence class a has an order if 1 appears in the row corresponding to a. Note, that there is no guarantee that 15 powers will be sufficient in all cases, so you may need to vary the second parameter of `seeallpows` to see the whole story.

◼ Research Question 1

Let n be a positive integer. Which integers a have an order?

◼ Benefits of Order

For the remainder of this notebook, we will restrict to integers a which have an order modulo n.

◼ A Sample Calculation

As mentioned above and used in the Prelab section, we can apply knowledge of the order of a modulo n and the first few powers of a to quickly compute very large powers modulo n. For example, we can see the first few powers of 7 modulo 10 by executing the next cell:

```
[ > seepows(7,10,20);
[ >
```

Suppose we wanted to know the final digit of 7^{12345}. (The final digit of a number is simply its remainder modulo 10.) We can see from the output above that the powers of 7 taken modulo 10 repeat with period 4. Moreover, $\mathrm{ord}_{10}(7) = 4$.) So, we can see that $7^4 \equiv 1 \pmod{10}$, $7^8 \equiv 1 \pmod{10}$, $7^{12} \equiv 1 \pmod{10}, \ldots, 7^{12344} \equiv 1 \pmod{10}$ because $12344 \equiv 0 \pmod 4$. Thus,

$$7^{12345} = 7^{(12344+1)} = 7^{12344} \cdot 7 \equiv 7 \pmod{10}.$$

We can check this result with Maple:

```
[ > modp(Power(7,12345), 10);
[ >
```

Or even by computing 7^{12345} and looking at the last digit:

```
[ > 7^12345;
[ >
```

As you can see, the last digit is, in fact, 7 (Yippee!). Clearly, our method using $\text{ord}_{10}(7)$ is simpler (and faster) than actually computing 7^{12345}.

General Properties

We would like to formalize the properties being used in the calculation above. Experiment with seepows and seeallpows to answer the following questions:

Research Question 2

Find a characterization, in terms of $\text{ord}(a)$, for *all* exponents i such that $a^i \equiv 1 \pmod{n}$.

Research Question 3

Generalize your conjecture for Research Question 2 to give a necessary and sufficient condition for j and k (in terms of $\text{ord}(a)$) so that $a^j \equiv a^k \pmod{n}$.

Since the answers to both of these questions are related to $\text{ord}_n(a)$, they tell us that the order of a modulo n is really the key to understanding all of the powers of a.

Limitations on Orders

In the last section, we studied the order of a modulo n and the behavior of the powers of a. Here we shall consider the following question:

Given n, what can we say about the possible orders of elements? That is, are all integers really candidates for values of $\text{ord}_n(a)$ as we keep n fixed and vary a?

To get a start on this question, defined below are two new functions to help investigate orders of integers modulo n. The first takes a and n as input, and returns the order of a modulo n:

```
> ord := proc(a, n)
   local current, count;
   if gcd(a, n)>1 then
    printf('%d has no order modulo %1d\n', a, n);
    0;
   else
    count := 1;
    current := modp(a, n);
    while current <> 1 do
      printf('%d^%d is congruent to %d modulo %d\n', a, count,
  current,n);
      count := count+1;
      current := modp(current*a, n);
    od;
    printf('%d^%d is congruent to %d modulo %d\n', a, count,
  current,n);
    printf('\nThe order of %d (mod %d) is %d',a,n,count);
   fi;
  end:
```

Try it out to compute $\text{ord}_7(2)$:

```
> ord(2,7);
>
```

Notice that `ord` prints out each power along the way. We can compute orders without this information being printed by using Maple's function `order`, which is loaded from the number theory library. We use it in our second function, `allords(n)`, which will give the order for each integer modulo n that has an order. Here's the definition:

```
> with(numtheory):
  allords := proc(n)
   local j;
   for j from 1 to n-1 do
    if gcd(j,n) = 1 then printf(`The order of %d (mod %d) is %d\n`,
  j, n, order(j,n))  fi; od;
  end:
>
```

If $n = 30$, here's what we get:

```
> allords(30);
>
```

Use `ord` and `allords` to collect enough data on the order of integers a modulo p, where p is a prime, to formulate a solution to the following question.

■ Research Question 4

What are the possible orders for an integer a modulo a fixed prime p?

Note: It will be *much* easier to prove your conjecture after you have completed the remaining research questions. So, do yourself a favor and hold off on this proof until after you've finished the last section. Even then, you might only be able to prove part of your conjecture. If there is some part you cannot prove, give numerical evidence to support your conjecture.

■ Fermat and Euler

Typically, we start with conjectures and try to prove them. In this case, we take a different approach. We will provide the basis for a proof, and you will need to see what sort of statement it leads to for the theorem. The theorems which result are known as *Fermat's Little Theorem* and Euler's generalization. Our goal is to answer the following question:

Given an integer n, can we produce another positive integer m such that $a^m \equiv 1 \pmod n$ for **all** integers a that have an order?

Note that this would imply that such an m is a period (although not necessarily minimal) for the powers of a mod n for **all** a that have an order.

Let's look at an example. We will work with powers of 2 modulo 15. In this computation, we will deal with the set of integers between 1 and n which are relatively prime to n. Since this set will appear in the course later, we will give its usual notation here:

Definition The *multiplicative group* modulo n, denoted $\mathbf{Z}_n^*$, is the set of elements of $\mathbf{Z}_n$ which are relatively prime to n. The number of integers in $\mathbf{Z}_n^*$ is denoted by $\phi(n)$.

To list the elements of $\mathbf{Z}_n^*$, we have the function `zstar`, which is defined below:

```
> zstar := proc(n)
    local zst,j;
    zst := NULL;
    for j from 1 to n do
      if gcd(j,n)=1 then zst := zst,j; fi;
    od;
    [zst]
  end:
```

Let's try it for $n = 15$:

```
> n := 15;
> zstar(n);
```

An interesting thing happens if we multiply every element of $\mathbf{Z}_n^*$ by a. Here's what we get:

```
> a := 2;
> a * zstar(n);
```

Since we are working modulo n, let's reduce the entire list modulo n at the same time:

```
> modp(a * zstar(n), n);
```

Looking closely, we recognize that these are the elements of $\mathbf{Z}_n^*$, but listed in a different order. It's easier to see this if we sort them:

```
> sort(modp(a * zstar(n), n));
>
```

As you can see, we can produce the elements of $\mathbf{Z}_n^*$ in a new way, namely by multiplying by a. We can exploit this observation by taking the product over both sets. Starting with the product of the elements of $\mathbf{Z}_n^*$ each multiplied by a ($= 2$ in this case), we have:

(1) $\qquad (2 \cdot 1) \cdot (2 \cdot 2) \cdot (2 \cdot 4) \cdot (2 \cdot 7) \cdot (2 \cdot 8) \cdot (2 \cdot 11) \cdot (2 \cdot 13) \cdot (2 \cdot 14).$

Rather than multiply everything out, we instead factor out a 2 from each term:

$$2^8 \cdot (1 \cdot 2 \cdot 4 \cdot 7 \cdot 8 \cdot 11 \cdot 13 \cdot 14).$$

On the other hand, we observed above that the terms of the product in equation (1) are congruent to the elements of $\mathbf{Z}_n^*$ (after reordering them). Hence it follows that

$$2^8 \cdot (1 \cdot 2 \cdot 4 \cdot 7 \cdot 8 \cdot 11 \cdot 13 \cdot 14) \equiv 1 \cdot 2 \cdot 4 \cdot 7 \cdot 8 \cdot 11 \cdot 13 \cdot 14 \pmod{15}.$$

We can cancel the equal terms from each side (why?), to find that

$$2^8 \equiv 1 \pmod{15}.$$

Let's check this last congruence:

```
[ > modp(2^8, 15);
[ >
```

Now, it is your turn. Try to follow the reasoning above with a different value of a. Remember that we are only interested in values of a that have an order.

As you try different values of a and keep $n = 15$, do you get the same exponent or does it vary? How can we predict what the exponent will be?

After you have a handle on $n = 15$, try different values for n.

■ Research Question 5

For $n = 15$ and any value of a relatively prime to n, find a value for m that satisfies the requirements given in the question at the beginning of this section. You should be able to model your proof after the example given above.

■ Research Question 6

Now generalize your solution to Research Question 5 to any positive integer n and any value of a that has an order.

■ Research Question 7

Specialize your solution to Research Question 6 to $n = p$, where p is prime, and any value of a that has an order modulo p. You should be able to be more specific about the value of "m" in this special case.

Once your proof of your conjecture for Research Question 7 is complete, you should be in a better position to prove your conjecture for Research Question 4. You may be able to prove all of that conjecture, or you may only be able to prove part of it. The proof of the best possible conjecture to Research Question 4 requires tools which we have not yet developed, so if you can't prove the whole thing, that may actually be a good sign!

The result of Research Question 7 is known as Fermat's Little Theorem. Research Question 6 is Euler's generalization of Fermat's Little Theorem to include composite values of n.

■ Why Is Fermat's Theorem Little?

The name Fermat's Little Theorem arose as a contrast to what is known as *Fermat's Last Theorem*:

If $n > 3$ and a, b, and c are integers such that

$$a^n + b^n = c^n,$$

then at least one of a, b, or c is zero.

Ironically, most mathematicians do not believe that Fermat proved this statement. It became famous years ago and is probably the most-studied mathematical problem in history. It was not considered a theorem until 1996 when Andrew Wiles completed its proof roughly 350 years after Fermat had conjectured it! Unfortunately, it has overshadowed the theorems which Fermat discovered and proved. So, the name Fermat's Little Theorem stuck as a way to distinguish it from Fermat's Last Theorem.

Mathematica Lab: Multiplicative Orders

■ Order, Order, Who Has an Order?

In the Prelab exercises, you started looking at powers of a modulo n (for fixed a and n). The first property these powers have is that they always repeat (eventually). The function **seepows** defined below allows us to see many powers of a modulo n at once to observe this.

```
seepows[a_, n_, highestpower_] :=
  Table[PowerMod[a, j, n], {j, 1, highestpower}]
```

♡ Mathematica Note

PowerMod is a special Mathematica function for computing powers modulo n very efficiently. **PowerMod[a,j,n]** computes the remainder of a^j divided by n.

Let's see it in action. To compute the remainders of 3^1, 3^2, 3^3, ... , 3^{30} when divided by 180, execute the following cell:

```
seepows[3, 180, 30]
```

Notice that the powers ultimately repeat. The repetition may not start with the very first term, as in the case shown. Recall that the terms before the repetition start are called the preperiod. In this case, the preperiod has a length of 1. After the preperiod, the sequence 9, 27, 81, 63 repeats from there on. The sequence is periodic starting with the second term and its minimal period is 4 (because the length of this repeating block is 4). Experiment with different values for a and n to see different combinations of periods and preperiods.

■ Exercise 1

Consider the sequence a^0, a^1, a^2, a^3, ... taken modulo n for a fixed integer n.

(a) Show that there exist distinct integers i and j such that $a^i \equiv a^j$ (mod n).

(b) Show that the sequence above is ultimately periodic. In other words, if we disregard the first few terms, there exists an integer $P > 0$ such that $a^k \equiv a^{k+P}$ (mod n) for all k sufficiently large.

The fact that powers of a fixed integer ultimately repeat modulo n makes the powers much more accessible. It makes it easy to compute very large powers of an integer mod n since one can extrapolate from the small powers.

We will now focus on powers of elements which have an *order*. As defined in the Prelab section, an integer a has order m if m is the least positive integer such that $a^m \equiv 1$ (mod n). So, an integer a has an order if some power of a is congruent to 1 modulo n. Our first order of business is to determine which elements have an order. To assist you in your investigation, we provide the function **seeallpows** defined below. This function shows the powers modulo n for every congruence class a.

```
seeallpows[n_, highestpower_] := TableForm[
      Table[seepows[a, n, highestpower],
          {a, 0, n - 1}], TableSpacing -> {0, 1}]
```

For example, to see the first 15 powers for each of the congruence classes modulo 6, execute the next cell:

```
seeallpows[6, 15]
```

Each row of the output corresponds to a different congruence class a. In a given row, we have a^1, a^2, a^3, a^4, ... , a^{15}. You can easily see the value of a for a given row by looking at the first entry, a^1.

A congruence class a has an order if 1 appears in the row corresponding to a. Note, that there is no guarantee that 15 powers will be sufficient in all cases, so you may need to vary the second parameter of **seeallpows** to see the whole story.

■ Research Question 1

Let n be a positive integer. Which integers a have an order?

■ Benefits of Order

For the remainder of this notebook, we will restrict to integers a which have an order modulo n.

■ A Sample Calculation

As mentioned above and used in the Prelab section, we can apply knowledge of the order of a modulo n and the first few powers of a to quickly compute very large powers modulo n. For example, we can see the first few powers of 7 modulo 10 by executing the next cell:

seepows[7, 10, 20]

Suppose we wanted to know the final digit of 7^{12345}. (The final digit of a number is simply its remainder modulo 10.) We can see from the output above that the powers of 7 taken modulo 10 repeat with period 4. (Moreover, $\operatorname{ord}_{10}(7) = 4$.) So, we can see that $7^4 \equiv 1 \pmod{10}$, $7^8 \equiv 1 \pmod{10}$, $7^{12} \equiv 1 \pmod{10}$, ... , $7^{12344} \equiv 1 \pmod{10}$ because $12344 \equiv 0 \pmod 4$. Thus,

$$7^{12345} = 7^{12344+1} = 7^{12344} \cdot 7^1 \equiv 7 \pmod{10}.$$

We can check this result with Mathematica:

PowerMod[7, 12345, 10]

Or even by computing 7^{12345} and looking at the last digit:

7 ^ 12345

As you can see, the last digit is, in fact, 7 (Yippee!). Clearly, our method using $\operatorname{ord}_{10}(7)$ is simpler (and faster) than actually computing 7^{12345}.

■ General Properties

We would like to formalize the properties being used in the calculation above. Experiment with **seepows** and **seeallpows** to answer the following questions:

■ Research Question 2

Find a characterization, in terms of ord(a), for *all* exponents i such that $a^i \equiv 1 \pmod n$.

■ **Research Question 3**

> Generalize your conjecture for Research Question 2 to give a necessary and sufficient condition for j and k (in terms of ord(a)) so that $a^j \equiv a^k \pmod{n}$.

Since the answers to both of these questions are related to $\text{ord}_n(a)$, they tell us that the order of a modulo n is really the key to understanding all of the powers of a.

■ Limitations on Orders

In the last section, we studied the order of a modulo n and the behavior of the powers of a. Here we shall consider the following question:

> Given n, what can we say about the possible orders of elements? That is, are all integers really candidates for values of $\text{ord}_n(a)$ as we keep n fixed and vary a?

To get a start on this question, defined below are two new functions to help investigate orders of integers modulo n. The first takes a and n as input, and returns the order of a modulo n:

```
ord[a_, n_, verbose_] := Module[{j = 1},
  If[GCD[a, n] > 1,
    Print[a, " has no order modulo ", n]; Return[],
        If[n > 1, While[PowerMod[a, j, n] != 1,
         If[verbose, Print[a, "^", j,
       " ≡ ", PowerMod[a, j, n], "   (mod ", n, ")"]];
         j = j + 1]];
        If[verbose, Print[a, "^", j,
      " ≡ ", PowerMod[a, j, n], "   (mod ", n, ")"]];
        Print["Order of ", a, " (mod ", n, ") is ", j, "."]]]

ord[a_, n_] := ord[a, n, True]
```

♡ Mathematica Note

Mathematica allows us to give two definitions for the same function. It can tell them apart by the number of arguments. The main function has a third input we call **verbose**. When this is **True**, the routine prints out messages to show what is happening in the calculation. The second definition of **ord** says that if we use only two arguments, then make **verbose=True** so we get the extra messages. This Mathematica trick allows us to define a function with an option, and set the default value for the option.

Try it out to compute $\text{ord}_7(2)$:

```
ord[2, 7]
```

If you want just the last line of output, add in the extra argument **False**:

```
ord[2, 7, False]
```

The second function is **allords[n]**, which will give the order for each integer modulo n that has an order. Here's the definition:

```
allords[n_] :=
    Do[If[GCD[j, n] == 1, ord[j, n, False]], {j, n - 1}];
```

If $n = 30$, here's what we get:

```
allords[30]
```

Use **ord** and **allords** to collect enough data on the order of integers a modulo p, where p is a prime, to formulate a solution to the following question.

■ Research Question 4

What are the possible orders for an integer a modulo a fixed prime p?

Note: It will be *much* easier to prove your conjecture after you have completed the remaining research questions. So, do yourself a favor and hold off on this proof until after you've finished the last section. Even then, you might only be able to prove part of your conjecture. If there is some part you cannot prove, give numerical evidence to support your conjecture.

■ Fermat and Euler

Typically, we start with conjectures and try to prove them. In this case, we take a different approach. We will provide the basis for a proof, and you will need to see what sort of statement it leads to for the theorem. The theorems which result are known as *Fermat's Little Theorem*, and Euler's generalization. Our goal is to answer the following question:

Given an integer n, can we produce another positive integer m such that $a^m \equiv 1 \pmod{n}$ for **all** integers a that have an order?

Note that this would imply that such an m is a period (although not necessarily minimal) for the powers of a mod n for **all** a that have an order.

Let's look at an example. We will work with powers of 2 modulo 15. In this computation, we will deal with the set of integers between 1 and n which are relatively prime to n. Since this set will appear in the course later, we will give its usual notation here:

Definition The *multiplicative group* modulo n, denoted $\mathbf{Z}_n^*$, is the set of is the set of elements of $\mathbf{Z}_n$ which are relatively prime to n. The number of integers in $\mathbf{Z}_n^*$ is denoted by $\phi(n)$.

To list the elements of $\mathbf{Z}_n^*$, we have the function **Zstar**, which is defined below:

```
Zstar[n_] := Select[Range[1, n - 1], GCD[#, n] == 1 &]
```

Let's try it for $n = 15$:

```
n = 15
Zstar[n]
```

An interesting thing happens if we multiply every element of $\mathbf{Z}_n^*$ by a. Here's what we get:

```
a = 2
a * Zstar[n]
```

Since we are working modulo n, let's reduce the entire list modulo n at the same time:

```
Mod[a * Zstar[n], n]
```

Looking closely, we recognize that these are the elements of $\mathbf{Z}_n^*$, but listed in a different order. It's easier to see this if we sort them:

```
Sort[Mod[a * Zstar[n], n]]
```

As you can see, we can produce the elements of $\mathbf{Z}_n^*$ in a new way, namely by multiplying by a. We can exploit this observation by taking the product over both sets. Starting with the product of the elements of $\mathbf{Z}_n^*$ each multiplied by a (= 2 in this case), we have:

$$(2 \cdot 1) \cdot (2 \cdot 2) \cdot (2 \cdot 4) \cdot (2 \cdot 7) \cdot (2 \cdot 8) \cdot (2 \cdot 11) \cdot (2 \cdot 13) \cdot (2 \cdot 14). \tag{1}$$

Rather than multiply everything out, we instead factor out a 2 from each term:

$$2^8 \cdot (1 \cdot 2 \cdot 4 \cdot 7 \cdot 8 \cdot 11 \cdot 13 \cdot 14).$$

On the other hand, we observed above that the terms of the product in equation (1) are congruent to the elements of $\mathbf{Z}_n^*$ (after reordering them). Hence it follows that

$$2^8 \cdot (1 \cdot 2 \cdot 4 \cdot 7 \cdot 8 \cdot 11 \cdot 13 \cdot 14) \equiv 1 \cdot 2 \cdot 4 \cdot 7 \cdot 8 \cdot 11 \cdot 13 \cdot 14 \pmod{15}.$$

We can cancel the equal terms from each side (why?), to find that

$$2^8 \equiv 1 \pmod{15}.$$

Let's check this last congruence:

```
Mod[2^8, 15]
```

Now, it is your turn. Try to follow the reasoning above with a different value of a. Remember that we are only interested in values of a that have an order.

As you try different values of a and keep $n = 15$, do you get the same exponent or does it vary? How can we predict what the exponent will be?

After you have a handle on $n = 15$, try different values for n.

■ Research Question 5

> For $n = 15$ and any value of a relatively prime to n, find a value for m that satisfies the requirements given in the question at the beginning of this section. You should be able to model your proof after the example given above.

■ Research Question 6

> Now generalize your solution to Research Question 5 to any positive integer n and any value of a that has an order.

■ Research Question 7

> Specialize your solution to Research Question 6 to $n = p$, where p is prime, and any value of a that has an order modulo p. You should be able to be more specific about the value of "m" in this special case.

Once your proof of your conjecture for Research Question 7 is complete, you should be in a better position to prove your conjecture for Research Question 4. You may be able to prove all of that conjecture, or you may only be able to prove part of it. The proof of the best possible conjecture to Research Question 4 requires tools which we have not yet developed, so if you can't prove the whole thing, that may actually be a good sign!

The result of Research Question 7 is known as *Fermat's Little Theorem*. Research Question 6 is Euler's generalization of Fermat's Little Theorem to include composite values of *n*.

■ Why Is Fermat's Theorem Little?

The name Fermat's Little Theorem arose as a contrast to what is known as *Fermat's Last Theorem*:

If $n > 3$ and a, b, and c are integers such that $a^n + b^n = c^n$, then at least one of a, b, or c is zero.

Ironically, most mathematicians do not believe that Fermat proved this statement. It became famous years ago and is probably the most-studied mathematical problem in history. It was not considered a theorem until 1996 when Andrew Wiles completed its proof roughly 350 years after Fermat had conjectured it! Unfortunately, it has overshadowed the theorems which Fermat discovered *and* proved. So, the name Fermat's Little Theorem stuck as a way to distinguish it from Fermat's Last Theorem.

Multiplicative Orders

8.1 Order, Order, Who Has an Order?

In the Prelab exercises, you started looking at powers of a modulo n (for fixed a and n). The first property these powers have is that they always repeat (eventually). The function below allows us to see many powers of a modulo n at once to observe this. To compute the remainders of 3^1, 3^2, 3^3, . . ., 3^{30}, when divided by 180, use the following applet.

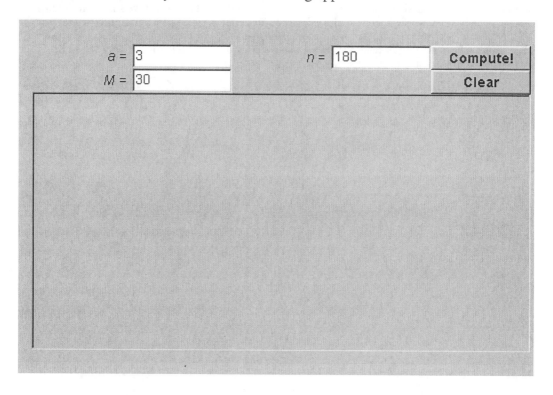

Notice that the powers ultimately repeat. The repetition may not start with the very first term, as in the case shown. Recall that the terms before the repetition start are called the preperiod. In this case, the preperiod has a length of 1. After the preperiod, the sequence 9, 27, 81, 63 repeats from there on. The sequence is periodic starting with the second term and its minimal period is 4 (because the length of this repeating block is 4). Experiment with different values for a and n to see different combinations of periods and preperiods.

> **Exercise 1**
>
> Consider the sequence a^0, a^1, a^2, a^3, ... taken modulo n for a fixed integer n.
>
> (a) Show that there exist distinct integers i and j such that $a^i \equiv a^j \pmod{n}$.
>
> (b) Show that the sequence above is ultimately periodic. (In other words, if we disregard the first few terms, there exists an integer $P > 0$ such that $a^k \equiv a^{k+P} \pmod{n}$ for all k sufficiently large.)

The fact that powers of a fixed integer ultimately repeat modulo n makes the powers much more accessible. It makes it easy to compute very large powers of an integer mod n since one can extrapolate from the small powers.

We will now focus on powers of elements which have an *order*. As defined in the Prelab section, an integer a has order m if m is the least positive integer such that $a^m \equiv 1 \pmod{n}$. So, an integer a has an order if some power of a is congruent to 1 modulo n. Our first order of business is to determine which elements have an order. To assist you in your investigation, we provide a function which shows the powers modulo n for every congruence class a.

For example, to see the first 15 powers for each of the congruence classes modulo 6, execute the next function:

Each row of the output corresponds to a different congruence class a. In a given row, we have $a^1, a^2, a^3, a^4, \ldots, a^{15}$. You can easily see the value of a for a given row by looking at the first entry, a^1.

A congruence class a has an order if 1 appears in the row corresponding to a. Note, that there is no guarantee that 15 powers will be sufficient in all cases, so you may need to vary the second parameter M to see the whole story.

Research Question 1

Let n be a positive integer. Which integers a have an order?

8.2 Benefits of Order

For the remainder of this chapter, we will restrict to integers a which have

8.2.1 A Sample Calculation

As mentioned above and used in the Prelab section, we can apply knowledge of the order of *a* modulo *n*) and the first few powers of *a* to quickly compute very large powers modulo *n*. For example, we can see the first few powers of 7 modulo 10 by executing the next function:

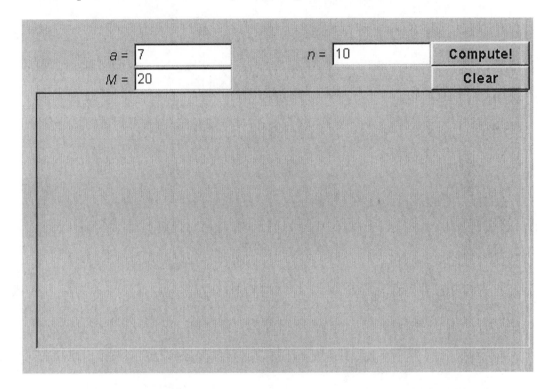

Suppose we wanted to know the final digit of 7^{12344}. (The final digit of a number is simply its remainder modulo 10.) We can see from the output above that the powers of 7 taken modulo 10 repeat with period 4. (Moreover, $\text{ord}_{10}(7) = 4$.) So, we can see that $7^4 \equiv 1 \pmod{10}$, $7^8 \equiv 1 \pmod{10}$, $7^{12} \equiv 1 \pmod{10}$, ..., $7^{12344} \equiv 1 \pmod{10}$ because $12344 \equiv 0 \pmod 4$. Thus,

$$7^{12345} = 7^{12344+1} = 7^{12344} \cdot 7 \equiv 7 \pmod{10}.$$

We can check this result by computing 7^{12345} and looking at the last digit (this may take a little while):

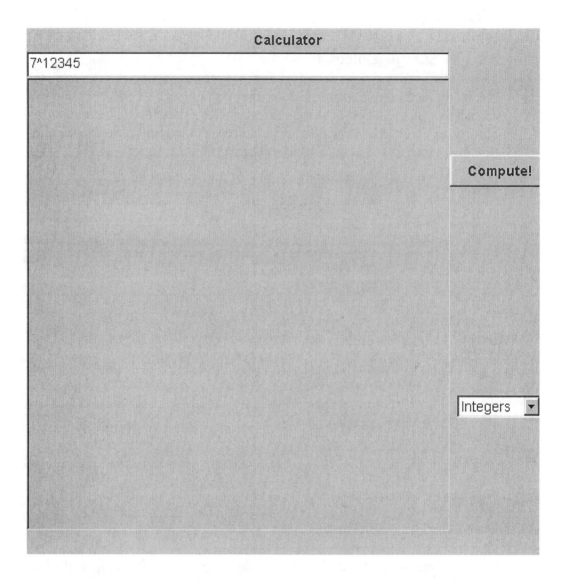

As you can see, the last digit is, in fact, 7 (Yippee!). Clearly, the method using $\mathrm{ord}_{10}(7)$ is simpler (and faster) than actually computing 7^{12345}.

8.2.2 General Properties

We would like to formalize the properties being used in the calculation above. Experiment with functions above to answer the following questions:

Research Question 2

Find a characterization, in terms of ord(a), for *all* exponents i such that $a^i \equiv 1 \pmod{n}$.

Research Question 3

Generalize your conjecture for Research Question 2 to give a necessary and sufficient condition for j and k (in terms of ord (a)) so that $a^j \equiv a^k \pmod{n}$.

Since the answers to both of these questions are related to $\mathrm{ord}_n(a)$, they tell us that the order of a modulo n is really the key to understanding all of the powers of a.

8.3 Limitations on Orders

In the last section, we studied the order of a modulo n and the behavior of the powers of a. Here we shall consider the following question:

Given n, what can we say about the possible orders of elements? That is, are all integers really candidates for values of $\mathrm{ord}_n(a)$ as we keep n fixed and vary a?

To get a start on this question, there are two new functions below to help investigate orders of integers modulo n. The first takes a and n as input, and returns the order of a modulo n. Try it out, taking $a = 2$ and $n = 7$:

Order of a modulo n		
$a =$ 2	$n =$ 7	Compute!
		Clear

The second function will give the order for each integer modulo n that has an order. If $n = 30$, here's what we get:

Use these two functions to collect enough data on the order of integers a modulo p, where p is a prime, to formulate a solution to the following question.

Research Question 4

What are the possible orders for an integer a modulo a fixed prime p?

Note: It will be *much* easier to prove your conjecture after you have completed the remaining research questions. So, do yourself a favor and hold off on this proof until after you've finished the last section. Even then, you might only be able to prove part of your conjecture. If there is some part you cannot prove, give numerical evidence to support your conjecture.

8.4 Fermat and Euler

Typically, we start with conjectures and try to prove them. In this case, we take a different approach. We will provide the basis for a proof, and you will need to see what sort of statement it leads to for the theorem. The theorems which result are known as *Fermat's Little Theorem* and Euler's generalization. Our goal is to answer the following question:

> Given an integer n, can we produce another positive integer m such that $a^m \equiv 1 \pmod{n}$ for **all** integers a that have an order?

Note that this would imply that such an m is a period (although not necessarily minimal) for the powers of a mod n for **all** a that have an order.

Let's look at an example. We will work with powers of 2 modulo 15. In this computation, we will deal with the set of integers between 1 and n which are relatively prime to n. Since this set will appear in the course later, we will give its usual notation here:

> **Definition** The *multiplicative group* modulo n, denoted $\mathbf{Z}_n^*$, is the set of elements of $\mathbf{Z}_n$ which are relatively prime to n. The number of integers in $\mathbf{Z}_n^*$ is denoted by $\phi(n)$.

In our example of $n = 15$, $\mathbf{Z}_{15}^*$ is

$$1, 2, 4, 7, 8, 11, 13, 14.$$

An interesting thing happens if we multiply every element of $\mathbf{Z}_n^*$ by a. Here's what we get with $n = 15$ and $a = 2$:

$$2, 4, 8, 14, 16, 22, 26, 28.$$

Since we are working modulo n, let's reduce the entire list modulo $n = 15$:

$$2, 4, 8, 14, 1, 7, 11, 13.$$

Looking closely, we recognize that these are the elements of $\mathbf{Z}_{15}^*$, but listed in a different order. It's easier to see this if we sort them:

$$1, 2, 4, 7, 8, 11, 13, 14.$$

As you can see, we can produce the elements of $\mathbf{Z}_n^*$ in a new way, namely by multiplying by a. We can exploit this observation by taking the product over both sets. Starting with the product of the elements of $\mathbf{Z}_{15}^*$ each multiplied by a ($= 2$ in this case), we have:

$$(1) \qquad (2 \cdot 1) \cdot (2 \cdot 2) \cdot (2 \cdot 4) \cdot (2 \cdot 7) \cdot (2 \cdot 8) \cdot (2 \cdot 11) \cdot (2 \cdot 13) \cdot (2 \cdot 14).$$

Rather than multiply everything out, we instead factor out a 2 from each term:

$$2^8 \cdot (1 \cdot 2 \cdot 4 \cdot 7 \cdot 8 \cdot 11 \cdot 13 \cdot 14).$$

On the other hand, we observed above that the terms of the product in equation (1) are congruent to the elements of $\mathbf{Z}_n^*$ (after reordering them). Hence it follows that

$$2^8 \cdot (1 \cdot 2 \cdot 4 \cdot 7 \cdot 8 \cdot 11 \cdot 13 \cdot 14) \equiv 1 \cdot 2 \cdot 4 \cdot 7 \cdot 8 \cdot 11 \cdot 13 \cdot 14 \pmod{15}.$$

We can cancel the equal terms from each side (why?), to find that

$$2^8 \equiv 1 \pmod{15}.$$

Let's check this last congruence:

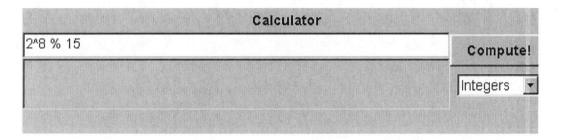

Now, it is your turn. Try to follow the reasoning above with a different value of a. Remember that we are only interested in values of a that have an order.

Here is a function which automates the computations we did above. You supply the values of a and n. It will list the elements of $\mathbf{Z}_n^*$, take that list and multiply each term by a, then give the list again after reducing modulo n, and finally sort the list.

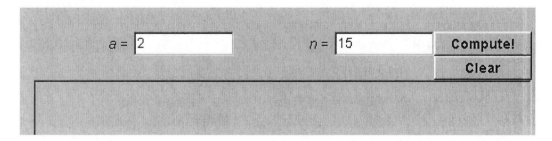

As you try different values of a and keep $n = 15$, do you get the same exponent or does it vary? How can we predict what the exponent will be?

After you have a handle on $n = 15$, try different values for n.

Research Question 5

For $n = 15$ and any value of a relatively prime to n, find a value for m that satisfies the requirements given in the question at the beginning of this section. You should be able to model your proof after the example given above.

Research Question 6

Now generalize your solution to Research Question 5 to any positive integer n and any value of a that has an order.

> **Research Question 7**
>
> Specialize your solution to Research Question 6 to $n = p$, where p is prime, and any value of a that has an order modulo p. You should be able to be more specific about the value of "m" in this special case.

Once your proof of your conjecture for Research Question 7 is complete, you should be in a better position to prove your conjecture for Research Question 4.

The result of Research Question 7 is known as *Fermat's Little Theorem*. Research Question 6 is Euler's generalization of Fermat's Little Theorem to include composite values of n.

8.4.1 Why Is Fermat's Theorem Little?

The name Fermat's Little Theorem arose as a contrast to what is known as *Fermat's Last Theorem*:

> If $n > 3$ and a, b, and c are integers such that $a^n + b^n = c^n$, then at least one of a, b, or c is zero.

Ironically, most mathematicians do not believe that Fermat proved this statement. It became famous years ago and is probably the most-studied mathematical problem in history. It was not considered a theorem until 1994 when Andrew Wiles completed its proof roughly 350 years after Fermat had conjectured it! Unfortunately, it has overshadowed the theorems which Fermat discovered *and* proved. So, the name Fermat's Little Theorem stuck as a way to distinguish it from Fermat's Last Theorem.

Homework

Throughout, n will denote a positive integer, and p will denote a prime number. Except where otherwise noted, you may use a computer to help with the computations.

1. The computations for this problem can be done by hand.

    (a) Compute $\text{ord}_{499}(140)$. (Hint: It is less than 10.)

    (b) Compute the remainder of $140^{1234567}$ when divided by 499.

2. Suppose that a is a positive integer. Suppose further that $\gcd(a, n) = 1$ and that $\phi(n) = 148$. On the basis of the results of this chapter, what are the possible values of $\text{ord}_n(a)$?

3. Suppose $a \in \mathbf{Z}$ and $p \nmid a$. Prove that a^{p-2} is a multiplicative inverse of a modulo p.

4. Suppose $a \in \mathbf{Z}$ and $p \nmid a$.

    (a) Show that $\left(1 - p^{\phi(a)}\right)/a$ is an integer.

    (b) Show that $\left(1 - p^{\phi(a)}\right)/a$ is a multiplicative inverse of a modulo p.

5. Prove that for any integer a, $a^p \equiv a \pmod{p}$. (Hint: Use Fermat's Little Theorem for most values of a.)

6. Suppose $a \in \mathbf{Z}$, p is an odd prime, and $p \nmid a$. Prove that $a^{(p-1)/2} \equiv \pm 1 \pmod{p}$. (Hint: What are the solutions to $x^2 \equiv 1 \pmod{p}$?)

7. Suppose $\gcd(a, n) = 1$. Show that a solution to $ax \equiv b \pmod{n}$ is given by $x = a^{\phi(n)-1}b$.

8. What is the value of $\sum_{j=1}^{p} j^{p-1} \% p$?

9. What is the value of $\sum_{j=1}^{p} j^p \% p$?

10. In this problem, we will find two proofs of the following statement:

    Let a be an integer relatively prime to n, and j and k be positive integers with $d = \gcd(j, k)$. If $a^j \equiv 1 \pmod{n}$ and $a^k \equiv 1 \pmod{n}$, then $a^d \equiv 1 \pmod{n}$.

    (a) Use the result of Research Question 2 to prove the statement above. (Hint: How are j, k, and d related to $\text{ord}_n(a)$?)

    (b) Use the GCD Trick to prove the statement above. (Recall that the GCD Trick states that there exists integers r and s such that $rj + sk = d$.)

11. Let n be a positive integer and $a \neq 0$.

    (a) Show that the sequence $a^0 \% n, a^1 \% n, a^2 \% n, \ldots$ is purely periodic if and only if $\gcd(a, n) = 1$.

    (b) Show that if $\gcd(a, n) = 1$, then $\operatorname{ord}_n(a)$ is the minimal period for the sequence in part (a).

12. Let n be a positive integer. Show that the sequence

$$2 \% n, 2^2 \% n, 2^{2^2} \% n, 2^{2^{2^2}} \% n, \ldots$$

    is ultimately constant. (This was a problem in the 20th USA Mathematical Olympiad.)

13. Show that if $a, n \in \mathbf{Z}$ (where $n > 0$) and $\gcd(a, n) = 1$, then there exists a positive integer j such that $n \mid (a^j - 1)$.

14. Suppose the $\gcd(a, n) = 1$ where $n > 0$. Let $d = \operatorname{ord}_n(a)$. Prove that if $\gcd(j, d) = 1$, then $\operatorname{ord}_n(a^j) = d$.

15. Show that if $a \in \mathbf{Z}$, then the congruence $x^7 \equiv a \pmod{277}$ always has a solution $x = a^{79}$. (Note, 277 is prime.)

16. Suppose p and q are primes, and $m = \operatorname{lcm}(p - 1, q - 1)$. If $\gcd(a, pq) = 1$, then $a^m \equiv 1 \pmod{pq}$. (Hint: Think about the congruence modulo p and then modulo q, and use the Chinese Remainder Theorem to put the results together.)

17. Prove that 2821 is a Carmichael number.

18. Prove that 6601 is a Carmichael number.

19. Prove that every positive integer can be written as a sum of distinct powers of 2 (including possibly 2^0).

20. Use the powering algorithm to compute $2^{32} \% 101$.

21. Use the powering algorithm to compute $2^{137} \% 101$.

22. Use the Fermat test to prove that 9017 is composite.

23. Use the Fermat test to prove that 9991 is composite.

9. The Euler ϕ-function

Prelab

In Chapter 8, you encountered the function $\phi(n)$: the number of congruence classes in $\mathbf{Z}_n^*$.[1] Stated formally,

Definition For each positive integer n, let $\phi(n)$ denote the number of integers m which satisfy $1 \leq m \leq n$ and $\gcd(m, n) = 1$.

This function is known as the *phi-function* or *Euler's phi-function*.[2] In this chapter, we will study ways of computing $\phi(n)$. A simple way to do this is by computing $\gcd(m, n)$ for each m between 1 and n, and then counting the numbers satisfying $\gcd(m, n) = 1$. For example, to compute $\phi(8)$ we look at the numbers between 1 and 8 (i.e., 1, 2, 3, 4, 5, 6, 7, and 8), and remove the ones which are not relatively prime to 8 (which leaves 1, 3, 5, and 7). There are 4 numbers left, and so $\phi(8) = 4$.

1. **Using the method described above, compute** $\phi(7)$, $\phi(10)$, $\phi(15)$, **and** $\phi(18)$.

Computing $\phi(n)$ by checking each integer less than or equal to n is fine for small values of n, but suppose that we want to compute $\phi(1000000000)$? Checking one integer at a time is not very practical in this case. Is there a better way? There is indeed, and by the end of this chapter you will have found it. To get things started, we look at the special case of $n = p$, where p is a prime number.

2. **Compute** $\phi(p)$ **for at least three primes** p **of your choosing.**

Maple electronic notebook `09-phi.mws`

Mathematica electronic notebook `09-phi.nb`

Web electronic notebook Start with the web page `index.html`

---

[1]Recall that this function is closely related to the orders of elements in $\mathbf{Z}_n^*$.

[2]Leonhard Euler was the first to consider this function. It initially appeared in his paper generalizing Fermat's Little Theorem.

Maple Lab: The Euler φ-function

■ A Comparison

We begin here by recalling that the function $\phi(n)$ counts the number of integers m such that $1 \leq m \leq n$ and $\gcd(m, n) = 1$. In the Prelab section of this chapter, you computed $\phi(n)$ for a few different values of n by systematically checking each m such that $1 \leq m \leq n$, and counting how many satisfied $\gcd(m, n) = 1$. Here's a Maple function to automatically do the same thing:

```
> firstphi := proc(n)
    local phi,m;
    phi := 0;
    for m from 1 to n do
     if gcd(m,n)=1 then phi := phi+1; fi;
    od;
    phi;
  end:
```

Here it is in action. (Compare the output to your responses to the Prelab exercises. Do they agree?)

```
[ > firstphi(7);
[ > firstphi(10);
[ > firstphi(15);
[ > firstphi(18);
```

Let's try it out on a larger number:

```
[ > firstphi(12345);
```

This one takes a bit longer, which shouldn't be surprising, because there are more values of m to check. Maple has a command called phi in its number theory library that also computes $\phi(n)$, but does so using a different method. First execute the next command to load the number theory library. Do not worry when Maple issues a warning about a new definition of order.

```
[ > with(numtheory):
```

Now, compute $\phi(12345)$ using Maple's function:

```
[ > phi(12345);
[ >
```

Here's a comparison of the time required for each function to compute $\phi(1000)$:

```
> start:=time():
    firstphi(1000);
    finish:=time():
    printf('%g seconds', finish - start);
> start:=time():
    phi(1000);
    finish:=time():
```

424

```
⌊   printf('%g seconds', finish - start);
[ >
```

The main reason that phi is so much faster than firstphi is that phi uses an algorithm that is more efficient than the one used in firstphi. One goal for this chapter is to find this more efficient method for computing $\phi(n)$.

▪ *Exercise 1*

(a) Use the code above to determine the time required to compute $\phi(10^4)$ using firstphi (i.e., replace 1000 by 10^4 and see how long it takes).

(b) On the basis of the results of part (a), extrapolate the number of years required for the command firstphi(10^20) to return an answer. (For simplicity, assume that all years have 365 days.)

(c) Use the code above to determine how long it takes phi to compute $\phi(10^{20})$.

▪ Computing $\phi(n)$: A Special Case

As has become our standard practice, we will work towards a general solution to our problem by starting with special cases that are easier to tackle. In this section, we consider the values of $\phi(n)$ when n is a power of a prime.

▪ Research Question 1

Find a formula for $\phi(p)$, where p is prime.

▪ Research Question 2

Find a formula for $\phi(p^a)$, where p is a prime and a is a positive integer.

Hint: You may find it easier to count the number of $m \le p^a$ such that $1 < \gcd(m, p^a)$, and subtract this from $n = p^a$.

▪ Computing $\phi(n)$: The General Case

In order to compute $\phi(n)$ for values of n that are more complicated than prime powers, we would like to develop a means for breaking the general computation into several simpler computations. One way to proceed is to determine combinations of m and n for which the statement

$$\phi(m \cdot n) = \phi(m) \cdot \phi(n)$$

is true. This leads us directly to the next Research Question.

▪ Research Question 3

Find a condition for pairs of integers m and n which guarantees that

$$\phi(m \cdot n) = \phi(m) \cdot \phi(n).$$

Note: The proof of your conjecture (assuming you find the one we are looking for) will be considered on your homework assignment. Unless you are told otherwise, you need not provide a proof in your lab report. However, since you are not mandated to include a proof for your conjecture, you should provide lots of numerical data to support your claim.

■ Research Question 4

Use your conjectures from Research Questions 2 and 3 to assist in finding a formula for $\phi(n)$, where $n = p^a q^b$, p and q are distinct primes, and a and b are positive integers.

We're almost there. Here's the last step:

■ Research Question 5

Find a formula for $\phi(n)$, where

$$n = p_1^{a_1} p_2^{a_2} \cdots p_k^{a_k},$$

$p_1, p_2, \ldots, p_k$ are distinct primes, and $a_1, a_2, \ldots, a_k$ are positive integers.

■ *Exercise 2*

Use your formula from Research Question 5 to compute $\phi(10^{20})$, explaining the steps. Compare the result with your answer to Exercise 1(c).

■ Going Farther: RSA

One of the best-known applications of number theory is to what is known as a *public key cryptosystem*. The idea behind a public key cryptosystem is that one person, let's call her Alice, can tell anyone and everyone enough information to encode a message to be sent to her, but only Alice can decode the message. This certainly runs counter to intuition. If one has enough information to encode a message, why can't they run the process in reverse to decode it?

Several different methods for implementing public key cryptosystems have been proposed. One of the first, and certainly the best known, was developed by Rivest, Shamir, and Adelman, and is known as *RSA*. Before describing the specific mechanics involved in implementing RSA, we first discuss a general scheme known as *text blocking* that is incorporated into most cryptosystems.

Before we get started, open the section titled "Cryptography Utility Functions" and execute the group it contains. This will load a few functions for converting between text and numbers. Some of these functions we have seen before, and others will be explained below in the section on text blocking.

■ Cryptography Utility Functions

■ Text Blocking

The crytosystems we have introduced thus far have been based on the idea of encoding messages a single letter at a time. Most cryptographic schemes of this type are not useful in practice because they will be vulnerable to frequency analysis. Recall that frequency analysis proceeds by counting the number of times each character occurs in an encoded message, and then making educated guesses about how to decode each character.

A simple way to work around frequency analysis is to organize the message text into blocks. Instead of representing each single character as a number, the text is taken in blocks of a certain length. Thus far, we have worked with a 95-character alphabet. We will now encode blocks of several characters of text as single numbers. For example, suppose we are going to convert the text of the quote

To be or not to be, that is the question. -- Hamlet

into numbers. If we take the text in blocks of four characters each, then the first four characters "To b" would be converted into a single number, the next four characters "e or" would be converted to a single number, and so on. The first step in converting each 4-character block to a number is to convert each individual character in the block to a number. For example:

```
> texttonums('T');
  texttonums('o');
  texttonums(' ');
  texttonums('b');
>
```

(Recall that `texttonums` converts a string to a list of numbers, each of which is in the range from 0 to 94.) We then combine these numbers together as shown:

$$52 \cdot 95^0 + 79 \cdot 95^1 + 0 \cdot 95^2 + 66 \cdot 95^3 = 56594307.$$

The number 56594307 is then encoded and transmitted. More generally, if we take blocks of size n, and if $a_1, a_2, \ldots, a_n$ are the numerical equivalents for each letter in a block, then the *block number* is computed using the formula

$$a_1 \cdot 95^0 + a_2 \cdot 95^1 + a_3 \cdot 95^2 + \ldots + a_n \cdot 95^{(n-1)}.$$

By blocking in this manner, we have effectively raised the size of our "alphabet" from 95 to 95^n, which makes frequency analysis much more difficult, if not completely infeasible. For example, if $n = 10$, then the size of our "alphabet" is $95^{10} = 59873693923837890625$.

To make life a little easier for you down the road, the function `texttoblocks` is provided (the code was in the Cryptography Utility Functions section). This function will take a message, break it into blocks, and then compute the numerical equivalent for each block.

```
> message := 'To be or not to be, that is the question. -- Hamlet';
  blocklength := 4;    # Set the block length here.

  blocks := texttoblocks(message,blocklength);
```

The counterpart to `texttoblocks` is `blockstotext`, which will convert a list of blocks back into text.

```
> blockstotext(blocks, blocklength);
>
```

Implementing RSA

The first step in implementing the RSA cryptosystem is to select two distinct primes p and q, and then set $M = p\, q$. If our messages are blocked into groups of n letters, then p and q must be large enough so that $95^n < M$. Taking logarithms (base 95), this translates to $n < \log_{95}(M)$. We will pick the block length n from this formula.

Once p and q have been selected, the next step is to select a positive integer e such that $\gcd(e, (p-1)(q-1)) = 1$. The integer e is called the encoding exponent, for reasons that will be clear shortly. Once suitable choices of M and e have been made, then we encode a text block number x to a new number c using the formula

$$c = x^e \ \% \ M.$$

The number c is then transmitted to the recipient. Note that the *only* information required to encode a message are the integers e and M. In particular, it is not necessary to know p and q in order to send a message.

To decode a message, we require a decoding exponent d, which is given by the formula

$$d = e^{(-1)} \ \% \ ((p-1)(q-1)).$$

Once d has been computed, the original number x may be recovered using the formula

$$x = c^d \ \% \ M.$$

Let's look at a specific example to see how this works. Here's our quote again:

```
> message := 'To be or not to be, that is the question. -- Hamlet';
```

We begin by selecting two primes p and q, and then setting $M = p\,q$.

```
> p := 12373:
  q := 32491:
  M := p*q;
```

We determine the block length by computing

```
> evalf(log[95](M));
```

Rounding down, we will end up using a block length of 4. The next step is to select an encoding exponent. A common choice is $e = 17$, but this is suitable only if $\gcd(e, (p-1)(q-1)) = 1$. Here's a check:

```
> e := 17:
  gcd(e, (p-1)*(q-1));
>
```

Now we're ready to encode. The function `RSAencode` takes the original message, e, and the modulus, M, as input, and returns a list of encoded blocks. It selects the block size by the formula we derived above. Here is the code.

```
> RSAencode := proc(message, e, M)
    local blocks, j, blocksize;
    blocksize := floor(log[95](M));
    blocks := texttoblocks(message, blocksize);
    [seq(modp(Power(blocks[j], e), M), j=1..nops(blocks))];
  end:
```

Here's the set of encoded blocks for our message.

```
> codedblocks := RSAencode(message, e, M);
```

```
[ >
```

In order to decode, we first need to compute the decoding exponent d. Recall that the formula for d is given by

$$d = e^{(-1)} \, \% \, ((p-1)(q-1)).$$

We can compute d as shown.

```
[ > d := modp(1/e, (p-1)*(q-1));
[ >
```

Once we have d, we can use the function RSAdecode to decode the set of encoded blocks. Here is the code for RSAdecode:

```
[ > RSAdecode := proc(codedblocks, d, M)
    local decodedblocks, j, blocklength;
    blocklength := floor(log[95](M));
    decodedblocks := [seq(modp(Power(codedblocks[j], d), M),
    j=1..nops(codedblocks))];
    blockstotext(decodedblocks, blocklength);
  end:
```

Now we can decode the encoded blocks.

```
[ > RSAdecode(codedblocks, d, M);
[ >
```

Exercise 3

In this exercise, we prove that the method given above for decoding messages using RSA will work. Suppose that $M = p\,q$ where p and q are distinct primes, $0 < x < M$, $c = x^e \, \% \, M$, and
$$d = e^{(-1)} \, \% \, ((p-1)(q-1)).$$
Show that
$$x = c^d \, \% \, M.$$

Exercise 4

Bill wants to send a message to Alice using RSA, so she tells him that her encoding exponent is $e = 17$ and her encoding modulus is $M = 12195377863$. Show Bill how to encode the message

```
[ > M := 12195377863:
    message := 'Hey Alice, isn't RSA cool!?  --Bill';
[ >
```

Exercise 5

Alice has received the encoded blocks shown below. Use the fact that $M = 104729 \cdot 116447$ to decode the message.

```
[ > encodedblocks :=
    [4048208982,7973981835,2232860160,1228506572,2392521799,235351
    8709,4207703255,3703820767,250037531,6326907769,3298628742];
[ >
```

Is RSA Secure?

It's clear from the decoding procedure that if one can compute d then one can decode a message. Recall that the formula for computing d is given by

$$d = e^{(-1)} \% ((p-1)(q-1)).$$

Thus computing d using this formula requires knowledge of the values of p and q. On the other hand, the only information known to anyone encoding messages are the values of e and M. Of course, we know that $M = p \cdot q$, so that even if we just know e and M, we can find p and q by factoring M. For example, in Exercises 4 and 5 above, Alice used the encoding modulus $M = 12195377863$. In practice, messages sent to her would not be secure because it is easy to factor this value of M. Below is the Maple command to do just that.

```
[ > ifactor(12195377863);
```

Exercise 6

Bill has decided he wants to have people send him messages encoded using RSA. He tells everyone that his encoding exponent is $e = 17$ and his encoding modulus is $M = 13776856397$. He has received his first set of encoded blocks, which is given below. Decode the message.

```
[ > M := 13776856397;
    encodedblocks := [4490149759, 9836048959, 4449780697,
    13166158759, 7302115749, 8442727016, 2486382600, 7344621302,
    5062584227, 1272601926, 12966640119];
[ >
```

At this point, you may be asking yourself "Why would anyone use this cryptosystem? It doesn't seem to be secure." Indeed, the examples that we have seen so far are not secure, because it has been possible to factor the modulus M.

The key to making RSA secure is to select primes p and q that are much larger, say on the order of 100 digits each. In this case, M has about 200 digits and would be impossible to factor without knowledge of p and q. (One has to be a bit careful when making the choice of p and q, but we won't worry about these details here.) Without knowing p and q, there is no known way to compute the decoding exponent d.

Additional Remarks:

- Our example of encoding exponent, $e = 17$, is commonly used in RSA. Implementations will typically fix a single encoding exponent for all users, and each user gives out his or her value of M as his or her public key.

- The decoding process can be streamlined computationally. However, all known methods of decoding are essentially equivalent to the example given above. In particular, they all rely on being able to factor M.

- There is no known *proof* that RSA is secure. To date, no one has discovered a simple way to factor large integers, but it is possible that such a discovery will occur. If this were to happen, then the security of RSA-based cryptosystems would be severely compromised. However, the general consensus among factoring experts is that such a discovery is unlikely to occur, and that no simple factoring method exists.

Mathematica Lab: The Euler ϕ–function

■ A Comparison

We begin here by recalling that the function $\phi(n)$ counts the number of integers m such that $1 \le m \le n$ and $\gcd(m, n) = 1$. In the Prelab section of this chapter, you computed $\phi(n)$ for a few different values of n by systematically checking each m such that $1 \le m \le n$, and counting how many satisfied $\gcd(m, n) = 1$. Here's a Mathematica function to automatically do the same thing:

```
firstphi[n_] := Module[{phi},
  phi = 0;
  Do[ If[GCD[m, n] == 1, phi = phi + 1],
    {m, 1, n}];
  Return[phi] ]
```

Here it is in action. (Compare the output to your responses in the Prelab exercises. Do they agree?)

```
firstphi[7]
```

```
firstphi[10]
```

```
firstphi[15]
```

```
firstphi[18]
```

Let's try it out on a larger number:

```
firstphi[12345]
```

This one takes a bit longer, which shouldn't be surprising, because there are more values of m to check. Mathematica has a built-in command called **EulerPhi** that also computes $\phi(n)$, but does so using a different method:

```
EulerPhi[12345]
```

Here's a comparison of the time required for each function to compute $\phi(1000)$:

```
Timing[firstphi[1000]]
Timing[EulerPhi[1000]]
```

Part of the difference in time is due to **EulerPhi** being built into Mathematica while **firstphi** is not. However, **EulerPhi** also uses an algorithm that is more efficient than the one used in **firstphi**. One goal for this chapter is to find this more efficient method for computing $\phi(n)$.

■ Exercise 1

(a) Use the **Timing** command to determine the time required to compute $\phi(10^4)$ using **firstphi**.
(b) On the basis of the results of part (a), extrapolate the number of years required for the command **firstphi[10^20]** to return an answer. (For simplicity, assume that all years have 365 days.)
(c) Use **Timing** to determine how long it takes **EulerPhi** to compute $\phi(10^{20})$.

431

■ Computing $\phi(n)$: A Special Case

As has become our standard practice, we will work towards a general solution to our problem by starting with special cases that are easier to tackle. In this section, we consider the values of $\phi(n)$ when n is a power of a prime.

■ Research Question 1

> Find a formula for $\phi(p)$, where p is prime.

■ Research Question 2

> Find a formula for $\phi(p^a)$, where p is a prime and a is a positive integer.
>
> **Hint:** You may find it easier to count the number of $m \le p^a$ such that $\gcd(m, p^a) > 1$, and subtract this from $n = p^a$.

■ Computing $\phi(n)$: The General Case

In order to compute $\phi(n)$ for values of n that are more complicated than prime powers, we would like to develop a means for breaking the general computation into several simpler computations. One way to proceed is to determine combinations of m and n for which the statement

$$\phi(m \cdot n) = \phi(m) \cdot \phi(n)$$

is true. This leads us directly to the next Research Question.

■ Research Question 3

> Find a condition for pairs of integers m and n which guarantees that
>
> $$\phi(m \cdot n) = \phi(m) \cdot \phi(n).$$
>
> **Note:** The proof of your conjecture (assuming you find the one we are looking for) will be considered on your homework assignment. You can provide a proof with the lab report if you have one, but this is not required. However, since you are not mandated to include a proof for your conjecture, you should provide *lots* of numerical data to support your claim.

■ Research Question 4

> Use your conjectures from Research Questions 2 and 3 to assist in finding a formula for $\phi(n)$, where $n = p^a q^b$, p and q are distinct primes, and a and b are positive integers.

We're almost there. Here's the last step:

■ Research Question 5

Find a formula for $\phi(n)$, where

$$n = p_1^{a_1}\, p_2^{a_2} \cdots p_k^{a_k},$$

$p_1, p_2, \ldots, p_k$ are distinct primes, and $a_1, a_2, \ldots, a_k$ are positive integers.

■ Exercise 2

Use your formula from Research Question 5 to compute $\phi(10^{20})$, explaining the steps. Compare the result with your answer to Exercise 1(c).

■ Going Farther: RSA

One of the best-known applications of number theory is to what is known as a *public key cryptosystem*. The idea behind a public key cryptosystem is that one person, let's call her Alice, can tell anyone and everyone enough information to encode a message to be sent to her, but only Alice can decode the message. This certainly runs counter to intuition. If one has enough information to encode a message, why can't they run the process in reverse to decode it?

Several different methods for implementing public key cryptosystems have been proposed. One of the first, and certainly the best known, was developed by Rivest, Shamir, and Adelman, and is known as *RSA*. Before describing the specific mechanics involved in implementing RSA, we first discuss a general scheme known as *text blocking* that is incorporated into most cryptosystems.

■ Text Blocking

The crytosystems we have introduced thus far have been based on the idea of encoding messages a single letter at a time. Most cryptographic schemes of this type are not useful in practice because they will be vulnerable to frequency analysis. Recall that frequency analysis proceeds by counting the number of times each character occurs in an encoded message, and then making educated guesses about how to decode each character.

A simple way to work around frequency analysis is to organize the message text into blocks. Instead of representing each single character as a number, the text is taken in blocks of a certain length. Thus far, we have worked with a 95–characteralphabet. We will now encode blocks of several characters of text as single numbers. For example, suppose we are going to convert the text of the quote

> To be or not to be, that is the question. --Hamlet

into numbers. If we take the text in blocks of four characters each, then the first four characters "To b" would be converted into a single number, the next four characters "e or" would be converted to a single number, and so on. The first step in converting each 4–characterblock to a number is to convert each individual character in the block to a number. For example:

```
texttonums["T"]
texttonums["o"]
texttonums[" "]
texttonums["b"]
```

(Recall that **texttonums** converts a string to a list of numbers, each of which is in the range from 0 to 94.) We then combine these numbers together as shown:

$$52 \cdot 95^0 + 79 \cdot 95^1 + 0 \cdot 95^2 + 66 \cdot 95^3 = 56594307.$$

The number 56594307 is then encoded and transmitted. More generally, if we take blocks of size n, and if $(a_1, a_2, \ldots, a_n)$ are the numerical equivalents for each letter in a block, then the *block number* is computed using the formula

$$a_1 \cdot 95^0 + a_2 \cdot 95^1 + a_3 \cdot 95^2 + \cdots + a_n \cdot 95^{n-1}.$$

By blocking in this manner, we have effectively raised the size of our "alphabet" from 95 to 95^n, which makes frequency analysis much more difficult, if not completely infeasible. For example, if $n = 10$, then the size of our "alphabet" is $95^{10} = 59873693923837890625$.

To make life a little easier for you down the road, the function **texttoblocks** is provided (the code was in the initialization section of the notebook). This function will take a message, break it into blocks, and then compute the numerical equivalent for each block.

```
message = "To be or not to be, that is the question. -- Hamlet";
blocklength = 4;   (* Set the block length here. *)

blocks = texttoblocks[message, blocklength]
```

The counterpart to **texttoblocks** is **blockstotext**, which will convert a list of blocks back into text.

```
blockstotext[blocks, blocklength]
```

■ Implementing RSA

The first step in implementing the RSA cryptosystem is to select two distinct primes p and q, and then set $M = p\,q$. If our messages are blocked into groups of n letters, then p and q must be large enough so that $M > 95^n$. Taking logarithms (base 95), this translates to $n < \log_{95}(M)$. We will pick the block length n by this formula.

Once p and q have been selected, the next step is to select a positive integer e such that $\gcd(e, (p-1)(q-1)) = 1$. The integer e is called the encoding exponent, for reasons that will be clear shortly. Once suitable choices of M and e have been made, then we encode a text block number x to a new number c using the formula

$$c = x^e \,\%\, M.$$

The number c is then transmitted to the recipient. Note that the **only** information required to encode a message are the integers e and M. In particular, it is not necessary to know p and q in order to send a message.

To decode a message, we require a decoding exponent d, which is given by the formula

$$d = e^{-1} \,\%\, ((p-1)(q-1)).$$

Once d has been computed, the original number x may be recovered using the formula

$$x = c^d \,\%\, M.$$

Let's look at a specific example to see how this works. Here's our quote again:

```
message = "To be or not to be, that is the question. -- Hamlet";
```

We begin by selecting two primes p and q, and then setting $M = p\,q$.

```
p = 12373;
q = 32491;
M = p * q
```

The block length will then be

```
N[Log[95, M]]
```

Rounding down, we will end up using a block length of 4. The next step is to select an encoding exponent. A common choice is $e = 17$, but this is suitable only if $\gcd(e, (p-1)(q-1)) = 1$. Here's a check:

```
e = 17;
GCD[e, (p - 1) * (q - 1)]
```

Now we're ready to encode. The function **RSAencode** takes the original message, e, and the modulus, M, as input, and returns a list of encoded blocks. It selects the block size by the formula we derived above. Here is the code.

```
RSAencode[string_, e_, M_] := Module[{blocklength},
        blocklength = Floor[Log[95, M]];
    PowerMod[texttoblocks[string, blocklength], e, M]]
```

Here's the set of encoded blocks for our message.

```
codedblocks = RSAencode[message, e, M]
```

In order to decode, we first need to compute the decoding exponent d. Recall that the formula for d is given by

$$d = e^{-1} \,\%\, ((p-1)(q-1)).$$

We can use the function **PowerMod** to compute d as shown.

```
d = PowerMod[e, -1, (p - 1) * (q - 1)]
```

Once we have d, we can use the function **RSAdecode** to decode the set of encoded blocks. Here is the code for **RSAdecode**:

```
RSAdecode[blocks_, d_, M_] := Module[{blocklength},
        blocklength = Floor[Log[95, M]];
    blockstotext[PowerMod[blocks, d, M], blocklength]]
```

Now we can decode the encoded blocks.

```
RSAdecode[codedblocks, d, M]
```

■ Exercise 3

In this exercise, we prove that the method given above for decoding messages using RSA will work. Suppose that $M = p\,q$ where p and q are distinct primes, $0 < x < M$, $c = x^e \,\%\, M$, and

$$d = e^{-1} \,\%\, ((p-1)(q-1)).$$

Show that

$$x = c^d \,\%\, M.$$

■ Exercise 4

Bill wants to send a message to Alice using RSA, so she tells him that her encoding exponent is $e = 17$ and her encoding modulus is $M = 12195377863$. Show Bill how to encode the message

```
M = 12195377863;
message = "Hey Alice, isn't RSA cool!?   --Bill";
```

■ Exercise 5

Alice has received the encoded blocks shown below. Use the fact that $M = 104729 \cdot 116447$ to decode the message.

```
encodedblocks = {4048208982, 7973981835, 2232860160,
     1228506572, 2392521799, 2353518709, 4207703255,
     3703820767, 250037531, 6326907769, 3298628742};
```

■ Is RSA secure?

It's clear from the decoding procedure that if one can compute d then one can decode a message. Recall that the formula for computing d is given by

$$d = e^{-1} \% ((p - 1)(q - 1)).$$

Thus computing d using this formula requires knowledge of the values of p and q. On the other hand, the only information known to anyone encoding messages are the values of e and M. Of course, we know that $M = p \cdot q$, so that even if we just know e and M, we can find p and q by factoring M. For example, in Exercises 4 and 5 above, Alice used the encoding modulus $M = 12195377863$. In practice, messages sent to her would not be secure because it is easy to factor this value of M. Below is the Mathematica to do just that.

```
FactorInteger[12195377863]
```

■ Exercise 6

Bill has decided he wants to have people send him messages encoded using RSA. He tells everyone that his encoding exponent is $e = 17$ and his encoding modulus is $M = 13776856397$. He has received his first set of encoded blocks, which is given below. Decode the message.

```
M = 13776856397
    encodedblocks = {4490149759, 9836048959, 4449780697,
   13166158759, 7302115749, 8442727016, 2486382600,
   7344621302, 5062584227, 1272601926, 12966640119}
```

At this point, you may be asking yourself "Why would anyone use this cryptosystem? It doesn't seem to be secure." Indeed, the examples that we have seen so far are not secure, because it has been possible to factor the modulus M.

The key to making RSA secure is to select primes p and q that are much larger, say on the order of 100 digits each. In this case, M has about 200 digits and would be impossible to factor without knowledge of p and q. (One has to be a bit careful when making the choice of p and q, but we won't worry about these details here.) Without knowing p and q, there is no known way to compute the decoding exponent d.

Additional Remarks:

- Our example of encoding exponent, $e = 17$, is commonly used in RSA. Implementations will typically fix a single encoding exponent for all users, and each user gives out his or her value of M as his or her public key.

- The decoding process can be streamlined computationally. However, all known methods of decoding are essentially equivalent to the example given above. In particular, they all rely on being able to factor M.

- There is no known *proof* that RSA is secure. To date, no one has discovered a simple way to factor large integers, but it is possible that such a discovery will occur. If this were to happen, then the security of RSA-based cryptosystems would be severely compromised. However, the general consensus among factoring experts is that such a discovery is unlikely to occur, and that no simple factoring method exists.

The Euler ϕ-function

9.1 A Comparison

We begin here by recalling that the function $\phi(n)$ counts the number of integers m such that $1 \leq m \leq n$ and $\gcd(m, n) = 1$. In the Prelab section of this chapter, you computed $\phi(n)$ for a few different values of n by systematically checking each m such that $1 \leq m \leq n$, and counting how many satisfied $\gcd(m, n) = 1$. Clicking on "First phi" in the applet below will automatically do the same thing. Try it out; click on "First phi" to evaluate $\phi(20)$.

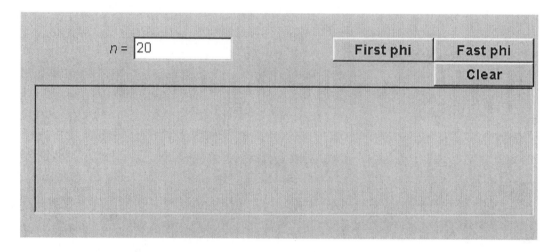

The output tells us that $\phi(20) = 8$. It also tells you how long it takes to perform the computation. Let's try out "First phi" on a larger number:

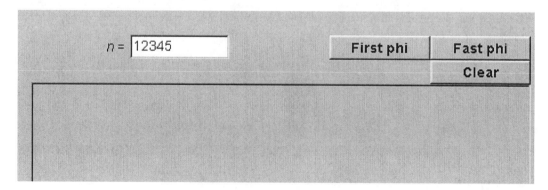

438

This one takes a bit longer, which shouldn't be surprising, because there are more values of m to check. The applet has a second built-in function called "Fast phi" that also computes $\phi(n)$, but does so using a method different from "First phi". Try out both "First phi" and "Fast phi" below to compare the computation times.

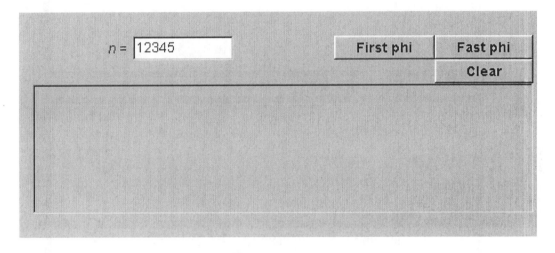

Clearly "Fast phi" is faster than "First phi". (Could we possibly expect a function called "Fast phi" to be *slower*?) The reason is that "Fast phi" uses an algorithm that is more efficient than the one used in "First phi". One goal for this chapter is to find this more efficient method for computing $\phi(n)$.

Exercise 1

(a) Use the applet to determine the time required to compute $\phi(10^4)$ using "First phi".

(b) On the basis of the results of part (a), extrapolate the number of years required for "First phi" to compute $\phi(10^{20})$. (For simplicity, assume that all years have 365 days.)

(c) Use the applet to determine how long it takes "Fast phi" to compute $\phi(10^{20})$.

9.2   Computing $\phi(n)$: A Special Case

As has become our standard practice, we will work towards a general solution to our problem by starting with special cases that are easier to tackle. In this section, we consider the values of $\phi(n)$ when n is a power of a prime.

Research Question 1

Find a formula for $\phi(p)$, where p is prime.

Research Question 2

Find a formula for $\phi(p^a)$, where p is a prime and a is a positive integer.

Hint: You may find it easier to count the number of $m \leq p^a$ such that $\gcd(m, p^a) > 1$, and subtract this from $n = p^a$.

9.3   Computing $\phi(n)$: The General Case

In order to compute $\phi(n)$ for values of n that are more complicated than prime powers, we would like to develop a means for breaking the general computation into several simpler computations. One way to proceed is to determine combinations of m and n for which the statement

$$\phi(m \cdot n) = \phi(m) \cdot \phi(n)$$

is true. This leads us directly to the next Research Question.

Research Question 3

Find a condition for pairs of integers m and n which guarantees that

$$\phi(mn) = \phi(m)\,\phi(n).$$

Note: The proof of your conjecture (assuming you find the conjecture we are looking for) will be considered on your homework assignment. You can provide a proof with the lab report if you have one, but this is not required. However, since you are not mandated to include a proof for your conjecture, you should provide *lots* of numerical data to support your claim.

Research Question 4

Use your conjectures from Research Questions 2 and 3 to assist in finding a formula for $\phi(n)$, where $n = p^a q^b$, p and q are distinct primes, and a and b are positive integers.

We're almost there. Here's the last step:

Research Question 5

Find a formula for $\phi(n)$, where

$$n = p_1^{a_1} p_2^{a_2} \cdots p_k^{a_k},$$

$p_1, p_2, \ldots, p_k$ are distinct primes, and $a_1, a_2, \ldots, a_k$ are positive integers.

Exercise 2

Use your formula from Research Question 5 to compute ϕ (10^{20}), explaining the steps. Compare the result with your answer to Exercise 1(c).

9.4   Going Farther: RSA

One of the best known applications of number theory is to what is known as a *public key cryptosystem*. The idea behind a public key cryptosystem is that one person, let's call her Alice, can tell anyone and everyone enough information to encode a message to be sent to her, but only Alice can decode the message. This certainly runs counter to intuition. If one has enough information to encode a message, why can't they run the process in reverse to decode it?

Several different methods for implementing public key cryptosystems have been proposed. One of the first, and certainly the best known, was developed by Rivest, Shamir, and Adelman, and is known as *RSA*. Before describing the specific mechanics involved in implementing RSA, we first discuss a general scheme known as *text blocking* that is incorporated into most cryptosystems.

9.4.1   Text Blocking

The crytosystems we have introduced thus far have been based on the idea of encoding messages a single letter at a time. Most cryptographic schemes of this type are not useful in practice because they will be vulnerable to frequency analysis. Recall that frequency analysis proceeds by counting the number of times each character occurs in an encoded message, and then making educated guesses about how to decode each character.

A simple way to work around frequency analysis is to organize the message text into blocks. Instead of representing each single character as a number, the text is taken in blocks of a certain length. Thus far, we have worked with a 95-character alphabet. We will now encode blocks of several characters of text as single numbers. For example, suppose we are

going to convert the text of the quote

To be or not to be, that is the question. -- Hamlet

into numbers. If we take the text in blocks of four characters each, then the first four characters "To b" would be converted into a single number, the next four characters "e or" would be converted to a single number, and so on. The first step in converting each 4-character block to a number is to convert each individual character in the block to a number. Use the applet below to convert "To b" to a list of numbers:

We then combine these numbers together as shown:

$$52 \cdot 95^0 + 79 \cdot 95^1 + 0 \cdot 95^2 + 66 \cdot 95^3 = 56594307.$$

The number 56594307 is then encoded and transmitted. More generally, if we take blocks of size n, and if $(a_1, a_1, \ldots, a_n)$ are the numerical equivalents for each letter in a block, then the *block number* is computed using the formula

$$a_1 \cdot 95^0 + a_2 \cdot 95^1 + a_3 \cdot 95^2 + \cdots + a_n \cdot 95^{n-1}.$$

By blocking in this manner, we have effectively raised the size of our "alphabet" from 95 to 95^n, which makes frequency analysis much more difficult, if not completely infeasible. For example, if $n = 10$, then the size of our "alphabet" is $95^{10} = 59873693923837890625$.

To make life a little easier for you down the road, the applet below is provided. This applet will take a message, break it into blocks, and then compute the numerical equivalent for each block. It also will reverse the process, converting blocks back into text.

Shift Cipher

> To be or not to be, that is the question. -- Hamlet

| Text->Blocks | Block Length: | 4 |

| Blocks->Text | Block Length: | 4 |

9.4.2 Implementing RSA

The first step in implementing the RSA cryptosystem is to select two distinct primes p and q, and then set $M = pq$. If our messages are blocked into groups of n letters, then p and q must be large enough so that $M > 95^n$. Taking logarithms (base 95), this translates to $n < \log_{95}(M)$. We will pick the block length n by this formula.

Once p and q have been selected, the next step is to select a positive integer e such that $\gcd(e, (p-1)(q-1)) = 1$. The integer e is called the *encoding exponent*, for reasons that will be clear shortly. Once suitable choices of M and e have been made, then we encode a text block number x to a new number c using the formula

$$c = x^e \,\%\, M.$$

The number c is then transmitted to the recipient. Note that the *only* information required to encode a message are the integers e and M. In particular, it is not necessary to know p and q in order to send a message.

To decode a message, we require a decoding exponent d, which is given by the formula

$$d = e^{-1} \,\%\, ((p{-}1)(q{-}1)).$$

Once d has been computed, the original number x may be recovered using the formula

$$x = c^d \,\%\, M.$$

Let's look at a specific example to see how this works. Here's our quote again:

> To be or not to be, that is the question. -- Hamlet

We begin by selecting two primes, say $p = 12373$ and $q = 32491$, and then setting $M = pq = 402011143$. The block length will then be

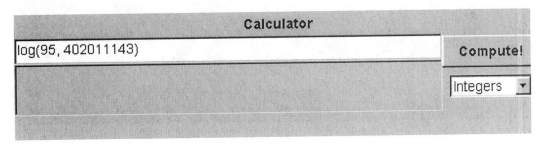

Recalling that the applet rounds down when in "Integers" mode, we will end up using a block length of 4. The next step is to select an encoding exponent. A common choice is $e = 17$, but this is suitable only if $\gcd(e, (p{-}1)(q{-}1)) = 1$. Here's a check:

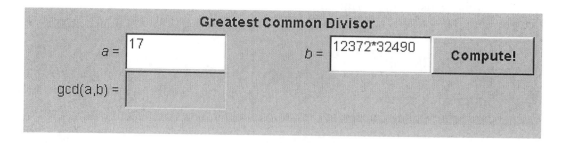

Now we're ready to encode. The applet below takes the original message, the encoding exponent e, and the modulus M as input, and returns a list of encoded blocks. It selects the block size by the formula we derived above. Here's the set of encoded blocks for our message.

```
Shift Cipher

To be or not to be, that is the question. -- Hamlet



RSA encode | Exponent: 17    Modulus: 402011143
```

In order to decode, we first need to compute the decoding exponent d. Recall that the formula for d is given by

$$d = e^{-1} \% ((p-1)(q-1)).$$

We can use the applet below to compute d as shown.

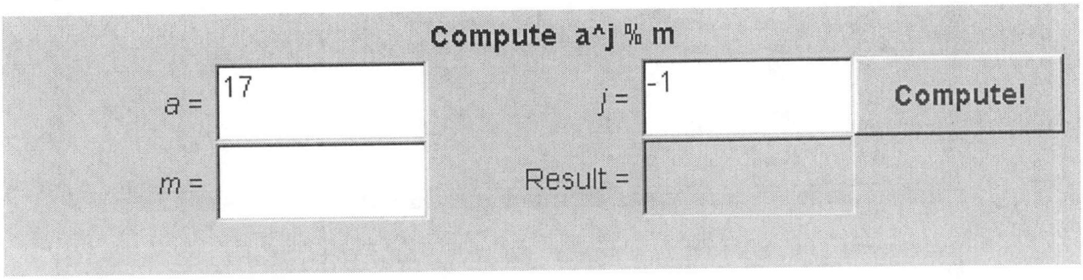

Compute a^j % m

$a =$ `17` $j =$ `-1`   **Compute!**

$m =$ ``            Result = ``

Once we have d, we can use the "RSA Decode" applet to decode the set of encoded blocks.

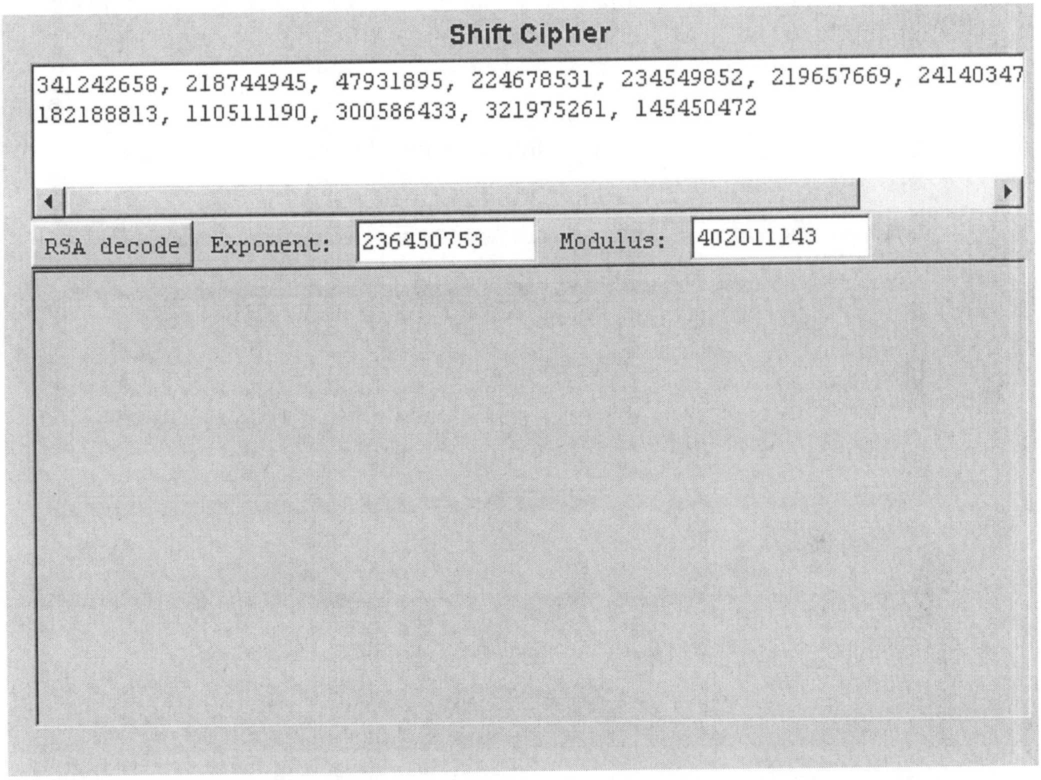

Shift Cipher

341242658, 218744945, 47931895, 224678531, 234549852, 219657669, 24140347
182188813, 110511190, 300586433, 321975261, 145450472

RSA decode   Exponent: `236450753`   Modulus: `402011143`

Exercise 3

In this exercise, we prove that the method given above for decoding messages using RSA will work. Suppose that $M = pq$ where p and q are distinct primes, $0 < x < M$, $c = x^e \% M$, and

$$d = e^{-1} \% ((p-1)(q-1)).$$

448 The Euler φ-function

Show that

$$x = c^d \% M.$$

Exercise 4

Bill wants to send a message to Alice using RSA, so she tells him that her encoding exponent is $e = 17$ and her encoding modulus is $M = 12195377863$. Show Bill how to encode the message

Hey Alice, isn't RSA cool!? --Bill

Exercise 5

Alice has received the encoded blocks shown below. Use the fact that $M = 104729 \cdot 116447$ to decode the message.

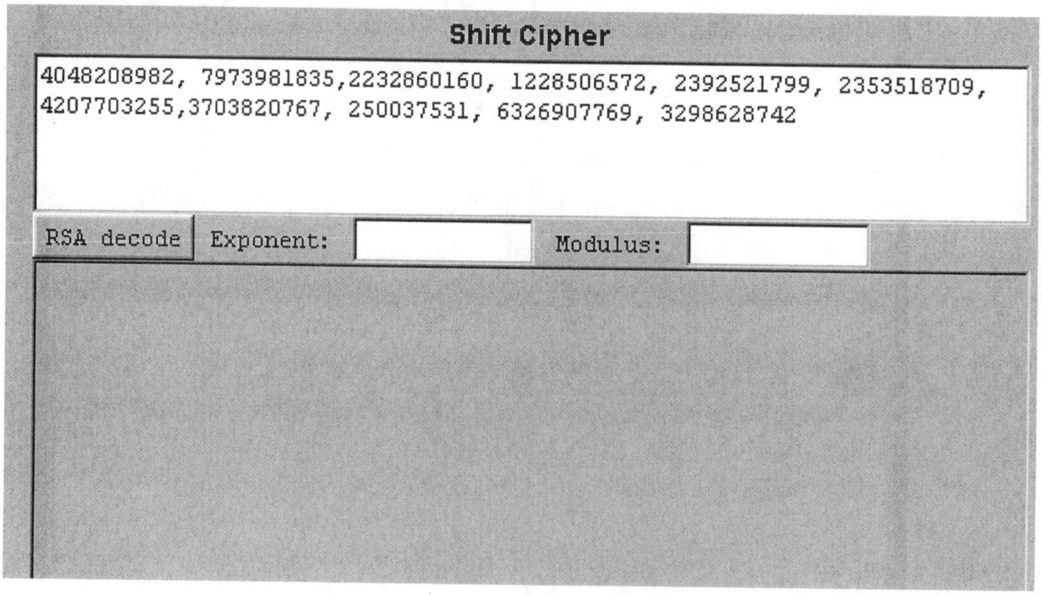

Shift Cipher

4048208982, 7973981835,2232860160, 1228506572, 2392521799, 2353518709,
4207703255,3703820767, 250037531, 6326907769, 3298628742

RSA decode | Exponent: [   ]   Modulus: [   ]

9.4.3 Is RSA secure?

It's clear from the decoding procedure that if one can compute d then one can decode a message. Recall that the formula for computing d is given by

$$d = e^{-1} \% ((p-1)(q-1)).$$

Thus computing d using this formula requires knowledge of the values of p and q. On the other hand, the only information known to anyone encoding messages are the values of e and M. Of course, we know that $M = p \cdot q$, so that even if we just know e and M, we can find p and q by factoring M. For example, in Exercises 4 and 5 above, Alice used the encoding modulus M = 12195377863. In practice, messages sent to her would not be secure because it is easy to factor this value of M. Below is an applet to do just that.

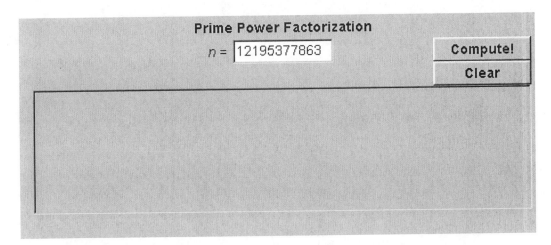

Exercise 6

Bill has decided he wants to have people send him messages encoded using RSA. He tells everyone that his encoding exponent is e = 17 and his encoding modulus is M =

> 13776856397. He has received his first set of encoded blocks, which is given below. Decode the message.

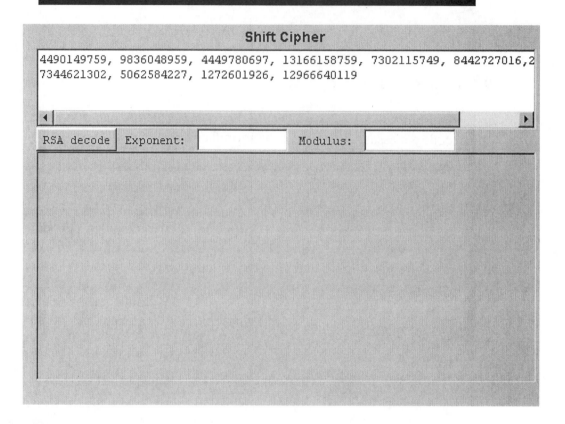

Shift Cipher

```
4490149759, 9836048959, 4449780697, 13166158759, 7302115749, 8442727016,2
7344621302, 5062584227, 1272601926, 12966640119
```

RSA decode | Exponent: [   ] | Modulus: [   ]

At this point, you may be asking yourself "Why would anyone use this cryptosystem? It doesn't seem to be secure." Indeed, the examples that we have seen so far are not secure, because it has been possible to factor the modulus M.

The key to making RSA secure is to select primes p and q that are much larger, say on the order of 100 digits each. In this case, M has about 200 digits and would be impossible to factor without knowledge of p and q. (One has to be a bit careful when making the choice of p and q, but we won't worry about these details here.) Without knowing p and q, there is no known way to compute the decoding exponent d.

Additional Remarks:

- Our example of encoding exponent, $e = 17$, is commonly used in RSA. Implementations will typically fix a single encoding exponent for all users, and each user gives out his or her value of *M* as his or her public key.

- The decoding process can be streamlined computationally. However, all known methods of decoding are essentially equivalent to the example given above. In particular, they all rely on being able to factor *M*.

- There is no known *proof* that RSA is secure. To date, no one has discovered a simple way to factor large integers, but it is possible that such a discovery will occur. If this were to happen, then the security of RSA-based cryptosystems would be severely compromised. However, the general consensus among factoring experts is that such a discovery is unlikely to occur, and that no simple factoring method exists.

Homework

The first part of these exercises is devoted to the unproved conjecture from the lab project. Recall the statement of this conjecture:

Conjecture *Suppose that m and n are positive integers. If $\gcd(m, n) = 1$, then*

$$\phi(mn) = \phi(m)\phi(n).$$

Since we have been recklessly applying this conjecture, we had better prove that it is actually true!

The underlying idea is to use the correspondence between the elements of the set $\mathbf{Z}_{mn}$ and the elements of the set $\mathbf{Z}_m \times \mathbf{Z}_n$ coming from the Chinese Remainder Theorem. In what follows, we will represent $\mathbf{Z}_n$ by $\{0, \ldots, n-1\}$. The subset of the integers in this range that are relatively prime to n is denoted by $\mathbf{Z}_n^*$.

In this discussion, $f \colon \mathbf{Z}_{mn} \to \mathbf{Z}_m \times \mathbf{Z}_n$ will be the function defined in the chapter summary for the Chinese Remainder Theorem given by $f(a) = (a \,\%\, m, a \,\%\, n)$. We already know that f is a bijective function.

The first two exercises illustrate the idea of the proof on a numerical level, and the next three outline the proof itself.

1. We consider the special case when $m = 3$ and $n = 4$.

   (a) Write down the correspondence between numbers in $\mathbf{Z}_{12}$ and pairs of integers in $\mathbf{Z}_3 \times \mathbf{Z}_4$ given by the function f. In other words, write out the 12 values $f(a)$ where $a \in \mathbf{Z}_{12}$.

   (b) For each value you computed above, circle the equations that correspond to $a \in \mathbf{Z}_{12}^*$.

   (c) How does this set of ordered pairs compare with $\mathbf{Z}_3^* \times \mathbf{Z}_4^*$?

2. We consider the special case when $m = 3$ and $n = 5$.

   (a) Find the explicit function $g \colon \mathbf{Z}_3 \times \mathbf{Z}_5 \to \mathbf{Z}_{15}$ from the Chinese Remainder Theorem chapter summary. (Recall that g is the inverse function of f.)

   (b) Write down all ordered pairs $(a, b) \in \mathbf{Z}_3^* \times \mathbf{Z}_5^*$.

   (c) Compute $g(a, b)$ for each ordered pair in part (b), reducing each answer to its remainder modulo 15.

   (d) Compare the list in part (c) with the integers modulo 15 which are relatively prime to 15.

Exercises 3 to 5 contain the steps of a formal proof.

3. If m and n are positive integers and k is any integer, show that $\gcd(k, mn) = 1$ if and only if $\gcd(k, m) = 1$ and $\gcd(k, n) = 1$.

4. Suppose $\gcd(m, n) = 1$. Prove that f establishes a bijection between $\mathbf{Z}^*_{mn}$ and $\mathbf{Z}^*_m \times \mathbf{Z}^*_n$. (In other words, if $f(k) = (a, b)$, then $\gcd(k, mn) = 1$ if and only if $\gcd(a, m) = 1$ and $\gcd(b, n) = 1$.)

5. Use the results of the preceeding two exercises to complete the proof of the conjecture.

The remaining exercises consist of applications of our formula for $\phi(n)$.

6. For each of the following values of n, find the prime factorization for n, and then use this factorization to compute $\phi(n)$.

   (a) $n = 400$

   (b) $n = 823543$

   (c) $n = 4443450741279$

7. For which positive integers n will $\phi(n)$ be an odd number?

8. Suppose n is a positive integer such that $\phi(n)$ is a power of 2. Show that n is of the form $n = 2^k \cdot p_1 p_2 \cdots p_r$ where the p_j are distinct Fermat primes. Recall that a Fermat prime is a prime number of the form $2^{2^j} + 1$ for some $j \geq 0$. (Note, this result is relevant to which polygons are *constructible*. Namely, a regular n-gon is constructible with compass and straightedge if and only if $\phi(n)$ is a power of 2.)

9. Suppose a is an odd positive integer. Show that $\phi(a) = \phi(2a)$.

10. Use the formula for $\phi(n)$ to find all positive integers n such that $\phi(n) = 4$.

11. Use the formula for $\phi(n)$ to find all positive integers n such that $\phi(n) = 6$.

12. Use the formula for $\phi(n)$ to find all positive integers n such that $\phi(n) = 8$.

13. Prove or disprove: There are no positive integers n such that $\phi(n) = 10$.

14. Prove or disprove: There are no positive integers n such that $\phi(n) = 12$.

15. Prove or disprove: There are no positive integers n such that $\phi(n) = 14$.

16. Prove that for every $k \in \mathbf{Z}$, there are only finitely many n such that $\phi(n) = k$.

17. Prove or disprove: If $m \mid n$, then $\phi(mn) = m\phi(n)$.

18. Prove or disprove: If $\phi(mn) = m\phi(n)$, then $m \mid n$.

19. Prove or disprove: If n is a positive integer, then $\phi(n^j) = n^{j-1}\phi(n)$.

20. Prove or disprove: If $m \mid n$, then $\phi(m) \mid \phi(n)$.

21. Prove or disprove: If $\phi(m) \mid \phi(n)$, then $m \mid n$.

22. Prove that

$$\phi(n) = n \prod_{p \mid n} \left(1 - \frac{1}{p}\right)$$

where the product runs over primes p dividing n.

23. For what values of n, if any, is $\phi(n) = n/2$?

24. For what values of n, if any, is $\phi(n) = n/3$?

25. For what values of n, if any, is $\phi(n) = n/5$?

26. Suppose that m is divisible by a perfect square bigger than 1. For what values of n, if any, is $\phi(n) = n/m$?

27. Prove: If $\phi(n) \mid n - 1$, then n is not divisible by a perfect square bigger than 1.

28. If n is a positive integer, the probability that a randomly chosen integer k is relatively prime to n is $\phi(n)/n$.

    (a) Determine $\lim_{k \to \infty} \phi(p_k)/p_k$ where p_k is the kth prime.

    (b) Determine $\lim_{k \to \infty} \phi(n_k)/n_k$ where n_k runs through the sequence of composite numbers which are not divisible by a perfect square.

    (c) Prove that $\lim_{n \to \infty} \phi(n)/n$ does not exist.

10. Primitive Roots

Prelab

We will continue our study of multiplicative orders modulo n. We start by reviewing a basic property of orders.

1. **Use the results contained in the Chapter 8 to show that $\mathrm{ord}_n(a)$ divides $\phi(n)$.**

On the basis of the above exercise, it's clear that the largest possible order of an integer modulo n is $\phi(n)$. An important question that we shall investigate in this chapter is: For which integers n does there exist an integer a of order *exactly equal to $\phi(n)$*? Such an integer a is called a *primitive root modulo n*.

2. **For each integer $n = 2$, 3, 4, ... , 15, determine if there exists an integer a such that $\mathrm{ord}_n(a) = \phi(n)$. (You do not have to show *all* of the computations leading to your answers. Please provide a brief description of how you computed the needed orders, and then make a table showing for each n, the values for n, $\phi(n)$, the largest $\mathrm{ord}_n(a)$ you found, the smallest positive value of a producing this largest order, and a "yes" or "no" indicating whether there is a primitive root modulo n.)**

Along with our study on the existence of primitive roots, we will consider how knowledge of the order of one integer (e.g., the existence of a primitive root) can imply information about the orders of other integers. For example, we will consider the following questions in the lab.

Suppose a is an integer relatively prime to n such that $\mathrm{ord}_n(a) = \phi(n)$. What are the other multiplicative orders which occur modulo n, and how many congruence classes take on each order?

Maple electronic notebook `10-primitive.mws`

Mathematica electronic notebook `10-primitive.nb`

Web electronic notebook Start with the web page `index.html`

Maple Lab: Primitive Roots

■ Orderly Orders

In the last chapter, we looked at multiplicative orders of integers a modulo n by considering different values of a separately. We can learn more by considering how the order of one integer might be related to the order of another integer.

Suppose we know the order of a modulo n. Then, we have a pretty good grasp on the powers of a when computed modulo n. For example, with $a = 3$ and $n = 10$, the first 30 powers a^j reduced modulo n are

```
> a := 3:
  n := 10:
  seq(modp(a^j, n), j=1..30);
>
```

We know that this sequence will be purely periodic, and the period, which is also $\mathrm{ord}_n(a)$, will be the first position which contains a 1. In this case, $\mathrm{ord}_{10}(3) = 4$. Moreover, we then know that $3^i \equiv 3^j \pmod{10}$ if and only if $i \equiv j \pmod{\mathrm{ord}_{10}(3)}$. In other words, $3^i \equiv 3^j \pmod{10}$ if and only if $i \equiv j \pmod 4$. Thus,

$$3^i \equiv 3 \quad \text{if} \quad i \equiv 1 \pmod 4,$$
$$3^i \equiv 9 \quad \text{if} \quad i \equiv 2 \pmod 4,$$
$$3^i \equiv 7 \quad \text{if} \quad i \equiv 3 \pmod 4,$$
$$3^i \equiv 1 \quad \text{if} \quad i \equiv 0 \pmod 4.$$

This gives us lots of information about the powers of 3 modulo 10. How can we really make use of this? If we look at the list of powers $3^j \bmod 10$ above, we see that $3^3 \equiv 7 \pmod{10}$. It then follows that

$$7^j \equiv (3^3)^j \equiv 3^{(3j)} \pmod{10}.$$

So, not only should the powers of 7 appear as powers of 3 (when each are reduced modulo 10), but we should be able to connect exactly which power of 7 matches with which power of 3. Let's look at the sequence of powers of 7 taken modulo 10:

```
> a := 7:
  n := 10:
  seq(modp(a^j, n), j=1..30);
>
```

If we had wanted to predict the precise value of $7^6 \% 10$, we could have reasoned that
$$7^6 \equiv 3^{18} \equiv 9 \pmod{10}$$
because $18 \equiv 2 \pmod 4$. Looking at the results above, we can see that this is right.

We now focus on $\mathrm{ord}_{10}(3^j)$ for different values of j. The next function takes a value for a and for n as input. It computes $\mathrm{ord}_n(a^j)$ for each j from 1 to $\mathrm{ord}_n(a)$. (Do not be alarmed if executing the next group causes Maple to issue a warning about a new definition of order. This is normal when loading Maple's

468

number theory library.)

```
> with(numtheory):

  powords := proc(a,n)
   local orda, j, pow;
   orda := order(a, n);
   for j from 1 to orda do
    pow := modp(Power(a, j), n);
    printf('The order of %1d^%1d modulo %1d is %1d.\n', a, j, n,
  order(pow, n));
   od;
  end:
```

Here it is in action when $a = 3$ and $n = 10$:

```
> powords(3,10);
>
```

Use `powords` to experiment with the next Research Question. (Some ideas for a proof are contained in the preceding discussion!)

■ Research Question 1

Suppose that a is relatively prime to n and $j > 0$. Find a formula for $\mathrm{ord}_n(a^j)$ in terms of $\mathrm{ord}_n(a)$ and j

■ Research Question 2

Suppose that a is relatively prime to n. What are the values of $\mathrm{ord}_n(a^j)$ for $j = 0, 1, 2, \ldots$?

■ Orders in the Presence of a Primitive Root

The existence of a primitive root simplifies working with the elements of $\mathbf{Z}_n^*$, the congruence classes relatively prime to n. Suppose that r is a primitive root modulo n. On the one hand, we know that there are $\phi(n)$ congruence classes relatively prime to n. On the other hand, since $\mathrm{ord}_n(r) = \phi(n)$, the powers r, $r^2, r^3, r^4, \ldots, r^{\phi(n)}$ give $\phi(n)$ distinct congruence classes relatively prime to n. Thus, every integer relatively prime to n is congruent to a power of r. This is a **big deal**, so much so that we repeat it below.

If r is a primitive root modulo n, then *every* integer m that is relatively prime to n is congruent to r^j for some integer j between 1 and $\phi(n)$.

Let's use Maple to check this out. First, we can observe this fact by looking at the powers of r in succession. You can tell that r is a primitive root if the powers of r hit every congruence class relatively prime to n. The function `seepows` shows each power of r reduced modulo n until it reaches a power of r which is congruent to 1.

```
> seepows := proc (r,n)
    local j;
    if gcd(r,n)<> 1 then print('%1d must be relatively prime to
  %1d',r,n);
```

```
      else
      j:=1:
      while modp(Power(r,j),n)<>1 do
        printf('%1d^%d is congruent to %1d (mod %1d)\n', r, j,
  modp(Power(r,j),n); n);
        j := j+1;
      od;
      printf('%1d^%1d is congruent to %1d (mod %1d)\n', r, j,
  modp(Power(r,j),n), n);
      fi;
      end:
```

For example, try it for the powers of 2 modulo 11:

```
[ > seepows(2,11);
```

Another way to observe this behavior is to try to write every element of $\mathbf{Z}_n^*$ as a power of r, where r is a primitive root modulo n. That is the purpose of the next function.

```
[ > writeaspowers:=proc(r,n)
      local pows, sortedpows,j;
      pows:=[seq([modp(Power(r,j), n),j], j=1..phi(n))];
      sortedpows:=sort(pows, (a,b)->evalb(a[1]<b[1]));
      for j to phi(n) do
        printf('%3d is congruent to %1d^%1d (mod %1d)\n',
      sortedpows[j][1], r, sortedpows[j,2], n);
      od;
      end:
```

Remember when applying this function that the value of r must be a primitive root modulo n. Let's look again at the case of powers of 2 modulo 11:

```
[ > writeaspowers(2,11);
```

If you try `writeaspowers(r,n)` where r is *not* a primitive root modulo n, weird things happen:

```
[ > writeaspowers(4,11);
[ >
```

If r is a primitive root modulo n, then every element of $\mathbf{Z}_n^*$ is congruent to r^j for some j. Furthermore, your formula given in response to Research Question 1 shows how to compute $\mathrm{ord}_n(a^j)$ from $\mathrm{ord}_n(a)$ and j.

Thus you have a formula for the order of *every* element of $\mathbf{Z}_n^*$. Pretty cool, eh? Keep your formula in mind as you tackle the next two Research Questions. As you work on these questions, you will want to collect data with the assistance of the functions defined above. To save you the trouble of hunting around for integers that have a primitive root, we provide below a list of all such integers $n \le 30$:

$n = 2, 3, 4, 5, 6, 7, 9, 10, 11, 13, 14, 17, 18, 19, 22, 23, 25, 26, 27, 29.$

■ Research Question 3

Suppose that d is the order of some element of $\mathbf{Z}_n^*$, and that r is a primitive root of n.

(a) What values of j satisfy $1 \le j \le \phi(n)$ and $\text{ord}_n(r^j) = d$?

(b) For what values of j is r^j a primitive root modulo n?

Research Question 4

Suppose that d is the order of some element of $\mathbf{Z}_n^{\;*}$, and that r is a primitive root of n.

(a) How many values of j satisfy $1 \le j \le \phi(n)$ and $\text{ord}_n(r^j) = d$?

(b) How many primitive roots modulo n are there?

When Is There a Primitive Root?

The Research Questions in the previous section assumed that an integer n had a primitive root. In this section, we shall consider the problem of determining which integers n have primitive roots, and which do not. A good place to start your investigation is with the function `allords` from the previous chapter. Here it is again:

```
> allords := proc(n)
    local a;
    printf('      a   ord a');
    for a to n-1 do
      if gcd(a,n)=1 then
         printf('\n%6d%6d',a,order(a,n))
      fi; od; end:
```

Recall what `allords(n)` does. It displays each element a that is relatively prime to the input n, along with the value of ord(a). Here is a sample of its output:

```
> allords(15);
```

Try it out, and watch for values of n for which there is an element of order $\phi(n)$. To make this task easier, you can compute $\phi(n)$ with `phi(n)` at the same time as executing `allords(n)`.

```
> n:= 15:  # Set n here, then execute this group

  printf('\nPhi of %1d is %1d\n\n', n, phi(n)):

  allords(n):
>
```

As you can see, the output confirms what you discovered when working the Prelab exercises: there are no primitive roots modulo 15. The positive integers $n \le 30$ that have primitive roots were given above; here they are again:

$n = 2, 3, 4, 5, 6, 7, 9, 10, 11, 13, 14, 17, 18, 19, 22, 23, 25, 26, 27, 29.$

Use the functions defined above to determine which other integers n have primitive roots until you have enough data to make a conjecture for the final Research Question of this chapter:

Research Question 5

Which positive integers n have primitive roots?

Note: The proof of this conjecture is quite difficult. It would be truly remarkable for a student to find a proof on his or her own. Unless told otherwise by your instructor, you should concentrate on making a good conjecture here and supporting it in your lab report with numerical evidence.

▪ Going Farther: Cryptography Using the "Lumberjack's Rock and Roll" Discretely

The function writeaspowers in the second section above has an application to cryptography. The output of writeaspowers gives all of the values of a certain function. If r is a primitive root modulo m, then writeaspowers(r,m) tells us: for each c, what is an exponent x such that $c \equiv r^x \pmod{m}$. Here is an example of writeaspowers (it is easy to check that 2 is a primitive root modulo 19):

```
> writeaspowers(2,19);
>
```

If we wanted to find an x such that $2^x \equiv 3 \pmod{19}$, we could easily read off that $x = 13$ works.

The next function lets us compute this value a little bit more directly. To solve the congruence $r^x \equiv c \pmod{m}$, we provide the base r, the value c, and the modulus m. The function will return the smallest nonnegative exponent x which satisfies the congruence.

```
> discretelog := proc(base, value, modulus)
    local expo, try, valmod;
    valmod := value mod modulus;
    expo := 0;
    try := 1;
    while try <> valmod do
     try := (try * base) mod modulus;
     expo := expo+1;
    od;
    expo;
   end:
>
```

This function is generally known as a *discrete logarithm* because its defining property is so similar to logarithm functions in calculus. We include the adjective *discrete* to distinguish it from its more familiar continuous counterparts. The function above works by trying successive values for x, starting with $x = 0$, until it finds one which works.

Let's try our new discretelog to answer the same question as above: find x such that $2^x \equiv 3 \pmod{19}$:

```
> discretelog(2,3,19);
>
```

Good news, we found the same answer (and with less output than writeaspowers).

The discrete logarithm is important to cryptography (thus the title of this section). Ironically, it is important because discrete logarithms are difficult to compute quickly. By contrast, the inverse of discrete logs (computing $b^n \% m$) can be accomplished quickly using the algorithm described in Chapter 8. On average, the time required to compute $c = (b^n \% m)$ is proportional to the number of digits in m whereas the time to undo this and compute n from b, c, and m is proportional to the value of m. When m is large enough, this makes the exponentiation direction reasonably fast and the discrete logarithm direction impossible for all practical purposes. For this reason, modular exponentiation is called a one-way function.

The previous paragraph oversimplifies the situation a little bit. There are faster methods for computing discrete logarithms, especially if $\phi(m)$ is divisible by only small primes. But the basic principle is right: if we choose m so that $\phi(m)$ is divisible by a very large prime, then the computation of discrete logs is impractical.

In order to have Maple use the fast modular exponentiation algorithm described in Chapter 8, we need to combine a couple of simple commands. Here is a special function `powermod` which invokes the right Maple syntax.

```
> powermod := (base, expo, modulus) ->
    modp(Power(base,expo),modulus):
```

We can use this to quickly compute $2^{135791} \bmod 1234577$:

```
> powermod(2, 135791, 1234577);
```

We can reverse this computation with our version of `discretelog`:

```
> discretelog(2, 189155, 1234577);
>
```

Below we look at two cryptographic applications of discrete logarithms. There are others, including a public key cryptosystem named El Gamal after its inventor. Before you enter the applications, open the next section, hit Enter in the execution group, and you will load a collection of basic crytography functions, most of which are from earlier chapters.

■ Cryptographic Utility Functions

■ Password Files

Many computers allow many different users to login by providing a login name and a password. The passwords are typically kept in a file on the system, and they are encrypted so that users cannot discover each other's password. The problem is to keep users from spending their time trying to decrypt each other's password.

The problem is solved by using a one-way function. The system can then encrypt passwords, but no one is able to decrypt them, not even the computer! Interestingly, this is sufficient for password checking. When a user attempts to login, the computer takes whatever the user types as his or her password and runs it through the encryption process. It can then check to see if the encrypted version of whatever the user typed matches the encrypted version of the password it has stored in a file.

We are almost ready. We will set the modulus and our primitive root which will be used for an example.

```
> modulus := 27449;
  base := 3;
  order(base, modulus);
```

The last line shows that our base is a primitive root for the modulus. To encode the password "Hi",
we convert it to a number, encrypt the number, and then convert the resulting number back to text.

```
> password := `Hi`:
  passwordasnum := texttonum(password);
  encryptednum := powermod(base, passwordasnum, modulus);
  final := numtotext(encryptednum);
```

To decode this, we reverse the process. We undo the `powermod` by taking a discrete log. As a
result, you may notice that decoding takes longer than encoding.

```
> encryptednum := texttonum(final);
  decryptednum := discretelog(base, encryptednum, modulus);
  numtotext(decryptednum);
>
```

In this example, we used a small modulus so that we could compute the required discrete logarithm in
our lifetimes. Because of the small modulus, we needed a very short password.

In practice, computer passwords are often cracked in a very nonmathematical way. The "mischievous
user" who gets other people's passwords often does it by guessing! Sometimes, this takes a certain
amount of trial and error.

■ *Exercise 1*

Louis Reasoner is very proud of the new password system he installed on his computer. It uses
the method we described above, but with a large modulus for better security. In fact, the next
group initializes Louis's modulus and base.

```
> modulus := 100000000000000003;
  base := 2;
```

Louis is so confident in his system that he brags "No one can crack my passwords. My system is
so secure that I do not even use capital letters or punctuation in my passwords. I also like to
change my password every day. I have one password for every day of the week!" When Louis
heads off to watch his favorite movie, Disney's *Snow White*, his friends try to guess Louis's
password. The encoded version of today's password is

```
> encoded := `vI41~,w](`;
>
```

What is Louis's password today?

■ Key Exchange: Creating Shared Secrets

Suppose Alice and Bob are communicating through the internet. They want to set up encryption for
sending messages, so they will need a key. All communications on the internet can be observed by an
outsider, so Alice cannot simply send a key to Bob.

One solution is for Alice and Bob to work out a "shared secret". The secret is just a large number
which Alice and Bob know, but nobody else knows including Oscar who is watching all of the
transmissions between Alice and Bob. The shared secret can then be used as the key for any of a
number of cryptosystems.

The Diffie-Hellman protocol was one of the first methods for creating shared secrets. The security of
Diffie-Hellman depends on discrete logarithms' being difficult to compute. Here is how it works.

Alice and Bob will be sending messages back and forth, and Oscar is eavesdropping. First, Alice sends a modulus *m* and and base *r* to Bob. Typically, *m* will be a large prime and *r* will be a primitive root modulo *m*. Now, everyone knows these two numbers. Execute the next group to set *m* and *r* for a working example.

```
> m := 27449;
  r := 3;
  order(r,m);
[ >
```

The last line checks that *r* is a primitive root modulo *m*. Alice and Bob secretly pick random numbers, which we will call *a* and *b* respectively. Then, Alice computes r^a % *m* and Bob computes r^b % *m*. These values are exchanged. Now, everyone knows *r*, *m*, r^a % *m*, and r^b % *m*. Only Alice and Bob know their secret random values, so we keep Maple from printing those values.

```
> a := rand() mod (m-1):
  b := rand() mod (m-1):
  ra := powermod(r, a, m);
  rb := powermod(r, b, m);
[ >
```

Finally, Alice takes r^b % *m*, which she got from Bob, and uses it to compute

$$\left(r^b\right)^a \% m = r^{(a\,b)} \% m$$

(only she can do this because only she knows the value of *a*). Similarly, Bob computes

$$\left(r^a\right)^b \% m = r^{(a\,b)} \% m$$

using *b*. In the end, both Alice and Bob know $r^{(a\,b)}$ % *m*.

```
> rab := powermod(ra, b, m);
  rba := powermod(rb, a, m);
[ >
```

Of course, these values should have worked out to be the same!

Now that Alice and Bob have their key, they will use it with some cryptosystem. Normally, people use a fairly sophisticated system such as the *Digital Encryption Standard*, or DES for short. DES is pretty secure and was designed to run especially fast on a computer. However, DES is not very interesting from a number theoretic point of view, so we will use a simple shift cipher instead (with text blocking). The encoding and decoding functions were contained in the pile of utility functions you executed above.

To encode a message, just send the message and key into the encoder:

```
> message := 'Did you solve exercise 1?';
  encoded := encode(message, rab);
```

Similarly, we can decode a message with the function decode:

```
[ > decode(encoded, rab);
[ >
```

Oscar could figure out this "secret" value $r^{(ab)}$ % m if he could compute discrete logarithms. He would take a discrete logarithm of r^a % m to find a, and then proceed as Alice did to find $r^{(ab)}$ % m. Our modulus is small, so we can try that now.

```
[ > discretelog(r, ra, m);
```

For comparison, here is the value of a we saved above:

```
[ > a;
[ >
```

No one knows of a better way for Oscar to proceed. If the modulus were bigger, Oscar would be stuck, so we assume that this exchange between Alice and Bob is secure.

We mentioned above that discrete logarithms are only difficult to compute if $\phi(m)$ is divisible by a very large prime. The next function implements an algorithm which computes discrete logarithms quickly if $\phi(m)$ is divisible by only small primes. If $\phi(m) = p_1^{a_1} p_2^{a_2} \bullet \bullet \bullet p_n^{a_n}$, then the running time of our first discretelog is proportional to $\phi(m) = p_1^{a_1} p_2^{a_2} \bullet \bullet \bullet p_n^{a_n}$. The running time for fastdiscretelog below is proportional to $\sum_{i=1}^{n} a_i p_i$. Do not try to decipher the code here. We will discuss the method in the chapter summary.

```
[ > fastdiscretelog := proc(base, val, modulus)
    local fim, phim, j, p, d, try, cmp, ans, cnt, k, ppart, ptry,
  ival, idem;
    phim := phi(modulus);
    fim := ifactors(phim)[2];
    ans := 0;
    ival := 1/val mod modulus;
    for j to nops(fim) do # Loop over primes dividing phi(modulus)
     p := fim[j,1];
     d := fim[j,2];
     idem := modp(p^d/phim, p^d);
     try := modp(Power(base,phim/p^d*idem),modulus);
     ptry := modp(Power(try, p^(d-1)), modulus);
     ppart := 0;
      for k to d do # Loop over powers of p
       cmp := modp(Power(ival, phim/p^k*idem), modulus) *
  modp(Power(try, ppart*p^(d-k)), modulus);
       cmp := cmp mod modulus;
       cnt := 0;
       while(cmp <> 1) do
         cmp := (cmp*ptry) mod modulus;
         cnt := cnt+1;
       od;
       ans := (ans + cnt*idem*phim/p^(d-k+1)) mod phim;
       ppart := ppart+cnt*p^(k-1);
      od;
    od;
```

```
    ans;
  end:
[ >
```

To see it in action, we start by carefully choosing *m*.

```
[ > m := 174636001;
    isprime(m);
    ifactor(m-1);
[ >
```

The second line of output verifies that our modulus is prime. The third line then shows that $\phi(m)$ has only small prime factors. After a little searching, it is not hard to discover that our modulus has a primitive root, namely 13.

```
[ > base := 13;
    order(base, m);
    phi(m);
[ >
```

We see that the order of 13 modulo *m* is the same as $\phi(m)$. Next we raise 13 to a moderately large power:

```
[ > bignum := powermod(base, 123456789, m);
[ >
```

Finally, we can undo this with our fast discrete log function:

```
[ > fastdiscretelog(base, bignum, m);
[ >
```

Computing this value using plain old `discretelog` would take more than an hour on most computers.

▰ *Exercise 2*

Alice and Bob cannot wait to try using the Diffie-Hellman protocol to establish encrypted communication with someone. They think that the only thing he needs to get a high level of security is a large modulus *m*. So, Alice picks the following modulus and primitive root *r*:

```
[ > m := 15*2^78+1;
    isprime(m);
    r := 11;
    order(r,m);
[ >
```

We can see that *m* is prime and that *r* is a primitve root modulo *m*.

We are able to observe the values exchanged by Alice and Bob.

```
[ > alicenum := 3268170018038853053202889;

    bobnum := 4168860017312113121130990;
[ >
```

This is enough to establish the key for Alice and Bob. Then Alice sends an encrypted message to Bob:

```
> alicetobob := [4287436560919001394387439,
   4379666011581448391288243, 4656180681496403797643909,
   4197584748237556580656485, 4367214525356700096276646,
   3981076401286882248383284, 1154612708230347541420200,
   1098901436212055158708132, 396432393592786066611479]:
>
```

Decode the message.

Mathematica Lab: Primitive Roots

■ Orderly Orders

In the last chapter, we looked at multiplicative orders of integers a modulo n by considering different values of a separately. We can learn more by considering how the order of one integer might be related to the order of another integer.

Suppose we know the order of a modulo n. Then, we have a pretty good grasp on the powers of a when computed modulo n. For example, with $a = 3$ and $n = 10$, the first 30 powers a^j reduced modulo n are

```
a = 3;
n = 10;
Table[Mod[a^j, n], {j, 30}]
```

We know that this sequence will be purely periodic, and the period, which is also $\operatorname{ord}_n(a)$, will be the first position which contains a 1. In this case, $\operatorname{ord}_{10}(3) = 4$. Moreover, we then know that $3^i \equiv 3^j \pmod{10}$ if and only if $i \equiv j \pmod{\operatorname{ord}_{10}(3)}$. In other words, $3^i \equiv 3^j \pmod{10}$ if and only if $i \equiv j \pmod 4$. Thus,

$$
\begin{aligned}
3^i \equiv 3 \quad &\text{if} \quad i \equiv 1 \pmod 4 \\
3^i \equiv 9 \quad &\text{if} \quad i \equiv 2 \pmod 4 \\
3^i \equiv 7 \quad &\text{if} \quad i \equiv 3 \pmod 4 \\
3^i \equiv 1 \quad &\text{if} \quad i \equiv 0 \pmod 4.
\end{aligned}
$$

This gives us lots of information about the powers of 3 modulo 10. How can we really make use of this? If we look at the list of powers 3^j mod 10 above, we see that $3^3 \equiv 7 \pmod{10}$. It then follows that

$$ 7^j \equiv (3^3)^j \equiv 3^{3j} \pmod{10}. $$

So, not only should the powers of 7 appear as powers of 3 (when each are reduced modulo 10), but we should be able to connect exactly which power of 7 matches with which power of 3. Let's look at the sequence of powers of 7 taken modulo 10:

```
a = 7;
n = 10;
Table[Mod[a^j, n], {j, 30}]
```

If we had wanted to predict the precise value of 7^6 % 10, we could have reasoned that

$$ 7^6 \equiv 3^{18} \equiv 9 \pmod{10} $$

because $18 \equiv 2 \pmod 4$. Looking at the results above, we can see that this is right.

We now focus on $\operatorname{ord}_{10}(3^j)$ for different values of j. The next function takes a value for a and for n as input. It computes $\operatorname{ord}_n(a^j)$ for each j from 1 to $\operatorname{ord}_n(a)$.

```
powords[a_, n_] := Module[{orda, j, pow},
       (* The function ord
   is defined in the Initialization section. *)
        orda = ord[a, n];
        Do[ pow = PowerMod[a, j, n];
            Print["The order of ",
   a, "^", j, " modulo ", n, " is ", ord[pow, n]],
            {j, orda}]]
```

Here it is in action when $a = 3$ and $n = 10$:

```
powords[3, 10]
```

Use **powords** to experiment with the next research question. (Some ideas for a proof are contained in the preceding discussion!)

■ Research Question 1

> Suppose that a is relatively prime to n and $j > 0$. Find a formula for $\mathrm{ord}_n(a^j)$ in terms of $\mathrm{ord}_n(a)$ and j.

■ Research Question 2

> Suppose that a is relatively prime to n. What are the values of $\mathrm{ord}_n(a^j)$ for $j = 0, 1, 2, \ldots$?

■ Orders in the Presence of a Primitive Root

The existence of a primitive root simplifies working with the elements of $\mathbf{Z}_n^*$, the congruence classes relatively prime to n. Suppose that r is a primitive root modulo n. On the one hand, we know that there are $\phi(n)$ congruence classes relatively prime to n. On the other hand, since $\mathrm{ord}_n(a) = \phi(n)$, the powers a^1, a^2, a^3, a^4, . . . , $a^{\phi(n)}$ give $\phi(n)$ distinct congruence classes relatively prime to n. Thus, *every* integer relatively prime to n is congruent to a power of r. This is a **big deal**, so much so that we repeat it below.

> If r is a primitive root modulo n, then every integer m that is relatively prime to n is congruent to r^j for some integer j between 1 and $\phi(n)$.

Let's use Mathematica to check this out. First, we can observe this fact by looking at the powers of r in succession. You can tell that r is a primitive root if the powers of r hit every congruence class relatively prime to n. The function **seepows** shows each power of r reduced modulo n until it reaches a power of r which is congruent to 1.

```
seepows[r_, n_] := Module[{j},
        If[GCD[r, n] != 1,
    Print[r, " must be relatively prime to ", n],
        j = 1;
            While[PowerMod[r, j, n] != 1,
                Print[r, "^", j,
    " is congruent to ", PowerMod[r, j, n], " (mod ", n, ")"];
                j = j + 1]; Print[r, "^", j,
    " is congruent to ", PowerMod[r, j, n], " (mod ", n, ")"]]]
```

For example, try it for the powers of 2 modulo 11:

```
seepows[2, 11]
```

Another way to observe this behavior is to try to write every element of $\mathbf{Z}_n^*$ as a power of r where r is a primitive root modulo n. That is the purpose of the next function:

```
writeaspowers[r_, n_] := Module[{pows, sortedpows},
          pows =
     Table[{PowerMod[r, j, n], j}, {j, 1, EulerPhi[n]}];
          sortedpows = Sort[pows, #1[[1]] < #2[[1]] &];
          Do[
              Print[sortedpows[[j, 1]], " is congruent to ",
     r, "^", sortedpows[[j, 2]], " (mod ", n, ")"],
              {j, Length[sortedpows]}];
          ]
```

Remember when applying this function that the value of a must be a primitive root modulo n. Let's look again at the case of powers of 2 modulo 11:

```
writeaspowers[2, 11]
```

Note, if you try **writeaspowers[r,n]** where r is not a primitive root modulo n, weird things happen:

```
writeaspowers[4, 11]
```

If r is a primitive root modulo n, then every element of $\mathbf{Z}_n^*$ is congruent to r^j for some j. Furthermore, your formula given in response to Research Question 1 shows how to compute $\operatorname{ord}_n(a^j)$ from $\operatorname{ord}_n(a)$ and j. Thus you have a formula for the order of every element of $\mathbf{Z}_n^*$. Pretty cool, eh? Keep your formula in mind as you tackle the next two Research Questions. As you work on these questions, you will want to collect data with the assistance of the functions defined above. To save you the trouble of hunting around for integers that have a primitive root, we provide below a list of all such integers $n \le 30$:

$$n = 2, 3, 4, 5, 6, 7, 9, 10, 11, 13, 14, 17, 18, 19, 22, 23, 25, 26, 27, 29.$$

■ Research Question 3

> Suppose that d is the order of some element of $\mathbf{Z}_n^*$, and that r is a primitive root of n.
> (a) What values of j satisfy $1 \le j \le \phi(n)$ and $\operatorname{ord}_n(r^j) = d$?
> (b) For what values of j is r^j a primitive root modulo n?

■ Research Question 4

> Suppose that d is the order of some element of $\mathbf{Z}_n^*$, and that r is a primitive root of n.
> (a) How many values of j satisfy $1 \le j \le \phi(n)$ and $\operatorname{ord}_n(r^j) = d$?
> (b) How many primitive roots modulo n are there?

■ When Is There a Primitive Root?

The Research Questions in the previous section assumed that an integer n had a primitive root. In this section, we shall consider the problem of determining which integers n have primitive roots, and which do not. A good place to start your investigation is with the function **allords** from the previous chapter. Here it is again:

```
allords[n_] :=
 Module[{buildoutput}, buildoutput = {{"a", "ord a"}};
 Do[If[GCD[j, n] == 1,
    buildoutput = Append[buildoutput, {j, ord[j, n, False]}]],
 {j, n-1}];
        DisplayForm[
   GridBox[buildoutput, ColumnAlignments -> Right]]]
```

Recall what **allords[n]** does. It displays each element *a* that is relatively prime to the input **n**, along with the value of ord(*a*). Here is a sample of its output:

```
allords[15]
```

Try it out, and watch for values of *n* for which there is an element of order $\phi(n)$. To make this task easier, you can compute $\phi(n)$ with **EulerPhi[n]** at the same time as executing **allords[n]**.

```
n = 15;  (* Set n here, then execute this cell *)

Print["Phi of ", n, " is ", EulerPhi[n]];
allords[n]
```

Try this for several values of *n*.

As you can see, the output confirms what you discovered when working the Prelab exercises: there are no primitive roots modulo 15. The positive integers $n \leq 30$ that have primitive roots were given above; here they are again:

$$n = 2, 3, 4, 5, 6, 7, 9, 10, 11, 13, 14, 17, 18, 19, 22, 23, 25, 26, 27, 29.$$

Use the functions defined above to determine which other integers *n* have primitive roots until you have enough data to make a conjecture for the final Research Question of this chapter:

■ **Research Question 5**

> Which positive integers *n* have primitive roots?
>
> **Note:** The proof of this conjecture is quite difficult. It would be truly remarkable for a student to find a proof on his or her own. Unless told otherwise by your instructor, you should concentrate on making a good conjecture here and supporting it in your lab report with numerical evidence.

■ Going Farther: Cryptography Using the "Lumberjack's Rock and Roll" Discretely

The function **writeaspowers** in the previous section has an application to cryptography. The output of **writeaspowers** gives all of the values of a certain function. If *r* is a primitive root modulo *m*, then **writeaspowers[r,m]** tells us: for each *c*, what is an exponent *x* such that $c \equiv r^x \pmod{m}$. Here is an example of **writeaspowers** (it is easy to check that 2 is a primitive root modulo 19):

```
writeaspowers[2, 19]
```

If we wanted to find an *x* such that $2^x \equiv 3 \pmod{19}$, we could easily read off that $x = 13$ works.

The next function lets us compute this value a little bit more directly. To solve the congruence $r^x \equiv c \pmod{m}$, we provide the base *r*, the value *c*, and the modulus *m*. The function will return the smallest nonnegative exponent *x* which satisfies the congruence.

```
discretelog[base_, value_, modulus_] :=
 Module[{expo, try, valmod},
        valmod = Mod[value, modulus];
        expo = 0;
        try = 1;
        While[try != valmod,
            try = Mod[try * base, modulus];
            expo = expo + 1];
        expo]
```

This function is generally known as a *discrete logarithm* because its defining property is so similar to logarithm functions in calculus. We include the adjective *discrete* to distinguish it from its more familiar continuous counterparts. The function above works by trying successive values for *x*, starting with $x = 0$, until it finds one which works.

Let's try our new **discretelog** to answer the same question as above: find *x* such that $2^x \equiv 3 \pmod{19}$:

```
discretelog[2, 3, 19]
```

Good news, we found the same answer (and with less output than **writeaspowers**).

The discrete logarithm is important to cryptography (thus the title of this section). Ironically, it is important because discrete logarithms are difficult to compute quickly. By contrast, the inverse of discrete logs (computing $b^n \% m$) can be accomplished quickly using the algorithm described in Chapter 8. On average, the time required to compute $c = (b^n \% m)$ is proportional to the number of digits in *m* whereas the time to undo this and compute *n* from *b*, *c*, and *m* is proportional to the value of *m*. When *m* is large enough, this makes the exponentiation direction reasonably fast and the discrete logarithm direction impossible for all practical purposes. For this reason, modular exponentiation is called a one-way function.

The previous paragraph oversimplifies the situation a little bit. There are faster methods for computing discrete logarithms, especially if $\phi(m)$ is divisible by only small primes. But the basic principle is right: if we choose *m* so that $\phi(m)$ is divisible by a very large prime, then the computation of discrete logs is impractical.

In order to have Mathematica use the fast modular exponentiation algorithm described in Chapter 8, we use the command **PowerMod**. We can use this to quickly compute $2^{135791} \% 1234577$:

```
PowerMod[2, 135791, 1234577]
```

We can reverse this computation with our version of **discretelog**:

```
discretelog[2, 189155, 1234577]
```

Below we look at two cryptographic applications of discrete logarithms. There are others, including a public key cryptosystem named El Gamal after its inventor. These sections use some cryptographic utilities which are not part of Mathematica. The definitions for these functions are in the initialization section of this notebook.

▪ Password Files

Many computers allow many different users to login by providing a login name and a password. The passwords are typically kept in a file on the system, and they are encrypted so that users cannot discover each other's password. The problem is to keep users from spending their time trying to decrypt each other's password.

The problem is solved by using a one-way function. The system can then encrypt passwords, but no one is able to decrypt them, not even the computer! Interestingly, this is sufficient for password checking. When a user attempts to login, the computer takes whatever the user types as his or her password and runs it through the encryption process. It can then check to see if the encrypted version of whatever the user typed matches the encrypted version of the password it has stored in a file.

We are almost ready. We will set the modulus and our primitive root which will be used for an example.

```
modulus = 27449
base = 3
ord[base, modulus]
```

The last line shows that our base is a primitive root for the modulus. To encode the password "Hi", we convert it to a number, encrypt the number, and then convert the resulting number back to text.

```
password = "Hi"
passwordasnum = texttonum[password]
encryptednum = PowerMod[base, passwordasnum, modulus]
final = numtotext[encryptednum]
```

To decode this, we reverse the process. We undo the **PowerMod** by taking a discrete log. As a result, you may notice that decoding takes longer than encoding.

```
encryptednum = texttonum[final]
decryptednum = discretelog[base, encryptednum, modulus]
numtotext[decryptednum]
```

In this example, we used a small modulus so that we could compute the required discrete logarithm in our lifetimes. Because of the small modulus, we needed a very short password.

In practice, computer passwords are often cracked in a very nonmathematical way. The "mischievous user" who gets other people's passwords often does it by guessing! Sometimes, this takes a certain amount of trial and error.

■ Exercise 1

Louis Reasoner is very proud of the new password system he installed on his computer. It uses the method we described above, but with a large modulus for better security. In fact, the next group initializes Louis's modulus and base.

```
modulus = 10000000000000003
base = 2
```

Louis is so confident in his system that he brags "No one can crack my passwords. My system is so secure that I do not even use capital letters or punctuation in my passwords. I also like to change my password every day. I have one password for every day of the week!" When Louis heads off to watch his favorite movie, Disney's *Snow White*, his friends try to guess Louis's password. The encoded version of today's password is

```
encoded := "vI41~,w](";
```

What is Louis's password today?

■ Key Exchange: Creating Shared Secrets

Suppose Alice and Bob are communicating through the internet. They want to set up encryption for sending messages, so they will need a key. All communications on the internet can be observed by an outsider, so Alice cannot simply send a key to Bob.

One solution is for Alice and Bob to work out a "shared secret". The secret is just a large number which Alice and Bob know, but nobody else knows including Oscar who is watching all of the transmissions between Alice and Bob. The shared secret can then be used as the key for any of a number cryptosystems.

The Diffie-Hellman protocol was one of the first methods for creating shared secrets. The security of Diffie-Hellman depends on discrete logarithms' being difficult to compute. Here is how it works.

Alice and Bob will be sending messages back and forth, and Oscar is eavesdropping. First, Alice sends a modulus m and and base b to Bob. Typically, m will be a large prime and r will be a primitive root modulo m. Now, everyone knows these two numbers. Execute the next group to set m and r for a working example.

```
m := 27449
r := 3
fastorder[r, m]
```

The last line checks that r is a primitive root modulo m. We will compute orders in this section using **fastorder**. As its name implies, it computes multiplicative orders more quickly than does **ord**. The function **fastorder** is defined in the initialization section of this notebook; its basic algorithm is explained there as well.

Alice and Bob secretly pick random numbers, which we will call a and b respectively. Then, Alice computes $r^a \% m$ and Bob computes $r^b \% m$. These values are exchanged. Now, everyone knows r, m, $r^a \% m$, and $r^b \% m$. Only Alice and Bob know their secret random values, so we keep Mathematica from printing those values.

```
a = Random[Integer, {0, m - 1}];
b = Random[Integer, {0, m - 1}];

ra = PowerMod[r, a, m]
rb = PowerMod[r, b, m]
```

Finally, Alice takes $r^b \% m$, which she got from Bob, and uses it to compute

$$(r^b)^a \% m = r^{ab} \% m$$

(only she can do this because only she knows the value of a). Similarly, Bob computes

$$(r^a)^b \% m = r^{ab} \% m$$

using b. In the end, both Alice and Bob know $r^{ab} \% m$.

```
rab = PowerMod[ra, b, m]
rba = PowerMod[rb, a, m]
```

Of course, these values should have worked out to be the same!

Now that Alice and Bob have their key, they will use it with some cryptosystem. Normally, people use a fairly sophisticated system such as the Digital Encryption Standard, or DES for short. DES is pretty secure and was designed to run especially fast on a computer. However, DES is not very interesting from a number theoretic point of view, so we will use a simple shift cipher instead (with text blocking). The encoding and decoding functions were contained in the pile of utility functions in the initialization section of the notebook.

To encode a message, just send the message and key into the encoder:

```
message = "Did you solve exercise 1?"
encoded = encode[message, rab]
```

Similarly, we can decode a message with the function decode:

```
decode[encoded, rab]
```

Oscar could figure out this "secret" value $r^{ab} \% m$ if he could compute discrete logarithms. He would take a discrete logarithm of $r^a \% m$ to find a, and then proceed as Alice did to find $r^{ab} \% m$. Our modulus is small, so we can try that now.

```
discretelog[r, ra, m]
```

For comparison, here is the value of a we saved above:

```
a
```

No one knows of a better way for Oscar to proceed. If the modulus were bigger, Oscar would be stuck, so we assume that this exchange between Alice and Bob is secure.

We mentioned above that discrete logarithms are only difficult to compute if $\phi(m)$ is divisible by a very large prime. The next function implements an algorithm which computes discrete logarithms quickly if $\phi(m)$ is divisible by only small primes. If $\phi(m) = p_1^{a_1} p_2^{a_2} \cdots p_n^{a_n}$, then the running time of our first **discretelog** is proportional to $\phi(m) = p_1^{a_1} p_2^{a_2} \cdots p_n^{a_n}$. The running time for **fastdiscretelog** below is proportional to $\sum_{i=1}^{n} a_i p_i$. Do not try to decipher the code here. We will discuss the method in the chapter summary.

```
fastdiscretelog[base_, val_, modulus_] :=
  Module[{fim, phim = EulerPhi[modulus], j, p, d, try,
    cmp, ans = 0, cnt, k, ppart, ptry, ival, idem},
      fim = FactorInteger[phim];
      ival = PowerMod[val, -1, modulus];
      (* Loop over primes dividing phi (modulus) *)
      Do[p = fim[[j, 1]];
        d = fim[[j, 2]];
        idem = PowerMod[phim / p^d, -1, p^d];
        try = PowerMod[base, phim / p^d * idem, modulus];
        ptry = PowerMod[try, p^(d - 1), modulus];
        ppart = 0;
        (* Loop over powers of p *)
        Do[
    cmp = PowerMod[ival, phim / p^k * idem, modulus] *
      PowerMod[try, ppart * p^(d - k), modulus];
          cmp = Mod[cmp, modulus];
          cnt = 0;
          While[
    cmp != 1, cmp = Mod[cmp * ptry, modulus];
            cnt = cnt + 1];
          ans =
    Mod[ans + cnt * idem * phim / p^(d - k + 1), phim];
          ppart = ppart + cnt * p^(k - 1)
        , {k, d}]
      , {j, Length[fim]}];
    ans]
```

To see it in action, we start by carefully choosing m.

```
m = 174636001
PrimeQ[m]
FactorInteger[m - 1]
```

The second line of output verifies that our modulus is prime. The third line then shows that $\phi(m)$ has only small prime factors. After a little searching, it is not hard to discover that our modulus has a primitive root, namely 13.

```
base = 13
fastorder[base, m]
EulerPhi[m]
```

We see that the order of 13 modulo *m* is the same as $\phi(m)$. Next we raise 13 to a moderately large power:

```
bignum = PowerMod[base, 123456789, m]
```

Finally, we can undo this with our fast discrete log function:

```
fastdiscretelog[base, bignum, m]
```

Computing this value using plain old **discretelog** would take more than an hour on most computers.

■ Exercise 2

Alice and Bob cannot wait to try using the Diffie-Hellman protocol to establish encrypted communication with someone. They think that the only thing he needs to get a high level of security is a large modulus *m*. So, Alice picks the following modulus and primitive root *r*.

```
m = 15 * 2 ^ 78 + 1
PrimeQ[m]
r = 11
fastorder[r, m]
```

We can see that *m* is prime and that *r* is a primitve root modulo *m*.

We are able to observe the values exchanged by Alice and Bob.

```
alicenum := 3268170018038853053202889;
bobnum := 4168860017312113121130990;
```

This is enough to establish the key for Alice and Bob. Then Alice sends an encrypted message to Bob:

```
alicetobob := {4287436560919001394387439,
    4379666011581448391288324, 4656180681496403797643909,
    4197584748237556580656485, 4367214525356700096276646,
    3981076401286882248383284, 1154612708230347541420200,
    1098901436212055158708132, 3964323935927860666114792};
```

Decode the message.

Primitive Roots

10.1 Orderly Orders

In the last chapter, we looked at multiplicative orders of integers a modulo n by considering different values of a separately. We can learn more by considering how the order of one integer might be related to the order of another integer.

Suppose we know the order of a modulo n. Then, we have a pretty good grasp on the powers of a when computed modulo n. For example, with $a = 3$ and $n = 10$, the first 30 powers a^j reduced modulo n are

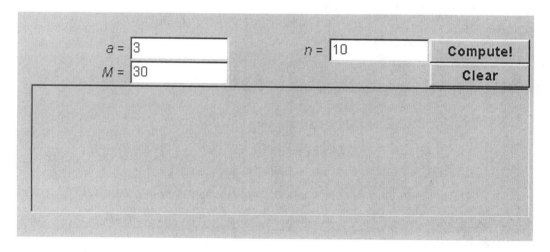

We know that this sequence will be purely periodic, and the period, which is also $\text{ord}_n(a)$, will be the first position which contains a 1. In this case, $\text{ord}_{10}(3) = 4$. Moreover, we then know that $3^i \equiv 3^j \pmod{10}$ if and only if $i \equiv j \pmod{\text{ord}_{10}(3)}$. In other words, $3^i \equiv 3^j \pmod{10}$ if and only if $i \equiv j \pmod 4$. Thus,

$$3^i \equiv 3 \pmod{10} \quad \text{if} \quad i \equiv 1 \pmod 4$$
$$3^i \equiv 9 \pmod{10} \quad \text{if} \quad i \equiv 2 \pmod 4$$
$$3^i \equiv 7 \pmod{10} \quad \text{if} \quad i \equiv 3 \pmod 4$$
$$3^i \equiv 1 \pmod{10} \quad \text{if} \quad i \equiv 0 \pmod 4.$$

This gives us lots of information about the powers of 3 modulo 10. How can we really make use of this? If we look at the list of powers 3^j mod 10 above, we see that $3^3 \equiv 7 \pmod{10}$. It then follows that

$$7^j \equiv (3^3)^j \equiv 3^{3j} \pmod{10}.$$

So, not only should the powers of 7 appear as powers of 3 (when each are reduced modulo 10), but we should be able to connect exactly which power of 7 matches with which power of 3. Let's look at the sequence of powers of 7 taken modulo 10:

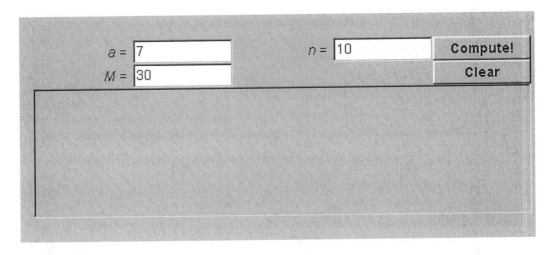

If we had wanted to predict the precise value of $7^6 \% 10$, we could have reasoned that

$$7^6 \equiv 3^{18} \equiv 9 \pmod{10}$$

because $18 \equiv 2 \pmod 4$. Looking at the results above, we can see that this is right.

We now focus on $\text{ord}_{10}(3^j)$ for different values of j. The next applet takes a value for a and for n as input. It computes $\text{ord}_n(a^j)$ for each j from 1 to ord_n

(*a*). Here it is in action when $a = 3$ and $n = 10$:

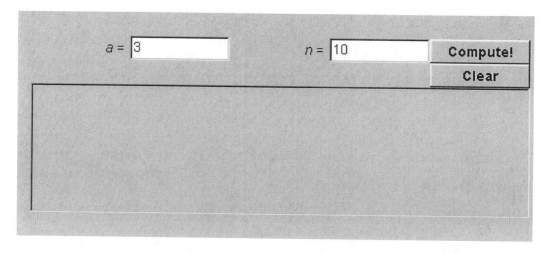

Use this applet to experiment with the next Research Question. (Some ideas for a proof are contained in the preceding discussion!)

Research Question 1

Suppose that a is relatively prime to n and $j > 0$. Find a formula for $\mathrm{ord}_n(a^j)$ in terms of $\mathrm{ord}_n(a)$ and j.

Research Question 2

Suppose that a is relatively prime to n. What are the values of $\mathrm{ord}_n(a^j)$ for $j = 0, 1, 2, \ldots$?

10.2 Orders in the Presence of a Primitive Root

The existence of a primitive root simplifies working with the elements of $\mathbf{Z}_n^*$, the congruence classes relatively prime to n. Suppose that r is a primitive root modulo n. On the one hand, we know that there are $\phi(n)$ congruence classes relatively prime to n. On the other hand, since $\mathrm{ord}_n(r) = \phi(n)$, the powers $r^1, r^2, r^3, \ldots, r^{\phi(n)}$ give $\phi(n)$ distinct congruence classes relatively prime to n. Thus, *every* integer relatively prime to n is congruent to a power of r. This is a **big deal**, so much so that we repeat it below.

> If r is a primitive root modulo n, then every integer m that is relatively prime to n is congruent to r^j for some integer j between 1 and $\phi(n)$.

Let's get some help checking this out. First, we can observe this fact by looking at the powers of r in succession. You can tell that r is a primitive root if the powers of r hit every congruence class relatively prime to n. The applet below shows each power of r reduced modulo n until it reaches a power of r which is congruent to 1. For example, here's what we get for the powers of 2 modulo 11:

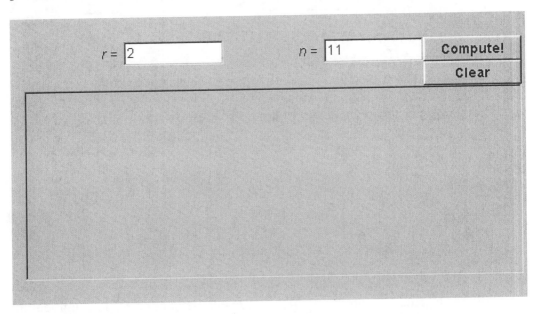

Another way to observe this behavior is to try to write every element of $\mathbf{Z}_n^*$ as a power of r where r is a primitive root modulo n. That is the purpose of the next applet. Let's look again at the case of powers of 2 modulo 11:

If you try this applet when r is not a primitive root modulo n, weird things happen:

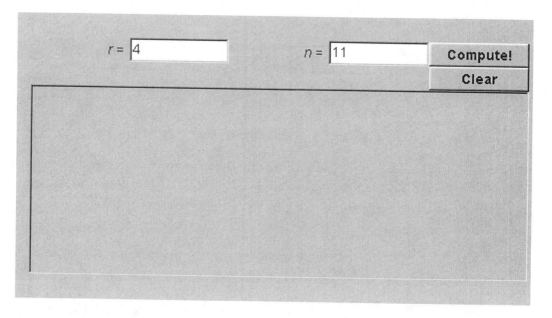

If r is a primitive root modulo n, then every element of $\mathbf{Z}_n^*$ is congruent to r^j for some j. Furthermore, your formula given in response to Research Question 1 shows how to compute $\mathrm{ord}_n(a^j)$ from $\mathrm{ord}_n(a)$ and j. Thus you have a formula for the order of every element of $\mathbf{Z}_n^*$. Pretty cool, eh? Keep your formula in mind as you tackle the next two Research Questions. As you work on these questions, you will want to collect data with the assistance of the applets defined above. To save you the trouble of hunting around for integers that have a primitive root, we provide below a list of all such integers $n \leq 30$:

$n = 2, 3, 4, 5, 6, 7, 9, 10, 11, 13, 14, 17, 18, 19, 22, 23, 25, 26, 27, 29.$

Research Question 3

Suppose that d is the order of some element of $\mathbf{Z}_n^*$, and that r is a primitive root of n.

(a) What values of j satisfy $1 \leq j \leq \phi(n)$ and $\mathrm{ord}_n(r^j) = d$?
(b) For what values of j is r^j a primitive root modulo n?

Research Question 4

Suppose that d is the order of some element of $\mathbf{Z}_n^*$, and that r is a primitive root of n.

(a) How many values of j satisfy $1 \leq j \leq \phi(n)$ and $\mathrm{ord}_n(r^j) = d$?
(b) How many primitive roots modulo n are there?

10.3   When Is There a Primitive Root?

The Research Questions in the previous section assumed that an integer n had a primitive root. In this section, we shall consider the problem of determining which integers n have primitive roots, and which do not. A good place to start your investigation is with an applet from a previous chapter. It displays each element a that is relatively prime to the input n, along with the value of $\mathrm{ord}_n(a)$. Here is a sample of its output:

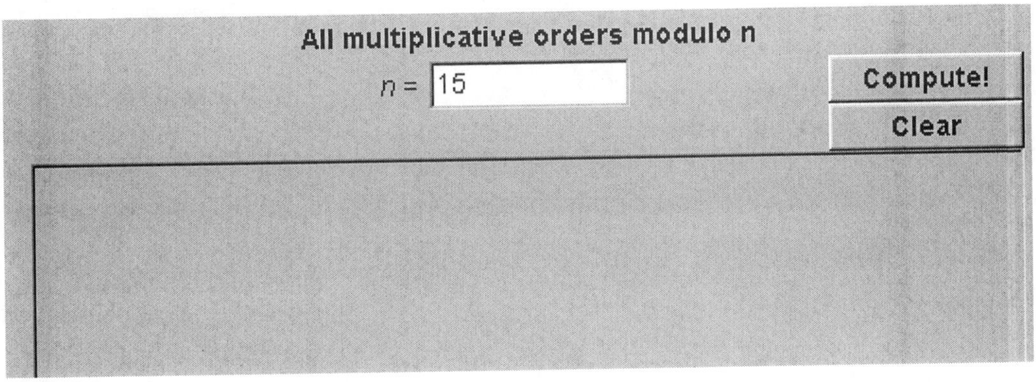

All multiplicative orders modulo n

$n =$ `15`   Compute!

Clear

Try it out, and watch for values of n for which there is an element of order $\phi(n)$. To make this task easier, you can use the following enhanced version of the previous applet, which includes a computation of $\phi(n)$ in the output:

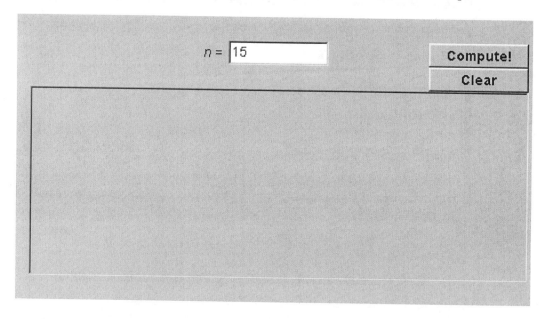

As you can see, the output confirms what you discovered when working the Prelab exercises: there are no primitive roots modulo 15. The positive integers $n \leq 30$ that have primitive roots were given in the previous section; here they are again:

$$n = 2, 3, 4, 5, 6, 7, 9, 10, 11, 13, 14, 17, 18, 19, 22, 23, 25, 26, 27, 29.$$

Use the functions defined above to determine which other integers n have primitive roots until you have enough data to make a conjecture for the final Research Question of this chapter:

Research Question 5

Which positive integers *n* have primitive roots?

Note: The proof of this conjecture is quite difficult. It would be truly remarkable for a student to find a proof on his or her own. Unless told otherwise by your instructor, you should concentrate on making a good conjecture here and supporting it in your lab report with numerical evidence.

10.4 Going Farther: Cryptography Using the "Lumberjack's Rock and Roll" Discretely

An applet in Section 10.2, which wrote every element of $\mathbf{Z}_n^*$ as a power of a primitive root, has an application to cryptography. The output gives all of the values of a certain function. If *r* is a primitive root modulo *m*, then the output of the applet tells us: for each *c*, what is an exponent *x* such that $c \equiv r^x \pmod{m}$. Here is an example (it is easy to check that 2 is a primitive root modulo 19):

If we wanted to find an x such that $2^x \equiv 3 \pmod{19}$, we could easily read off that $x = 13$ works.

The next applet lets us compute this value a little bit more directly. For the congruence $r^x \equiv c \pmod{m}$, the applet will return the smallest nonnegative exponent x which satisfies the congruence.

This function is generally known as a *discrete logarithm* because its defining property is so similar to logarithm functions in calculus. We include the adjective *discrete* to distinguish it from its more familiar continuous counterparts. The applet above works by trying successive values for x, starting with $x = 0$, until it finds one which works.

Let's try our new discrete log applet to find the same value as above: find x such that $2^x \equiv 3 \pmod{19}$:

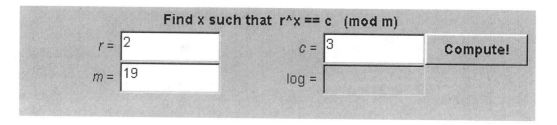

Good news, we found the same answer (and with less output than the applet above).

The discrete logarithm is important to cryptography (thus the title of this section). Ironically, it is important because discrete logarithms are difficult to compute quickly. By contrast, the inverse of discrete logs (computing $b^n \% m$) can be accomplished quickly using the algorithm described in Chapter 8. On average, the time required to compute $c = (b^n \% m)$ is proportional to the number of digits in m whereas the time to undo this and compute n from b, c, and m is proportional to the value of m. When m is large enough, this makes the exponentiation direction reasonably fast and the discrete logarithm direction impossible for all practical purposes. For this reason, modular exponentiation is called a one-way function.

The previous paragraph oversimplifies the situation a little bit. There are faster methods for computing discrete logarithms, especially if $\phi(m)$ is divisible by only small primes. But the basic principle is right: if we choose m so that $\phi(m)$ is divisible by a very large prime, then the computation of discrete logs is impractical.

We have seen an applet before which uses the fast modular exponentiation algorithm described in Chapter 8. We can use this to quickly compute 2^{135791} mod 1234577:

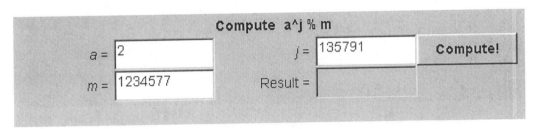

We can reverse this computation with our applet for computing discrete logs:

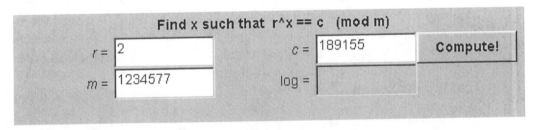

Below we look at two cryptographic applications of discrete logarithms. There are others, including a public key cryptosystem named El Gamal after its inventor.

10.4.1 Password Files

Many computers allow many different users to login by providing a login name and a password. The passwords are typically kept in a file on the system, and they are encrypted so that users cannot discover each other's password. The problem is to keep users from spending their time trying to decrypt each other's password.

The problem is solved by using a one-way function. The system can then encrypt passwords, but no one is able to decrypt them, not even the computer! Interestingly, this is sufficient for password checking. When a user attempts to login, the computer takes whatever the user types as his or her password and runs it through the encryption process. It can then check to see if the encrypted version of whatever the user typed matches the encrypted version of the password it has stored in a file.

We are almost ready. We will use $m = 27449$ as our modulus and $r = 3$ as our primitive root for a running example. Running the next applet will compute the order of 3 modulo 27449, which verifies that 3 is in fact a primitive root.

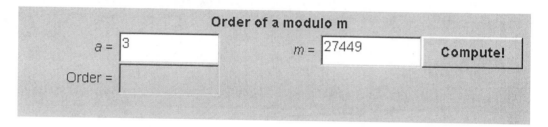

To encode the password "Hi", we convert it to a number, encrypt the number, and then convert the resulting number back to text. The next applet lets us do this one step at a time. Click on the buttons in succession starting with the first one (Text->Num). Note that the Text->Num button will convert the text to a single number using text blocking (where the entire text is treated as one block). Similarly the Num->Text button will convert a number into a single block of text.

To decode this, we reverse the process. We undo the exponentiation step by taking a discrete log. As a result, you may notice that the discrete log step of decoding takes longer than encoding. As above, you have to click on the sequence of buttons to take the text through the sequence of steps for decoding.

In this example, we used a small modulus so that we could compute the required discrete logarithm in our lifetimes. Because of the small modulus, we needed a very short password.

In practice, computer passwords are often cracked in a very nonmathematical way. The "mischievous user" who gets other people's passwords often does it by guessing! Sometimes, this takes a certain amount of trial and error.

Exercise 1

Louis Reasoner is very proud of the new password system he installed on his computer. It uses the method we described above, but with a large modulus for better security. In fact, Louis's modulus is $10^{17} + 3$ and his base is 2.

Louis is so confident in his system that he brags "No one can crack my passwords. My system is so secure that I do not even use capital letters or punctuation in my passwords. I also like to change my password every day. I have one password for every day of the week!" When Louis heads off to watch his favorite movie, Disney's *Snow White*, his friends try to guess Louis's password. The encoded version of today's password is "vI41~,w](" (not including the quotation marks).

Use (one of) the applets above to determine what Louis's password is today.

10.4.2 Key Exchange: Creating Shared Secrets

Suppose Alice and Bob are communicating through the internet. They want to set up encryption for sending messages, so they will need a key. All communications on the internet can be observed by an outsider, so Alice cannot simply send a key to Bob.

One solution is for Alice and Bob to work out a "shared secret". The secret is just a large number which Alice and Bob know, but nobody else knows including Oscar who is watching all of the transmissions between Alice and Bob. The shared secret can then be used as the key for any of a number cryptosystems.

The Diffie-Hellman protocol was one of the first methods for creating shared secrets. The security of Diffie-Hellman depends on discrete logarithms' being difficult to compute. Here is how it works.

Alice and Bob will be sending messages back and forth, and Oscar is eavesdropping. First, Alice sends a modulus m and and base r to Bob. Typically, m will be a large prime and r will be a primitive root modulo m. Now, everyone knows these two numbers. Here we will set

$$m = 27449 \quad \text{and} \quad r = 3$$

for a working example.

We check that r is a primitive root modulo m in the next applet.

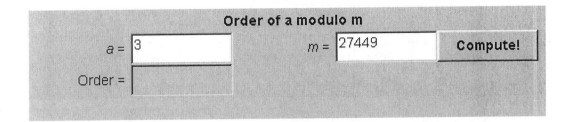

Alice and Bob secretly pick random numbers, which we will call a and b respectively. Then, Alice computes r^a % m and Bob computes r^b % m. These values are exchanged. Now, everyone knows r, m, r^a % m, and r^b % m. Only Alice and Bob know their secret random values. To simulate how this looks to Oscar, we have picked values for a and b for this example, but we are not telling them to you. We will say that

$$r^a \text{ \% } m = 955 \quad \text{and} \quad r^b \text{ \% } m = 11859.$$

Finally, Alice takes r^b % m, which she got from Bob, and uses it to compute

$$(r^b)^a \text{ \% } m = r^{ab} \text{ \% } m$$

(only she can do this because only she knows the value of a). Similarly, Bob computes

$$(r^a)^b \text{ \% } m = r^{ab} \text{ \% } m$$

using b. In the end, both Alice and Bob know r^{ab} % m. In our example, this turns out to be 8081.

Now that Alice and Bob have their key, they will use it with some cryptosystem. Normally, people use a fairly sophisticated system such as the Digital Encryption Standard, or DES for short. DES is pretty secure and was designed to run especially fast on a computer. However, DES is not very interesting from a number theoretic point of view, so we will use a simple shift cipher instead (with text blocking).

To encode a message, just run the message and key through the following applet:

Shift Cipher

Did you solve exercise 1?

| Encode | Key: | 8081 |

Similarly, we can decode a message with the a corresponding decoding applet:

Shift Cipher

6027, 8149, 6650, 8166, 6644, 7302, 8150, 7485, 6915, 6058, 5694,671, 8112

| Decode | Key: | 8081 |

Oscar could figure out this "secret" value $r^{ab} \% m$ if he could compute discrete logarithms. He would take a discrete logarithm of $r^a \% m$ to find a, and then proceed as Alice did to find $r^{ab} \% m$. Our modulus is small, so we can try that now.

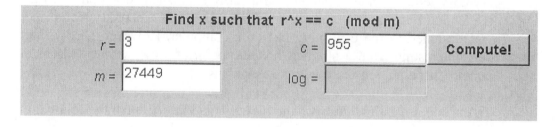

Find x such that r^x == c (mod m)

$r =$ 3 $c =$ 955   **Compute!**

$m =$ 27449   log =

For comparison, the value of a we used above was 11478.

No one knows of a better way for Oscar to proceed. If the modulus were bigger, Oscar would be stuck, so we assume that this exchange between Alice and Bob is secure.

We mentioned above that discrete logarithms are only difficult to compute if $\phi(m)$ is divisible by a very large prime. The next applet implements an algorithm which computes discrete logarithms quickly if $\phi(m)$ is divisible by only small primes. If $\phi(m) = p_1^{a_1} p_2^{a_2} \cdots p_n^{a_n}$, then the running time of our first discrete log applet is proportional to $\phi(m) = p_1^{a_1} p_2^{a_2} \cdots p_n^{a_n}$. The running time for the faster discrete log applet below is proportional to $\sum_{i=1}^{n} a_i p_i$. We will discuss the method used by this faster applet in the chapter summary.

To get set up, we start by carefully choosing m. Here we will use $m = 174636001$. First, we check that m is prime.

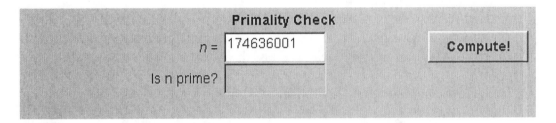

Next we check that $\phi(m) = m-1$ has only small prime factors.

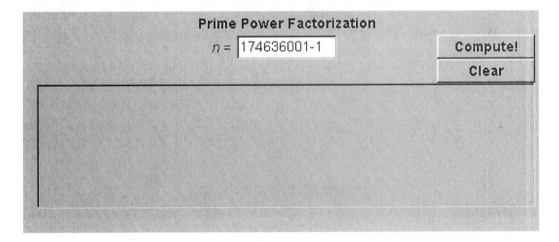

After a little searching, it is not hard to discover that our modulus has a primitive root, namely 13. We check that here by computing ord$_m$(13):

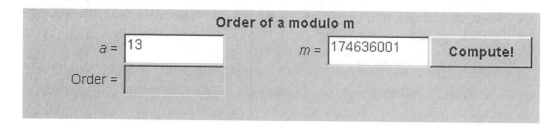

We see that the order of 13 modulo m is the same as $\phi(m) = m-1$. Next we raise 13 to a moderately large power:

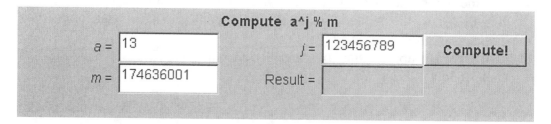

Finally, we can undo this with our fast discrete log applet:

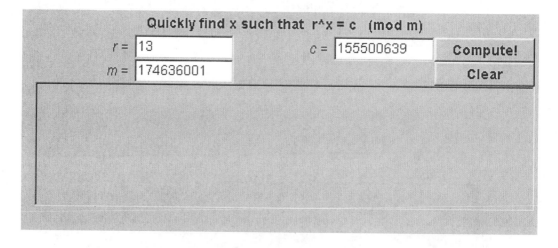

Computing this value using our original discrete log applet would take more than an hour on most computers.

Exercise 2

Alice and Bob cannot wait to try using the Diffie-Hellman protocol to establish encrypted communication with someone. They think that the only thing he needs to get a high level of security is a large modulus m. So, Alice picks the following modulus m and primitive root r.

$$m = 15 * 2^{78} + 1 \qquad \text{and} \qquad r = 11.$$

We can easily check that m is prime (see below).

We are able to observe the values exchanged by Alice and Bob.

Alice sends: 3268170018038853053202889

Bob sends: 4168860017312113121130990

This is enough to establish the key for Alice and Bob. Then Alice sends an encrypted message to Bob:
4287436560919001394387433,
4379666011581448391288343,
4656180681496403797643992,
4197584748237556580656485,
4367214525356700096276646,
3981076401286882248383284,
1154612708230347541420204,
1098901436212055158708139,
3964323935927860666114792

Decode the message. Note, since some of the numbers are very large in this problem, it may take the Java applets a few minutes to do some of the computations. Also, input and output areas may not look large enough to hold some of the numbers. They will hold the numbers, but you will not be able to see the whole number at once. By clicking in the area, you can use the arrow keys to move left and right within the number. Finally, you may want to use your mouse to copy-and-paste numbers from one place to another (if that is possible with your browser).

Primality Check

$n =$ `15*2^78+1`

Is n prime?

Compute!

Shift Cipher

```
237556580656485, 4367214525356700096276646, 3981076401286882248383284,
2303475414202004, 10989014362120551587081329, 3964323935927860666114792
```

Decode   Key:

Homework

1. Suppose d and n are integers greater than 1 such that $d \mid n$. If a is an integer relatively prime to n, show that $\text{ord}_d(a) \mid \text{ord}_n(a)$.

2. Suppose that n is odd and a is a primitive root modulo n.

   (a) Show that there exists an integer b such that $a \equiv b \pmod{n}$ and $\gcd(b, 2n) = 1$.

   (b) Show that b is a primitive root modulo $2n$. (Hint: Use exercise 1 to relate $\text{ord}_{2n}(b)$ and $\text{ord}_n(b)$.)

3. Let d and n be integers greater than 1 such that $d \mid n$. Suppose that a is a primitive root modulo n. Prove that a is a primitive root modulo d.

4. Use exercise 3 to prove that if $n = 8k$ where $k \geq 1$, then there is no primitive root modulo n. (Hint: What would a primitive root modulo n imply about primitive roots modulo 8?)

5. Assume r and s are relatively prime positive integers and that $n = rs$. Let $m = \text{lcm}(\phi(r), \phi(s))$, and assume that $\gcd(a, n) = 1$. Prove:

   (a) $a^m \equiv 1 \pmod{r}$ and $a^m \equiv 1 \pmod{s}$.

   (b) $a^m \equiv 1 \pmod{n}$. (Hint: Use part (a) and CRT.)

6. Suppose $n = 2^k b$, where $k \geq 2$ and b is odd. Let $\gcd(a, n) = 1$. Use exercise 5b to show that $\text{ord}_n(a) < \phi(n)$.

7. Suppose that $n = n_1 p^u q^v$, where p and q are distinct odd primes, $u \geq 1$, $v \geq 1$, and $\gcd(n_1, pq) = 1$. Use exercise 5b to show that $\text{ord}_n(a) < \phi(n)$.

8. Determine the number of primitive roots modulo n for each given value of n.

   (a) $n = 29$ (c) $n = 27$
   (b) $n = 71$ (d) $n = 625$

9. For each value of n given below, find the smallest positive primitive root modulo n, or say that no such primitive root exists.

   (a) $n = 7$ (c) $n = 9$ (e) $n = 11$
   (b) $n = 8$ (d) $n = 10$ (f) $n = 12$

10. For each value of n given below, the given value of r is a primitive root modulo n. In each part, find all of the primitive roots modulo n.

    (a) $n = 13$, $r = 2$ (c) $n = 19$, $r = 2$
    (b) $n = 17$, $r = 3$ (d) $n = 23$, $r = 5$

519

11. Given that 2 is a primitive root modulo 19, find an integer modulo 19 with each of the following orders:

    (a) Order 2   (c) Order 6
    (b) Order 3   (d) Order 9

12. In this exercise we derive an alternate proof of

    Wilson's Theorem If p is an odd prime, then $(p-1)! \equiv -1 \pmod{p}$.

    (a) Show that if a is a primitive root modulo p, then $(p-1)! \equiv a^{p(p-1)/2} \pmod{p}$.

    (b) Show that $x = a^{p(p-1)/2}$ is a solution to the equation $x^2 \equiv 1 \pmod{p}$.

    (c) Show that $a^{p(p-1)/2} \not\equiv 1 \pmod{p}$.

    (d) Conclude that $a^{p(p-1)/2} \equiv -1 \pmod{p}$, so that $(p-1)! \equiv -1 \pmod{p}$.

13. Prove that if p is a prime such that $p \equiv 1 \pmod 4$, then there exists an integer a such that $a^2 \equiv -1 \pmod{p}$.

14. Prove that if p is a prime such that $p \equiv 3 \pmod 3$, r is a primative root modulo p, and $a \equiv r^{(p-1)/3}$, then 1, a, and a^2 give solutions to $x^3 \equiv 1 \pmod{p}$ which are distinct modulo p.

The next few exercises prove results on Carmichael numbers. Recall that n is a Carmichael number if n is composite and for all $a \in \mathbf{Z}$ with $\gcd(a, n) = 1$ we have that $a^{n-1} \equiv 1 \pmod{n}$.

15. Prove that if n is a Carmichael number, then for every $d \mid n$ and every $a \in \mathbf{Z}$ such that $\gcd(a, d) = 1$ we have that $\operatorname{ord}_d(a) \mid (n-1)$.

16. Let $n = p_1^{e_1} p_2^{e_2} \cdots p_k^{e_k}$ be the prime-power factorization of a composite integer n. Suppose that for each i and every $a \in \mathbf{Z}$ such that $\gcd(a, p_i) = 1$ we have that $\operatorname{ord}_{p_i^{e_i}}(a) \mid (n-1)$. Prove that if n is a Carmichael number.

17. Let $n = p_1^{e_1} p_2^{e_2} \cdots p_k^{e_k}$ be the prime power factorization of a composite integer n. Prove that n is a Carmichael number if and only if for each i, $p_i^{e_i-1}(p_i-1) \mid n-1$.

18. Prove that if p is a prime such that $p^2 \mid n$, then n is not a Carmichael number.

19. Prove that a composite integer n is a Carmichael number if and only if n is a product of distinct primes and for each prime $p \mid n$ we have that $p - 1 \mid n - 1$.

20. Prove that every Carmichael number is odd.

21. Prove that if p and q are distinct primes, then pq is not a Carmichael number. (Hint: If $p < q$, what can you determine about $(pq - 1) \% (q - 1)$?)

22. Prove that if k is a positive integer such that $6k + 1$, $12k + 1$, and $18k + 1$ are all prime, then $(6k + 1)(12k + 1)(18k + 1)$ is a Carmichael number.

11. Quadratic Congruences

Prelab

In Chapter 5, we completely determined all solutions to the linear equation

$$ax + b \equiv 0 \pmod{m}.$$

In this chapter, we shall consider the problem of finding solutions to the quadratic equation

$$ax^2 + bx + c \equiv 0 \pmod{m}.$$

Our first step towards solving this problem will be to study the solutions to the less general equation

$$x^2 \equiv a \pmod{m}.$$

Note that a solution to this equation can be regarded as a square root of a. After all, when working with real numbers,

$$b = \sqrt{a} \Rightarrow b^2 = a.$$

That is, b is a solution to the equation $x^2 = a$. If we restrict to integers, then some, such as 1, 4, 9, and 16, are perfect squares, while others are not. The same is true when considering integers modulo m. In the modular case, we shall confine our attention to those equations

$$x^2 \equiv a \pmod{m}$$

such that $\gcd(a, m) = 1$.

Definition The *quadratic residues* modulo m are the integers a relatively prime to m which are congruent to a perfect square modulo m. An integer relatively prime to m which is not a quadratic residue is called a *quadratic nonresidue*.

Thus, the quadratic residues and nonresidues are distinguished by whether or not the equation $x^2 \equiv a \pmod{m}$ has a solution.

How do we determine if a specific integer is a quadratic residue or nonresidue modulo m? One way is by good old-fashioned brute force. We simply take each integer modulo m, compute the square, and check if our integer is among the list of squares. For example, to determine if 6 is a quadratic residue modulo 13, we would compute

$$1^2 \equiv 1 \qquad 2^2 \equiv 4 \qquad 3^2 \equiv 9 \qquad 4^2 \equiv 3$$
$$5^2 \equiv 12 \qquad 6^2 \equiv 10 \qquad 7^2 \equiv 10 \qquad 8^2 \equiv 12$$
$$9^2 \equiv 3 \qquad 10^2 \equiv 9 \qquad 11^2 \equiv 4 \qquad 12^2 \equiv 1$$

with all congruences being modulo 13. We see from the above results that 6 is a quadratic nonresidue modulo 13. In fact, we have learned much more. Modulo 13, we have

1, 3, 4, 9, 10, and 12 are the quadratic residues;
2, 5, 6, 7, 8, and 11 are the quadratic nonresidues.

Later in this chapter, we'll learn a more efficient method for determining if an integer a is a quadratic residue or nonresidue modulo m. For now, use the method illustrated above to answer the first four exercises.

1. **Find the quadratic residues modulo 3.**

2. **Find the quadratic residues modulo 5.**

3. **Find the quadratic residues modulo 7.**

4. **Find the quadratic residues modulo 11.**

Based on your results for the above problems, you may be ready to start forming a conjecture that addresses the following question: if p is prime, how many quadratic residues are there modulo p? We will explore this question further in the lab.

We next introduce a piece of useful notation.

Notation Let p be a prime. The *Legendre symbol* for an integer a is defined as follows:

$$\left(\frac{a}{p}\right) = \begin{cases} 1 & \text{if } a \text{ is a quadratic residue mod } p, \\ -1 & \text{if } a \text{ is a quadratic nonresidue mod } p, \\ 0 & \text{if } (a, p) > 1. \end{cases}$$

Thus, the Legendre symbol provides a convenient way to express whether or not a is a quadratic residue mod p.

5. **Do the three cases in the definition of the Legendre symbol cover all possibilities? Do they overlap? Explain.**

6. **Suppose that p is prime, and that $\gcd(a, p) = \gcd(b, p) = 1$. Prove:**

(a) $\left(\dfrac{1}{p}\right) = 1.$

(b) $\left(\dfrac{a^2}{p}\right) = 1.$

(c) $\left(\dfrac{a^2 b}{p}\right) = \left(\dfrac{b}{p}\right).$

(d) If $a \equiv b \pmod{p}$, **then** $\left(\dfrac{a}{p}\right) = \left(\dfrac{b}{p}\right).$

Your calculations for earlier questions, together with the identies proved in the previous exercise, can be used to assist in completing the remaining problems.

7. **Find** $\left(\dfrac{1}{3}\right)$, $\left(\dfrac{2}{3}\right)$, $\left(\dfrac{3}{3}\right)$, $\left(\dfrac{4}{3}\right)$, **and** $\left(\dfrac{5}{3}\right)$.

8. **Find** $\left(\dfrac{2}{7}\right)$, $\left(\dfrac{9}{7}\right)$, **and** $\left(\dfrac{-5}{7}\right)$.

9. **Find** $\left(\dfrac{110}{11}\right)$, $\left(\dfrac{111}{11}\right)$, **and** $\left(\dfrac{112}{11}\right)$.

Maple electronic notebook `11-quadratic.mws`

Mathematica electronic notebook `11-quadratic.nb`

Web electronic notebook Start with the web page `index.html`

Maple Lab: Quadratic Congruences

■ Definitions

In the Prelab section of this chapter, we introduced the notions of quadratic residue and quadratic nonresidue. We repeat the definitions here:

The quadratic residues modulo m are the integers relatively prime to m which are congruent to squares modulo m.

The integers relatively prime to m which are not congruent to a square modulo m are the quadratic nonresidues.

■ Counting Quadratic Residues

It is easy to have Maple produce a list of quadratic residues by automating the method used in the Prelab section. Specifically, we compute $a^2 \% n$ for each value of a satisfying $\gcd(a, n) = 1$ and $1 \le a \le n$. The resulting list contains the quadratic residues. Here's the code to carry out the dirty work:

```
> qr := proc(n)
   local answer, j;
   answer := NULL;
   for j from 1 to n do
    if gcd(j,n)=1 then
     answer := answer,modp(j^2,n);
    fi;
   od;
  # we sort and remove duplicates
   sort(convert({answer},list));
   end:
```

Try it out:

```
> qr(11);
```

If you want to count the number of quadratic residues modulo a prime, you can apply the `nops` command to the output of `qr`:

```
> qrlist := qr(17);
  nops(qrlist);
>
```

■ Research Question 1

If p is an odd prime, how many quadratic residues are there mod p?

Hint: The content of the next few sections may be helpful in proving your conjecture.

■ Counting Square Roots

A question that is related to Research Question 1 is the following:

526

How many square roots mod p can an integer have?

We say that m is a square root of a modulo p if $a \equiv m^2 \pmod{p}$. The function `modsqrt(a,p)` returns a list of all square roots of a mod p.

```
> modsqrt := proc(a,p)
    local answer, j;
    answer := NULL;
    for j from 1 to p do
     if modp(j^2,p) = modp(a,p) then
      answer := answer,j;
     fi;
    od;
    [answer];
  end:
```

For example, here are the square roots of 3 modulo 11:

```
[ > modsqrt(3,11);
```

And here's a check:

```
[ > modp(5^2,11);
[ > modp(6^2,11);
[ >
```

◼ Research Question 2

Let p be an odd prime and a an integer that is not divisible by p. How many square roots can a have modulo p?

Once you have completed Research Question 2, you may find that you have enough information to prove your conjecture for Research Question 1. Go back, and give it a try. If you have success, great! If not, don't despair. Go on to the next section, and return to Research Question 1 later.

◼ Quadratic Residues and Primitive Roots

Recall that if we fix a prime p, then a *primitive root* modulo p is an integer r such that every element of $\mathbf{Z}_p^*$ (the residue classes prime to p) is congruent to a power of r. As you may have discovered in the previous chapter, every prime has a primitive root. Thus, if r is a primitive root modulo p, then every nonzero residue class a modulo p can be written as

$$a \equiv r^j \pmod{p}$$

for a unique value of j between 0 and $p - 1$. In this section, we shall determine which values of j produce quadratic residues, and which values produce quadratic nonresidues.

To aid you in your explorations, several functions are provided that can be used in conjunction with `qr(p)` defined earlier. The function `primitive(p)` will return the smallest primitive root mod p. Do not be concerned if Maple warns you about a new definition for order when you execute the next execution group.

```
[ > # The numbtheory package defines the order function and the
```

```
legendre symbol.

with(numtheory):

# Find an element mod p of order p-1
primitive := proc(p)
 local try;
 try := 2;
 while order(try,p) <> p-1 do
  try := try + 1;
 od;
 try;
end:
```

For example, here's the smallest primitive root mod 23:

```
[ > primitive(23);
[ >
```

The function primpowers (p) takes a prime p as input, and provides the following output: a list of the quadratic residues modulo p; the smallest primitive root r modulo p; and a table of the values of r, r^2, r^3, . . . , $r^{(p-1)}$, each reduced modulo p. Here's the definition of the function:

```
[ > primpowers := proc(p)
   local prim, j;
   prim := primitive(p);
   printf('The quadratic residues are %a', qr(p));
   printf('\nA primitive root: r = %d\n\n',prim);
   for j from 1 to p-1 do
    printf('r^%-2d ', j);
   od;
   printf('\n');
   for j from 1 to p-1 do
    printf(' %-3d ', modp(Power(prim,j),p));
   od;
  end:
```

Go ahead -- try it out. (You know you want to!)

```
[ > primpowers(17);
[ >
```

Note that the table of values r, r^2, r^3, . . . , may be too wide to fit in the window, in which case Maple will wrap the lines at the right edge. You should be able to use data collected from primpowers to tackle the next Research Question.

■ Research Question 3

Let p be a prime and r a primitive root mod p. Characterize the exponents j such that r^j is a quadratic residue mod p.

Once you have completed Research Question 3, you should be able to finish off Research Question 1. If you haven't already done so, go back and complete your proof.

We close out this section with an exercise that gives a result that is useful, in certain situations, for determining whether an integer is a quadratic residue modulo a prime. The result is known as *Euler's Criterion*, and is stated in the exercise below.

Exercise 1

Prove the following:

Euler's Criterion: Suppose that p is an odd prime and that a is an integer not divisible by p. If a is a quadratic residue modulo p, then

$$a^{\left(\frac{p-1}{2}\right)} \equiv 1 \pmod{p}$$

and if a is a quadratic nonresidue modulo p, then

$$a^{\left(\frac{p-1}{2}\right)} \equiv -1 \pmod{p}.$$

Quadratic Reciprocity

Basics

We will be interested in comparing when p is congruent to a square mod q with whether q is a square mod p, where p and q are odd primes. In other words, we will be comparing the Legendre symbols $\left(\dfrac{p}{q}\right)$ and $\left(\dfrac{q}{p}\right)$. Maple computes Legendre symbols with the function `legendre`. To compare them, we will use the function `seeboth(p,q)`.

```
> seeboth := (p,q)->
    [legendre(p,q),legendre(q,p)]:
```

Let's see how it works:

```
> seeboth(3,5);
>
```

The resulting output tells us that 3 is not congruent to a square mod 5, and 5 is not congruent to a square mod 3. Try different combinations of odd primes with `seeboth`.

Research Question 4

Let p and q be odd primes. Conjecture a relationship between $\left(\dfrac{p}{q}\right)$ and $\left(\dfrac{q}{p}\right)$. Specifically, find conditions for p and q that will determine when $\left(\dfrac{p}{q}\right) = \left(\dfrac{q}{p}\right)$ and when $\left(\dfrac{p}{q}\right) = -\left(\dfrac{q}{p}\right)$.

Hint: This result, which is known as the Quadratic Reciprocity Theorem, is a tough one. You may find the suggestions given below helpful.

Tabulating Data

To make it easier to spot patterns, we will keep one prime fixed (say p) and use lots of primes q. The following function allows us to fix one of the primes, and compare it with lots of other odd primes (

howmany of them, to be precise). The output shows one row for each prime q. After displaying q, it shows the output from seeboth (for the fixed value of p and the q for that line).

```
> seelots := proc(p, howmany)
   local q, j;
   q := 2;
   for j from 1 to howmany do
    q := nextprime(q);
    printf('q =%3d  %3d %3d\n', q, legendre(p,q), legendre(q,p));
   od;
  end:
```

If we keep the prime $p = 3$ fixed and consider 20 primes q, we compute

```
> seelots(3,20);
>
```

◼ Additional Tips

Try different values of p with the function showlots. The relationship between $\left(\dfrac{p}{q}\right)$ and $\left(\dfrac{q}{p}\right)$ is subtle and you may not get it all in one shot. Try to find some values of p for which the relationship is simple. Once you can find some primes p for which you are comfortable conjecturing the relationship between $\left(\dfrac{p}{q}\right)$ and $\left(\dfrac{q}{p}\right)$ for varying q, try to pin down exactly which primes p are easy (i.e., go from a list of "easy values of p" to a conjecture about easy primes).

While it would be wonderful if you are able to prove your conjectures in this section, proofs are somewhat hard to come by. Ironically, there may be more proofs of quadratic reciprocity than of any other theorem in the course. However, they are all significantly more difficult that the theorems we have covered thus far.

◼ What's Quadratic Reciprocity Good For?

A reasonable question. One use is to efficiently compute values of $\left(\dfrac{p}{q}\right)$. For example, suppose that you wish to evaluate

$$\left(\frac{3}{1000003}\right)$$

Determining if the equation $x^2 \equiv 3 \pmod{1000003}$ has solutions is a tall order. However, the Quadratic Reciprocity Theorem will tell us that

$$\left(\frac{3}{1000003}\right) = -\left(\frac{1000003}{3}\right)$$

and since $1000003 \equiv 1 \pmod 3$, we see instantly that

$$\left(\frac{1000003}{3}\right) = \left(\frac{1}{1000003}\right) = 1.$$

Putting this together, we see that $\left(\dfrac{3}{1000003}\right) = -1$, and so $x^2 \equiv 3 \pmod{1000003}$ has no solutions.

Thus by applying the Quadratic Reciprocity Theorem, we will be able to quickly simplify a complicated calculation. We will return to applications of the Quadratic Reciprocity Theorem in the chapter summary.

Special Examples of Legendre Symbols

In practice, there are two cases of Legendre symbols which can be computed easily, but are not covered by the previous section. In both cases, they involve $\left(\dfrac{a}{p}\right)$ with a fixed, and the prime p varying. Let's look back at an earlier example of Legendre symbols of this type.

The value of $\left(\dfrac{5}{p}\right)$ depends only on the congruence class of p modulo 5. This is somewhat surprising! In fact, it may not be clear yet that this is the case. Let's compute some Legendre symbols to check it out. First, we have a function to display the value of $\left(\dfrac{5}{p}\right)$ together with the remainder of p modulo n, for a value of n of your choosing.

```
> checkem := proc(a, n, howmany)
    local i, p;
    for i from 2 to howmany+1 do
     p := ithprime(i);
     printf('%2d %% %d = %2d, LS(%d, %2d) = %2d\n', p, n, modp(p, n),
    a, p, legendre(a,p));
    od;
  end:
```

The function checkem(a, n, howmany) will display a table, with each line providing the value of a % n together with the value of $\left(\dfrac{a}{p_i}\right)$ where p_i is the ith prime. The value of i runs from 2 (thus we start with $p_2 = 3$) and ends at howmany+1. In the output, we write LS for Legendre symbol. In the example below, we set $a = 5$ and $n = 3$.

```
> checkem(5,3,20);
```

Examining the output, we don't see any obvious patterns. Let's try setting $n = 5$:

```
> checkem(5,5,20);
>
```

This time a pattern does emerge. Upon close inspection, we see that $\left(\dfrac{5}{p}\right) = 1$ when p is congruent to 1 or 4 modulo 5, and $\left(\dfrac{5}{p}\right) = -1$ when p is congruent to 2 or 3 modulo 5.

In the next two subsections, we examine two cases similar to the above illustration. For each, your task is to find a pattern similar to the one we spotted for $a = 5$.

Computing $\left(\dfrac{-1}{p}\right)$

Let p be an odd prime and we want to know when -1 is congruent to a square mod p. As above, use the function `checkem` with different values of n to look for patterns. Here is what we get if $n = 2$.

```
[ > checkem(-1, 2, 10);
[ >
```

That may have not been too helpful! Try other values of n until you find a pattern.

■ Research Question 5

For which odd primes p is $\left(\dfrac{-1}{p}\right) = 1$?

Hint: Once you have a good conjecture, consider Euler's Criterion when searching for a proof.

■ Computing $\left(\dfrac{2}{p}\right)$

Same song, second verse. Use the same approach as above to tackle Research Question 6. By now you should be an expert at using `checkem`, so we leave all of the entries for you to fill in.

```
[ > checkem(?, ?, ?);
[ >
```

■ Research Question 6

For which odd primes p is $\left(\dfrac{2}{p}\right) = 1$?

Note: We are primarily looking for a good conjecture for this problem. A proof is fairly difficult.

Mathematica Lab: Quadratic Congruences

■ Definitions

In the Prelab section of this chapter, we introduced the notions of quadratic residue and quadratic nonresidue. We repeat the definitions here:

> The *quadratic residues* modulo *m* are the integers relatively prime to *m* which are congruent to squares modulo *m*.

> The integers relatively prime to *m* which are not congruent to a square modulo *m* are the *quadratic nonresidues*.

■ Counting Quadratic Residues

It is easy to have Mathematica produce a list of quadratic residues by automating the method used in the Prelab section. Specifically, we compute a^2 % n for each value of a satisfying gcd(a, n) = 1 and $1 \le a \le n$. The resulting list contains the quadratic residues. Here's the code to carry out the dirty work:

```
qr[n_] := Module[{answer = {}},
(* Compute all of the squares of relatively prime integers *)
Do[If[GCD[j, n] == 1, AppendTo[answer, Mod[j^2, n]]],
   {j, 1, n - 1}];
(* Remove duplicates and sort *)
answer = Sort[Union[answer]];
answer]
```

Try it out:

```
qr[17]
```

If you want to count the number of quadratic residues modulo a prime, you can apply the **Length** command to the output of **qr**:

```
qr[17]
Length[qr[17]]
```

■ Research Question 1

> If p is an odd prime, how many quadratic residues are there mod p?
>
> **Hint:** The content of the next few sections may be helpful in proving your conjecture.

■ Counting Square Roots

A question that is related to Research Question 1 is the following:

How many square roots mod p can an integer have?

We say that m is a square root of a modulo p if $a \equiv m^2$ mod p. The function **sqrt[a,p]** returns a list of all square roots of a mod p.

```
sqrt[a_, p_] :=
            (* Find the square roots by trying everything *)
      Select[Range[1, p], Mod[#^2 - a, p] == 0 &]
```

For example, here are the square roots of 3 modulo 11:

```
sqrt[3, 11]
```

And here's a check:

```
Mod[5^2, 11]
Mod[6^2, 11]
```

■ Research Question 2

> Let p be an odd prime and a an integer that is not divisible by p. How many square roots can a have modulo p?

Once you have completed Research Question 2, you may find that you have enough information to prove your conjecture for Research Question 1. Go back, and give it a try. If you have success, great! If not, don't despair. Go on to the next section, and return to Research Question 1 later.

■ Quadratic Residues and Primitive Roots

Recall that if we fix a prime p, then a *primitive root* modulo p is an integer r such that every element of $\mathbf{Z}_p^*$ (the residue classes relatively prime to p) is congruent to a power of r. As you may have discovered in the previous chapter, every prime has a primitive root. Thus, if r is a primitive root modulo p, then every nonzero residue class a modulo p can be written as

$$a \equiv r^j \pmod{p}$$

for a unique value of j between 0 and $p - 1$. In this section, we shall determine which values of j produce quadratic residues, and which values produce quadratic nonresidues.

To aid you in your explorations, several functions are provided that can be used in conjunction with **qr[p]** defined earlier. The function **primitive[p]** will return the smallest primitive root mod p.

```
(* Find an element mod p of order p-1 *)
primitive[p_] :=
  Module[{try = 2}, While [ord[try, p] != p - 1, try = try + 1]; try]
```

For example, here's the smallest primitive root mod 23:

```
primitive[23]
```

The function **primpowers[p]** takes a prime p as input, and provides the following output: a list of the quadratic residues modulo p; the smallest primitive root r modulo p; and a table of the values of r^1, r^2, r^3, ..., r^{p-1}, each reduced modulo p. Here's the definition of the function:

```
primpowers[p_] := Module[{prim, mytable, header},
      prim = primitive[p];  (* Find a primitive root *)
  (* Make a table of the powers in order *)
   mytable = Table[PowerMod[prim, j, p], {j, 1, p - 1}];
      (* Make a header for this spiffy table *)
   header =
  Table[ToString[r^j, TraditionalForm], {j, 1, p - 1}];
      (* Some information for reference *)
   Print["The quadratic residues are: ",
  qr[p], "\nA primitive root: r = ", prim, "."];
      (* Now for the data *)
   Nicetable[{header, mytable}]
   ]
```

Go ahead —try it out. (You know you want to!)

```
primpowers[17]
```

Note that the table of values $r^1, r^2, r^3, \ldots$ may scroll off of the page to the right; use the scrollbar at the bottom to see the whole list. You should be able to use data collected from **primpowers** to tackle the next research question.

■ Research Question 3

> Let p be a prime and r a primitive root mod p. Characterize the exponents j such that r^j is a quadratic residue mod p.

Once you have completed Research Question 3, you should be able to finish off Research Question 1. If you haven't already done so, go back and complete your proof.

We close out this section with an exercise that gives a result that is useful, in certain situations, for determining whether or not an integer is a quadratic residue modulo a prime. The result is known as *Euler's Criterion*, and is stated in the exercise below.

■ Exercise 1

Prove the following:

Euler's Criterion: Suppose that p is an odd prime and that a is an integer not divisible by p. If a is a quadratic residue modulo p, then

$$a^{(p-1)/2} \equiv 1 \pmod{p},$$

and if a is a quadratic nonresidue modulo p, then

$$a^{(p-1)/2} \equiv -1 \pmod{p}.$$

■ Quadratic Reciprocity

■ Basics

We will be interested in comparing when p is congruent to a square mod q with whether q is a square mod p, where p and q are odd primes. In other words, we will be comparing the Legendre symbols $\left(\frac{p}{q}\right)$ and $\left(\frac{q}{p}\right)$. Mathematica computes Legendre symbols with the function **JacobiSymbol** (which works in slightly more generality than a Legendre symbol). To compare them, we will use the function **seeboth[p,q]**.

```
seeboth[p_, q_] := {JacobiSymbol[p, q], JacobiSymbol[q, p]}
```

Let's see how it works:

```
seeboth[3, 5]
```

The resulting output tells us that 3 is not congruent to a square mod 5, and 5 is not congruent to a square mod 3. Try different combinations of odd primes with **seeboth**.

■ Research Question 4

Let p and q be odd primes. Conjecture a relationship between $\left(\frac{p}{q}\right)$ and $\left(\frac{q}{p}\right)$. Specifically, find conditions for p and q that will determine when $\left(\frac{p}{q}\right) = \left(\frac{q}{p}\right)$ and when $\left(\frac{p}{q}\right) = -\left(\frac{q}{p}\right)$.

Hint: This result, which is known as the Quadratic Reciprocity Theorem, is a tough one. You may find the suggestions given below helpful.

■ Tabulating Data

To make it easier to spot patterns, we will keep one prime fixed (say p) and use lots of primes q. The following function allows us to fix one of the primes, and compare it with lots of other odd primes (**howmany** of them, to be precise). The output shows one row for each prime q. After displaying q, it shows the output from **seeboth** (for the fixed value of p and the q for that line).

```
seelots[p_, howmany_] := TableForm[
   Table[Flatten[{"q =", Prime[j], seeboth[p, Prime[j]]}],
    {j, 2, howmany + 1}],
   TableSpacing -> {0, 1}, TableAlignments -> Right]
```

If we keep the prime $p = 3$ fixed and consider 20 primes q, we compute

```
seelots[3, 20]
```

■ Additional Tips

Try different values of p with the function **showlots**. The relationship between $\left(\frac{p}{q}\right)$ and $\left(\frac{q}{p}\right)$ is subtle and you may not get it all in one shot. Try to find some values of p for which the relationship is simple. Once you can find some primes p for which you are comfortable conjecturing the relationship between $\left(\frac{p}{q}\right)$ and $\left(\frac{q}{p}\right)$ for varying q, try to pin down exactly which primes p are easy (i.e., go from a list of "easy values of p" to a conjecture about easy primes).

While it would be wonderful if you are able to prove your conjectures in this section, proofs are somewhat hard to come by. Ironically, there may be more proofs of quadratic reciprocity than of any other theorem in the course. However, they are all significantly more difficult that the theorems we have covered thus far.

■ What's Quadratic Reciprocity Good For?

A reasonable question. One use is to efficiently compute values of $\left(\frac{p}{q}\right)$. For example, suppose that you wish to evaluate

$$\left(\tfrac{3}{1000003}\right).$$

Determining if the equation $x^2 \equiv 3 \pmod{1000003}$ has solutions is a tall order. However, the Quadratic Reciprocity Theorem will tell us that

$$\left(\tfrac{3}{1000003}\right) = -\left(\tfrac{1000003}{3}\right),$$

and since $1000003 \equiv 1 \pmod 3$, we see instantly that

$$\left(\tfrac{1000003}{3}\right) = \left(\tfrac{1}{3}\right) = 1.$$

Putting this together, we see that $\left(\tfrac{3}{1000003}\right) = -1$, and so $x^2 \equiv 3 \pmod{1000003}$ has no solutions.

Thus by applying the Quadratic Reciprocity Theorem, we will be able to quickly simplify a complicated calculation. We will return to applications of the Quadratic Reciprocity Theorem in the chapter summary.

■ Special Examples of Legendre Symbols

In practice, there are two cases of Legendre symbols which can be computed easily, but are not covered by the previous section. In both cases, they involve $\left(\frac{a}{p}\right)$ with a fixed, and the prime p varying. Let's look back at an earlier example of Legendre symbols of this type.

The value of $\left(\frac{5}{p}\right)$ depends only on the congruence class of p modulo a specific integer. This is somewhat surprising! In fact, it may not be clear yet that this is the case. Let's compute some Legendre symbols to check it out. First, we define a function to display the value of $\left(\frac{5}{p}\right)$ together with the remainder of p modulo n for a value of n of your choosing.

```
checkem[a_, n_, howmany_] :=
TableForm[
        Table[{Prime[i], " % ", n, " = ",
     Mod[Prime[i], n], ";  LS(", a, ",", Prime[i], ") = ",
     JacobiSymbol[a, Prime[i]]}, {i, 2, howmany + 1}],
        TableSpacing -> {0, 0}, TableAlignments -> Right];
```

The function **checkem[a, n, howmany]** will display a table, with each line providing the value of a % n together with the value of $\left(\frac{a}{p_i}\right)$, where p_i is the ith prime. The value of i runs from 2 (thus we start with $p_2 = 3$) and ends at **howmany+1**. In the output, we write LS for Legendre Symbol. In the example below, we set $a = 5$ and $n = 3$.

```
checkem[5, 3, 20]
```

Examining the output, we don't see any obvious patterns. Let's try setting $n = 5$:

```
checkem[5, 5, 20]
```

This time a pattern does emerge. Upon close inspection, we see that $\left(\frac{5}{p}\right) = 1$ when p is congruent to 1 or 4 modulo 5, and $\left(\frac{5}{p}\right) = -1$ when p is congruent to 2 or 3 modulo 5.

In the next two subsections, we examine two cases similar to the above illustration. For each, your task is to find a pattern similar to the one we spotted for $a = 5$.

■ Computing $\left(\frac{-1}{p}\right)$

Let p be an odd prime. We want to know when -1 is congruent to a square mod p. As above, use the function **checkem** with different values of n to look for patterns. Here's what we get if $n = 2$.

    checkem[-1, 2, 20]

That may have not been too helpful! Try other values of n until you find a pattern.

■ Research Question 5

> For which odd primes p is $\left(\frac{-1}{p}\right) = 1$?
>
> **Hint:** Once you have a good conjecture, consider Euler's Criterion when searching for a proof.

■ Computing $\left(\frac{2}{p}\right)$

Same song, second verse. Use the same approach as above to tackle Research Question 6. By now you should be an expert at using **checkem**, so we leave all of the entries for you to fill in.

    checkem[?,?,?]

■ Research Question 6

> For which odd primes p is $\left(\frac{2}{p}\right) = 1$?
>
> **Note:** We are primarily looking for a good conjecture for this problem. A proof is fairly difficult.

Quadratic Congruences

11.1 Definitions

In the Prelab section of this chapter, we introduced the notions of quadratic residue and quadratic nonresidue. We repeat the definitions here:

> The *quadratic residues* modulo m are the integers relatively prime to m which are congruent to squares modulo m.

> The integers relatively prime to m which are not congruent to a square modulo m are the *quadratic nonresidues*.

11.2 Counting Quadratic Residues

It is easy to have Java produce a list of quadratic residues by automating the method used in the Prelab section. Specifically, the applet below computes $a^2 \% n$ for each value of a satisfying $\gcd(a, n) = 1$ and $1 \leq a \leq n$. The resulting output contains the quadratic residues. Try it out:

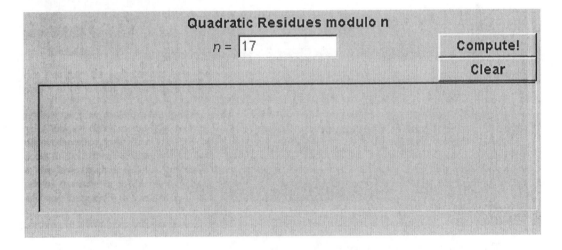

In the very near future, you will want to count the number of quadratic residues modulo a prime. You can do so using a modified version of the previous applet that includes the number of quadratic residues in the list, thus avoiding the tedium of counting them yourself.

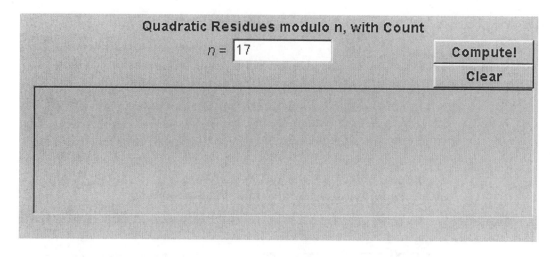

Research Question 1

If p is an odd prime, how many quadratic residues are there mod p?

Hint: The content of the next few sections my be helpful in proving your conjecture.

11.3 Counting Square Roots

A question that is related to Research Question 1 is the following:

How many square roots mod p can an integer have?

We say that m is a square root of a modulo p if $a \equiv m^2$ mod p. The applet below returns a list of all square roots of a mod p. For example, here are the square roots of 3 modulo 11:

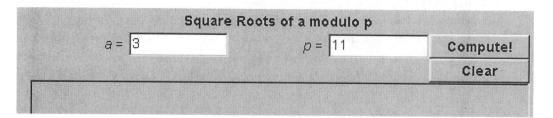

You can check the results using the Java calculator:

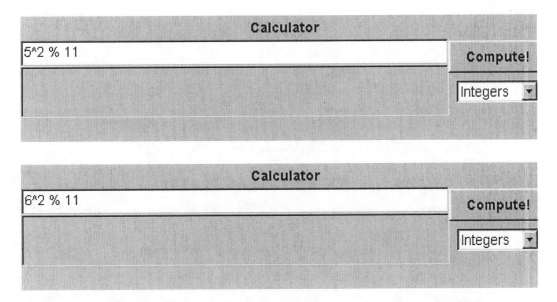

Research Question 2

Let p be an odd prime and a an integer that is not divisible by p. How many square roots can a have modulo p?

Once you have completed Research Question 2, you may find that you have enough information to prove your conjecture for Research Question 1. Go back, and give it a try. If you have success, great! If not, don't despair. Go on to the next section, and return to Research Question 1 later.

11.4 Quadratic Residues and Primitive Roots

Recall that if we fix a prime p, then a *primitive root* modulo p is an integer r such that every element of $\mathbf{Z}_p^*$ (the residue classes relatively prime to p) is congruent to a power of r. As you may have discovered in the previous chapter, every prime has a primitive root. Thus, if r is a primitive root modulo p, then every nonzero residue class a modulo p can be written as

$$a \equiv r^j \pmod{p}$$

for a unique value of j between 0 and $p - 1$. In this section, we shall determine which values of j produce quadratic residues, and which values produce quadratic nonresidues.

To aid you in your explorations, several applets are provided for your use. We begin with one you saw in a previous section:

The next applet will return the smallest primitive root mod p For example, here's the smallest primitive root mod 23:

The third applet takes a prime p as input, and provides the following output: a list of the quadratic residues modulo p; the smallest primitive root *r* modulo *p*; and, a table of the values of $r^1, r^2, r^3, \ldots, r^{p-1}$, each reduced modulo *p*. Go ahead – try it out. (You know you want to!)

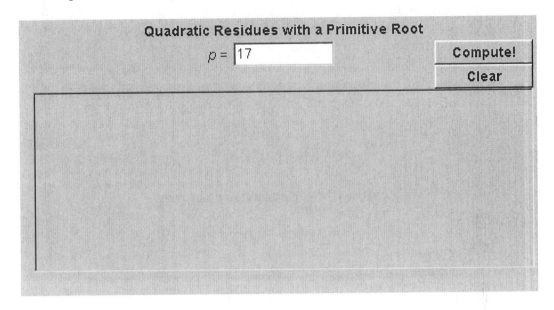

Note that the table of values $r^1, r^2, r^3, \ldots$ may scroll off of the page to the right; use the scrollbar at the bottom to see the whole list. You should be able to use data collected from this applet to tackle the next research question.

Research Question 3

Let *p* be a prime and *r* a primitive root mod *p*. Characterize the exponents *j* such that r^j is a quadratic residue mod *p*.

Once you have completed Research Question 3, you should be able to finish off Research Question 1. If you haven't already done so, go back and complete your proof.

We close out this section with an exercise that gives a result that is useful, in certain situations, for determining whether an integer is a quadratic residue modulo a prime. The result is known as *Euler's Criterion*, and is stated in the exercise below.

Exercise 1

Prove the following:

Euler's Criterion: Suppose that p is an odd prime and that a is an integer not divisible by p. If a is a quadratic residue modulo p, then

$$a^{(p-1)/2} \equiv 1 \ (\text{mod } p),$$

and if a is a quadratic nonresidue modulo p, then

$$a^{(p-1)/2} \equiv -1 \ (\text{mod } p).$$

11.5   Quadratic Reciprocity

Note: The limitations of typesetting mathematics in web pages makes the use of the usual version of the Legendre symbol impractical. Instead, we shall write $\text{LS}(p, q)$ to indicate whether p is a quadratic residue modulo q.

11.5.1   Basics

We will be interested in comparing when p is congruent to a square mod q with whether q is a square mod p, where p and q are odd primes. In other words, we will be comparing the Legendre symbols $\text{LS}(p, q)$ and $\text{LS}(q, p)$.

The applet below allows us to easily compare the values of $\text{LS}(p, q)$ and $\text{LS}(q, p)$. Try it out:

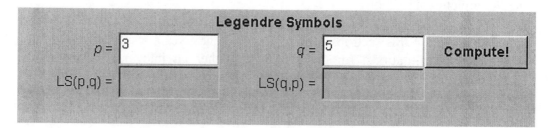

The resulting output tells us that 3 is not congruent to a square mod 5, and 5 is not congruent to a square mod 3. Try different combinations of odd primes with this applet.

Research Question 4

Let p and q be odd primes. Conjecture a relationship between $LS(p, q)$ and $LS(q, p)$. Specifically, find conditions for p and q that will determine when $LS(p, q) = LS(q, p)$ and when $LS(p, q) = -LS(q, p)$.

Hint: This result, which is known as the Quadratic Reciprocity Theorem, is a tough one. You may find the suggestions given below helpful.

11.5.2 Tabulating Data

To make it easier to spot patterns, we will keep one prime fixed (say p) and use lots of primes q. The following applet allows us to fix the prime p, and compare it with the first B odd primes. The output shows one row for each prime q. After displaying q, it shows the values of $LS(p, q)$ and $LS(q, p)$ for the fixed value of p and the q for that line. For example, if we keep the prime $p = 3$ fixed and take the first 20 primes for q, we get

11.5.3 Additional Tips

Try different values of p with the preceding applet. The relationship between LS(p, q) and LS(q, p) is subtle and you may not get it all in one shot. Try to find some values of p for which the relationship is simple. Once you can find some primes p for which you are comfortable conjecturing the relationship between LS(p, q) and LS(q, p) for varying q, try to pin down exactly which primes p are easy (i.e., go from a list of "easy values of p" to a conjecture about easy primes).

While it would be wonderful if you are able to prove your conjectures in this section, proofs are somewhat hard to come by. Ironically, there may be more proofs of quadratic reciprocity than of any other theorem in the course. However, they are all significantly more difficult that the theorems we have covered thus far.

11.5.4 What's Quadratic Reciprocity Good For?

A reasonable question. One use is to efficiently compute values of LS(p, q). For example, suppose that you wish to evaluate

$$LS(3, 1000003).$$

Determining if the equation $x^2 \equiv 3 \pmod{1000003}$ has solutions is a tall order. However, the Quadratic Reciprocity Theorem will tell us that

$$LS(3, 1000003) = -LS(1000003, 3)$$

and since $1000003 \equiv 1 \pmod 3$, we see instantly that

$$LS(1000003, 3) = LS(1, 3) = 1.$$

Putting this together, we see that LS(3, 1000003) = -1, and so $x^2 \equiv 3 \pmod{1000003}$ has no solutions.

Thus by applying the Quadratic Reciprocity Theorem, we will be able to quickly simplify a complicated calculation. We will return to applications of the Quadratic Reciprocity Theorem in the chapter summary.

11.6   Special Examples of Legendre Symbols

In practice, there are two cases of Legendre symbols which can be computed easily, but are not covered by the previous section. In both cases, they involve $LS(a, p)$ with a fixed, and the prime p varying. Let's look back at an earlier example of Legendre symbols of this type.

The value of $LS(5, p)$ depends only on the congruence class of p modulo a specific integer. This is somewhat surprising! In fact, it may not be clear yet that this is the case. Let's compute some Legendre symbols to check it out.

The applet below will display a table, with each line providing the remainder of a modulo n together with the value of $LS(a, p_i)$, where p_i is the ith prime. The value of i runs from 2 (thus we start with $p_2 = 3$) and ends at $B + 1$. In the example below, we set $a = 5$ and $n = 3$.

Legendre Symbols (a/p) with p modulo n

| $a =$ | 5 | $n =$ | 3 | Compute! |
| $B =$ | 10 | | | Clear |

Examining the output, we don't see any obvious patterns. Let's try setting $n = 5$:

Legendre Symbols (a/p) with p modulo n

| $a =$ | 5 | $n =$ | 5 | Compute! |
| $B =$ | 10 | | | Clear |

This time a pattern does emerge. Upon close inspection, we see that LS(5, p) = 1 when p is congruent to 1 or 4 modulo 5, and LS(5, p) = −1 when p is congruent to 2 or 3 modulo 5.

In the next two subsections, we examine two cases similar to the above illustration. For each, your task is to find a pattern similar to the one we spotted for $a = 5$.

11.6.1 Computing LS(−1, p)

Let p be an odd prime. We want to know when −1 is congruent to a square mod p. As above, use the applet with different values of n to look for patterns. Here's what we get if $n = 2$.

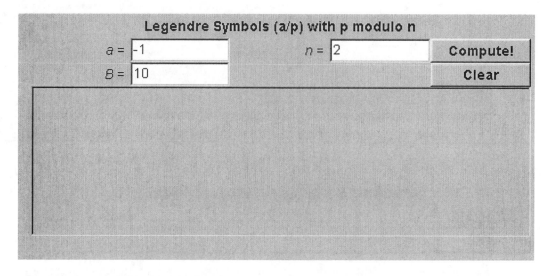

That may have not been too helpful! Try other values of n until you find a pattern.

Research Question 5

For which odd primes p is $LS(-1, p) = 1$?

Hint: Once you have a good conjecture, consider Euler's Criterion when searching for a proof.

11.6.2 Computing $LS(2, p)$

Same song, second verse. Use the same approach as above to tackle Research Question 6. By now you should be an expert at using the applet, so we leave all of the entries for you to fill in.

Legendre Symbols (a/p) with p modulo n

$a =$		$n =$		Compute!
$B =$				Clear

Research Question 6

For which odd primes p is $LS(2, p) = 1$?

Note: We are primarily looking for a good conjecture for this problem. A proof is fairly difficult.

Homework

1. Let p be an odd prime. Complete the proof of Research Question 6 by showing that $\left(\dfrac{2}{p}\right) = -1$ if $p \equiv 5 \pmod 8$ and that $\left(\dfrac{2}{p}\right) = 1$ if $p \equiv 7 \pmod 8$.

2. Let p be an odd prime. Prove that if $p \nmid a$ and $p \nmid b$, then

$$\left(\frac{ab}{p}\right) = \left(\frac{a}{p}\right)\left(\frac{b}{p}\right).$$

   (Hint: Euler's Criterion.)

3. Evaluate each of the following:

   (a) $\left(\dfrac{78}{97}\right)$

   (b) $\left(\dfrac{221}{881}\right)$

   (c) $\left(\dfrac{855}{1009}\right)$

4. Determine which of the following quadratic congruences has solutions.

   (a) $3x^2 + 7x + 5 \equiv 0 \pmod{13}$

   (b) $7x^2 + 13x - 12 \equiv 0 \pmod{29}$

   (c) $16x^2 + 5x + 1 \equiv 0 \pmod{31}$

5. Some of the theorems in this chapter can be formulated in very compact ways. In this exercise, you will deduce the compact version from the version stated in the section. In each case, p and q are odd primes and a is an integer.

   (a) $\left(\dfrac{a}{p}\right) \equiv a^{(p-1)/2} \pmod p$

   (b) $\left(\dfrac{-1}{p}\right) = (-1)^{(p-1)/2}$

   (c) $\left(\dfrac{2}{p}\right) = (-1)^{(p^2-1)/8}$

   (d) $\left(\dfrac{p}{q}\right)\left(\dfrac{q}{p}\right) = (-1)^{(p-1)(q-1)/4}$

6. What are the quadratic residues modulo 19? (Hint: 2 is primitive modulo 19.)

7. If p is an odd prime, what is the value of

$$\sum_{a=1}^{p-1} \left(\frac{a}{p}\right)?$$

561

8. Let p be an odd prime. For what values of p does $\left(\dfrac{-2}{p}\right) = 1$, and for what values of p does $\left(\dfrac{-2}{p}\right) = -1$?

9. Suppose that a is a quadratic residue modulo an odd prime p. Show that if $p \equiv 1$ (mod 4), then $p - a$ is a quadratic residue modulo p, and if $p \equiv 3$ (mod 4), then $p - a$ is a quadratic nonresidue modulo p.

10. Let p be an odd prime. For what values of p does $\left(\dfrac{3}{p}\right) = 1$, and for what values of p does $\left(\dfrac{3}{p}\right) = -1$?

11. Prove that if $p \equiv 3$ (mod 4) and q is an odd prime, then

$$\left(\frac{-p}{q}\right) = \left(\frac{q}{p}\right).$$

12. Let p be an odd prime. Show that the quadratic congruence equation $ax^2 + bx + c \equiv 0$ (mod p) has a solution if and only if $b^2 - 4ac \equiv 0$ (mod p) or $b^2 - 4ac$ is a quadratic residue modulo p.

13. Suppose that $p = 2^k + 1$ is a prime. Prove that if a is a quadratic nonresidue modulo p, then a is a primitive root modulo p.

The next two exercises give the beginnings of an efficient method for solving congruences

$$x^2 \equiv a \quad (\text{mod } p). \tag{11.4}$$

We have already seen how quadratic reciprocity can be used to efficiently determine if there are any solutions, but finding the solutions is a different matter. These exercises cover all prime moduli p except when $p \equiv 1$ (mod 8). The general algorithm increases in complexity with the power of 2 dividing $p - 1$.

14. Suppose that p is a prime such that $p \equiv 3$ (mod 4), and that a is a quadratic residue modulo p. Show that $x \equiv a^{(p+1)/4}$ (mod p) is a solution to equation (11.4).

15. Suppose p is a prime such that $p \equiv 5$ (mod 8), and a is a quadratic residue modulo p.

    (a) Prove that $a^{(p-1)/4} \equiv \pm 1$ (mod p).

(b) Show that if $a^{(p-1)/4} \equiv 1 \pmod p$, then $x \equiv a^{(p+3)/8} \pmod p$ is a solution to equation (11.4).

(c) Show that $2^{(p-1)/2} \equiv -1 \pmod p$.

(d) Show that if $a^{(p-1)/4} \equiv -1 \pmod p$, then $x \equiv 2a(4a)^{(p-5)/8} \pmod p$ is a solution to equation (11.4).

16. Use exercise 14 to find all solutions (if there are any) for each of the following congruence equations.

   (a) $x^2 \equiv 7 \pmod{19}$

   (b) $x^2 \equiv 2 \pmod{23}$

   (c) $x^2 \equiv 13 \pmod{47}$

17. Use exercise 14 to help you solve the following congruences.

   (a) $x^2 + 11x + 2 \equiv 0 \pmod{23}$

   (b) $x^2 + 11x + 2 \equiv 0 \pmod{31}$

12. Representation Problems

Prelab

Historically, people have studied many questions which come down to writing one type of integer as a sum of integers of another type. For example, suppose we wanted to see which integers can be written as the sum of two cubes. In other words, for which $n \in \mathbf{Z}$ do there exist $x, y \in \mathbf{Z}$ such that

$$n = x^3 + y^3?$$

It is not too hard to see that not every integer is the sum of two cubes. If $n = x^3 + y^3$, then $n \equiv x^3 + y^3 \pmod 9$. By cubing one integer from each congruence class modulo 9 we find that for any $x \in \mathbf{Z}$, $x^3 \equiv 0$, 1, or 8 (mod 9). So, the only possibilities are that $x^3 + y^3 \equiv 0$, 1, 2, 7, or 8 (mod 9). Thus integers congruent to 3, 4, 5, and 6 modulo 9 cannot be equal to the sum of two cubes.

1. **Show that there are infinitely many integers which are not the sum of 3 cubes.**

In this chapter, we will focus on representation problems which involve squares. In particular, we will consider when integers are the sum of two squares. For example, 5 is the sum of two squares because $5 = 1^2 + 2^2$.

2. **For each prime $p < 20$, find all ways (if any) of writing $p = x^2 + y^2$ with $0 \le x \le y$ and $x, y \in \mathbf{Z}$. You do not need to give justification for your results here.**

Maple electronic notebook `12-represent.mws`

Mathematica electronic notebook `12-represent.nb`

Web electronic notebook Start with the web page `index.html`

Maple Lab: Representation Problems

Introduction

In this lab, we will look at some representation problems where we try to express certain integers as sums of other special types of integers. The problems are of varying difficulty. In some cases you may be able to formulate strong conjectures and provide complete proofs. At other times, finding a good conjecture may be a big challenge for some Research Questions (and proofs may be very hard to come by).

When finding the right conjecture is difficult, remember that it is often useful to consider prime values before working on more general integers.

Differences of Two Squares

The basic question here is: which integers can be written as the difference of two squares? In other words, we want to know the integer solutions to the equation

$$n = x^2 - y^2.$$

Since the equation is already solved for n, a natural way to run experiments is to substitute lots of pairs (x, y) into the right-hand side and see what comes out. Since x and y are squared, we will only use nonnegative values for them. Here we compute all values of $x^2 - y^2$ with $0 \le x \le 10$ and $0 \le y \le 10$:

```
[ > seq(seq(x^2-y^2, x=0..10), y=0..10);
[ >
```

OK, so that is pretty much a jumbled mess. We can take several steps to clean it up a bit. First, we will sort the output. Next, we observe that whenever $x = y$, we will get 0, so we omit those combinations. Finally, we notice that switching the roles of x and y will negate n. So, it suffices to take $y < x$. The result will then be that we will only see the positive values of n which are the difference of two squares. Here is the resulting function:

```
[ > diffofsquares := bnd -> sort([seq(seq(x^2-y^2, x=y+1..bnd),
[   y=0..bnd)]):
```

Now we can try it out. The one parameter bnd is a bound determining how large values we should try for x and y. We use the same bound as above.

```
[ > diffofsquares(10);
[ >
```

Notice that we have not removed duplicates; some numbers appear more than once. This will give you an indication of how many ways certain integers can be written as the sum of two squares. Keep in mind that some numbers may not appear on this list because the bound bnd is too low. Similarly, increasing the value for bnd may uncover more ways of writing a given integer as the difference of two squares.

Research Question 1

Which positive integers n can be written as the difference of two squares, and in how many ways can it be done?

Sums of Two Squares

The idea here is just like the previous section, but for the sum of two squares. In other words, we want to find integer solutions to the equation

$$n = x^2 + y^2.$$

This time we will state the Research Question and then give some useful Maple functions. We will only ask the question: *how many* ways can an integer n be written as the sum of two squares for prime values of n?

Research Question 2

Which primes p can be written as the sum of two squares, and in how many ways can it be done?

Note: This is is a case where you will only be able to prove part of your conjecture. The complete proof of an optimal conjecture is beyond the scope of the chapter summary.

Research Question 3

Which positive integers n can be written as the sum of two squares?

Hint: It is probably a good idea to use the standard progression for investigating problems which we have used earlier in the course. Research Question 2 takes care of the case when n is prime. Try products of distinct primes and prime powers before trying to jump to the complete conjecture. Also keep in mind that your answer here needs to be consistent with your conjecture for Research Question 2.

We start with a function analogous to the one above.

```
> sumofsquares := bnd -> sort([seq(seq(x^2+y^2, x=0..y), y=1..bnd)]):
```

It will compute all values of $x^2 + y^2$ with x and y nonnegative and ranging up to some bound. This time, switching the values of x and y leaves $x^2 + y^2$ unchanged, so we take $x \leq y$. Here is a sample.

```
> sumofsquares(10);
```

For the problem of sums of squares, if we have a fixed value of n and we want to determine if $n = x^2 + y^2$ for some x and y, there are clearly only finitely many x and y we have to try. In fact, if we take $x \leq y$, then we can be sure that $x \leq \sqrt{\dfrac{n}{2}}$. The next function takes a single value of n as input and uses this observation to search for all ways of writing n as a sum of two squares x and y with $x \leq y$.

```
> assumofsquares := proc(n)
    local ans, bound, j;
    ans := NULL;
    bound := floor(sqrt(n/2));
    for j to bound do
     if(type(sqrt(n-j^2), integer)) then
      ans := ans, [j, sqrt(n-j^2)];
     fi;
    od;
```

```
    if ans = NULL then 'None' else ans fi;
  end:
```

For example, the earlier output indicated that 29 could be written as the sum of two squares. Here are the *x* and *y* that do it:

```
> assumofsquares(29);
```

On the other hand, 50 can be written as a sum of two squares in two different ways:

```
> assumofsquares(50);
```

So, $50 = 1^2 + 7^2 = 5^2 + 5^2$. From the earlier output, it appeared that 15 was *not* the sum of two squares. We can see that here because `assumofsquares(15)` produces the word None.

```
> assumofsquares(15);
>
```

■ Pythagorean Triples

Finding Pythagorean triples can be thought of as a special case of determining when an integer is the sum of two squares; a Pythagorean triple consists of three positive integers *x*, *y*, and *z* such that

$$x^2 + y^2 = z^2.$$

You may recognize this equation as being the relation satisfied by the lengths of the sides of a right triangle. Historically, interest in Pythagorean triples started from this geometric connection.

We can search for Pythagorean triples with Maple. For the following function, you provide a bound and Maple will find all Pythagorean triples with *x* and *y* less than or equal to the bound.

```
> pythag := proc(bnd)
    local x, y, ans;
    ans := NULL;
    for x from 1 to bnd do
     for y from x to bnd do
       if(type(sqrt(x^2+y^2), integer)) then
       ans := ans, [x, y, sqrt(x^2+y^2)];
       fi;
     od;
    od;
    ans;
  end:
```

Let's try it out.

```
> pythag(50);
>
```

As you start to look at these triples, you may notice that some are related to others. For example, we have the first triple, (3, 4, 5), and also (6, 8, 10) which can be gotten from (3, 4, 5) by multiplying each number by 2. We also have (9, 12, 15), which can be gotten by multiplying each term of (3, 4, 5) by 3.

More generally, if (*x*, *y*, *z*) satisfies $x^2 + y^2 = z^2$ and *k* is an integer, then

$$(k\,x)^2 + (k\,y)^2 = k^2\,(x^2 + y^2) = k^2\,z^2 = (k\,z)^2.$$

So, $(k\,x, k\,y, k\,z)$ satisfies the same equation. By reversing this argument, we see that if (x, y, z) is a Pythagorean triple, then we can cancel a factor of $\gcd(x, y, z)$ to get a triple with $\gcd(x, y, z) = 1$. A Pythagorean triple with $\gcd(x, y, z) = 1$ is called a *primitive* Pythagorean triple. Every Pythagorean triple is a multiple of a primitive triple.

The next function works just like `pythag` above, but it only outputs the primitive triples. If you decide to examine the code, you will see that `primpythag` tests whether a triple is primitive by checking only $\gcd(x, y) = 1$, which differs from our definition of primitive. A homework exercise at the end of the chapter proves that this test is nonetheless correct.

```
> primpythag := proc(bnd)
    local x, y, ans;
    ans := NULL;
    for x from 1 to bnd do
     for y from x to bnd do
      if gcd(x,y)=1 then
       if type(sqrt(x^2+y^2), integer) then
        ans := ans, [x, y, sqrt(x^2+y^2)]
       fi;
      fi;
     od;
    od;
    ans;
   end:
```

Here we will try `primpythag` with the same initial bound as above.

```
> primpythag(50);
>
```

As you can see, there are considerably fewer primitive triples in this range. We are now ready for our next Research Question.

■ Research Question 4

Discover as much as you can about primitive Pythagorean triples.

Research Question 4 is more open ended than any previous Research Question in this course. Do as much as you can with it. We will give you two hints:

1. The strongest known conjecture here is connected to the first two parts of the lab! You may want to start by looking at the numbers which come out as hypotenuses.

2. You may want to consider sums and/or differences between the legs and hypotenuse of a Pythagorean triangle.

■ Sums of Many Squares

We have seen that some positive integers can be written as the sum of two squares, but some cannot. Can all positive integers be written as the sum of three squares, or as the sum of four squares? How many squares would it take to express every integer? Maybe there is no fixed value m so that every positive integer is a sum of *m* squares.

First, we should clarify what we mean by writing a positive integer as the sum of say three squares. We allow using $0^2 = 0$, so we would say that 5 is the sum of three squares because $5 = 1^2 + 2^2 + 0^2$.

The next function will help in our investigations. The programming is more complicated than usual here, so we will focus on the result the function produces. You get to supply a bound; let's call the bound *B*. The function tries to write each integer from 1 to *B* as a sum of as few squares as possible. As it goes along, it prints out information saying how many it was *not* able to write as a sum of two squares, as a sum of three squares, and so on. Finally, it returns a list of *B* integers saying how many squares were needed for expressing that integer. Here is the definition of the function:

```
> sumofmany := proc(B)
   local j, k, vals, numsqrs, notdone;
   vals := array(1..B);
   for j to B do vals[j] := 0; od;
  # Fill in the squares
   for j to floor(sqrt(B)) do
    vals[j^2]:=1;
   od;
  # Now we start to do the rest
   numsqrs := 1;
   notdone:=1; # This is just to get into the loop
   while(notdone>0) do
    notdone := 0;
    for j to B do
     if(vals[j] = 0) then notdone := notdone +1; fi;
    od;
    printf('%d out of %d are not the sum of %d squares\n', notdone,
  B, numsqrs);
    numsqrs := numsqrs+1;
    for j from B to 1 by -1 do
     if(vals[j] = numsqrs-1) then
       for k to floor(sqrt(B-j)) do
        if vals[j+k^2] = 0 then
         vals[j+k^2] := numsqrs;
        fi;
       od;
     fi;
    od;
   od;
   print(vals);
  end:
```

■ Maple Note (read this only if you really want to know how this function works)
OK brave soul, here we go.

The function makes an array of length B called `vals`, where B is the bound supplied by the user. First all of the entries are initialized with the value of 0. We want to fill the *j*th entry with the smallest number of squares for expressing the integer *j*. So, we fill the entries where *j* itself is a square with the value of 1.

Now we head for a `while` loop. We will keep looping until there are no zeros left in the array. The first time through the `while` loop, we are marking the entries which are a sum of two squares. The next time we mark the integers which are a sum of three squares, and so on. We keep track of how many times we have been through the loop in the variable `numsqrs`.

Each time through the loop, we count how many entries are still zero, and report this to the user. Suppose we want to mark entries which are the sum of five squares. Then, we search through the list for entries which are the sum of four (but no fewer) squares. For each of these, we add all possible squares (keeping below the bound B). If we hit an entry which is zero, then that entry corresponds to an integer which is not the sum of four squares, but is the sum of five squares. So we mark that entry with a 5.

When all of the entries are nonzero, we are done with the `while` loop. All that is left is to print our list `vals`.

Now we try it with a bound of 50:

```
[ > sumofmany(50);
```

The output tells us that some of the integers could not be written as a sum of three squares, but all of the integers from 1 to 50 could be written as a sum of four squares. In that final list of numbers, we can see a 3 in the sixth position. This means that 6 is the sum of three squares, but is not the sum of two squares.

Use `sumofmany` to try to answer the following question:

■ Research Question 5

How many squares are needed to write every integer, or are there integers which need arbitrarily large numbers of squares?

(You should be able to formulate a conjecture here, but the proof of the "right" conjecture is very difficult. In fact, both Fermat and Euler knew the right conjecture, but neither was able to prove it!)

Mathematica Lab: Representation Problems

■ Introduction

In this lab, we will look at some representation problems where we try to express certain integers as sums of other special types of integers. The problems are of varying difficulty. In some cases you may be able to formulate strong conjectures and provide complete proofs. At other times, finding a good conjecture may be a big challenge for some Research Questions (and proofs may be very hard to come by).

When finding the right conjecture is difficult, remember that it is often useful to consider prime values before working on more general integers.

■ Differences of Two Squares

The basic question here is: which integers can be written as the difference of two squares? In other words, we want to know the integer solutions to the equation

$$n = x^2 - y^2.$$

Since the equation is already solved for n, a natural way to run experiments is to substitute lots of pairs (x, y) into the right-hand side and see what comes out. Since x and y are squared, we will only use non-negative values for them. Here we compute all values of $x^2 - y^2$ with $0 \le x \le 10$ and $0 \le y \le 10$:

```
Flatten[Table[x^2 - y^2, {y, 0, 10}, {x, 0, 10}]]
```

OK, so that is pretty much a jumbled mess. We can take several steps to clean it up a bit. First, we will sort the output. Next, we observe that whenever $x = y$, we will get 0, so we omit those combinations. Finally, we notice that switching the roles of x and y will negate n. So, it suffices to take $x > y$. The result will then be that we will only see the positive values of n which are the difference of two squares. Here is the resulting function:

```
diffofsquares[bnd_] :=
  Sort[Flatten[Table[x^2 - y^2, {y, 0, bnd}, {x, y + 1, bnd}]]]
```

Now we can try it out. The one parameter **bnd** is a bound determining how large values we should try for x and y. We use the same bound as above.

```
diffofsquares[10]
```

Notice that we have not removed duplicates; some numbers appear more than once. This will give you an indication of how many ways certain integers can be written as the sum of two squares. Keep in mind that some numbers may not appear on this list because the bound **bnd** is too low. Similarly, increasing the value for **bnd** may uncover more ways of writing a given integer as the difference of two squares.

■ Research Question 1

> Which positive integers n can be written as the difference of two squares, and in how many ways can it be done?

■ Sums of Two Squares

The idea here is just like the previous section, but for the sum of two squares. In other words, we want to find integer solutions to the equation

$$n = x^2 + y^2.$$

This time we will state the Research Question and then give some useful Mathematica functions. We will only ask the question: *how many* ways can an integer n be written as the sum of two squares for prime values of n?

■ Research Question 2

Which primes p can be written as the sum of two squares, and in how many ways can it be done?

Note: This is a case where you will only be able to prove part of your conjecture. The complete proof of an optimal conjecture is beyond the scope of the chapter summary.

■ Research Question 3

Which positive integers n can be written as the sum of two squares?

Hint: It is probably a good idea to use the standard progression for investigating problems which we have used earlier in the course. Research Question 2 takes care of the case when n is prime. Try products of distinct primes and prime powers before trying to jump to the complete conjecture. Also keep in mind that your answer here needs to be consistent with your conjecture for Research Question 2.

We start with a function analogous to the one above.

```
sumofsquares[bnd_] :=
   Sort[Flatten[Table[x^2 + y^2, {y, 1, bnd}, {x, 0, y}]]]
```

It will compute all values of $x^2 + y^2$ with x and y nonnegative and ranging up to some bound. This time, switching the values of x and y leaves $x^2 + y^2$ unchanged, so we take $x \le y$. Here is a sample.

```
sumofsquares[10]
```

For the problem of sums of squares, if we have a fixed value of n and we want to determine if $n = x^2 + y^2$ for some x and y, there are clearly only finitely many x and y we have to try. In fact, if we take $x \le y$, then we can be sure that $x \le \sqrt{n/2}$. The next function takes a single value of n as input and uses this observation to search for all ways of writing n as a sum of two squares x and y with $x \le y$.

```
assumofsquares[n_] := Module[{ans, bound, j},
        ans = {};
        bound = Floor[Sqrt[n / 2]];
        Do[
            If[IntegerQ[Sqrt[n - j^2]],
    ans = AppendTo[ans, {j, Sqrt[n - j^2]}]],
            {j, bound}];
        If[ans == {}, None, ans]]
```

For example, the earlier output indicated that 29 could be written as the sum of two squares. Here are the x and y that do it:

```
assumofsquares[29]
```

On the other hand, 50 can be written as a sum of two squares in two different ways:

```
assumofsquares[50]
```

So, $50 = 1^2 + 7^2 = 5^2 + 5^2$. From the earlier output, it appeared that 15 was not the sum of two squares. We can see that here because **assumofsquares[15]** produces the word None.

```
assumofsquares[15]
```

■ Pythagorean Triples

Finding Pythagorean triples can be thought of as a special case of determining when an integer is the sum of two squares; a Pythagorean triple consists of three positive integers x, y, and z such that

$$x^2 + y^2 = z^2.$$

You may recognize this equation as being the relation satisfied by the lengths of the sides of a right triangle. Historically, interest in Pythagorean triples started from this geometric connection.

We can search for Pythagorean triples with Mathematica. For the following function, you provide a bound and Mathematica will find all Pythagorean triples with x and y less than or equal to the bound.

```
pythag[bnd_] := Module[{x, y, ans},
        ans = {};
        Do[
            Do[
                If[IntegerQ[Sqrt[x^2 + y^2]],
        ans = AppendTo[ans, {x, y, Sqrt[x^2 + y^2]}]],
                {y, x, bnd}],
            {x, bnd}];
        ans]
```

Let's try it out.

```
pythag[50]
```

As you start to look at these triples, you may notice that some are related to others. For example, we have the first triple, (3, 4, 5), and also (6, 8, 10) which can be gotten from (3, 4, 5) by multiplying each number by 2. We also have (9, 12, 15), which can be gotten by multiplying each term of (3, 4, 5) by 3.

More generally, if (x, y, z) satisfies $x^2 + y^2 = z^2$ and k is an integer, then

$$(k\,x)^2 + (k\,y)^2 \ = \ k^2\,(x^2 + y^2) \ = \ k^2\,z^2 \ = \ (k\,z)^2.$$

So, $(k\,x, k\,y, k\,z)$ satisfies the same equation. By reversing this argument, we see that if (x, y, z) is a Pythagorean triple, then we can cancel a factor of $\gcd(x, y, z)$ to get a triple with $\gcd(x, y, z) = 1$. A Pythagorean triple with $\gcd(x, y, z) = 1$ is called a *primitive* Pythagorean triple. Every Pythagorean triple is a multiple of a primitive triple.

The next function works just like **pythag** above, but it only outputs the primitive triples. If you decide to examine the code, you will see that **primpythag** tests whether a triple is primitive by checking only $\gcd(x, y) = 1$, which differs from our definition of primitive. A homework exercise at the end of the chapter proves that this test is nonetheless correct.

```
primpythag[bnd_] := Module[{x, y, ans},
        ans = {};
        Do[
            Do[
                If[GCD[x, y] == 1,
                    If[IntegerQ[Sqrt[x^2 + y^2]],
        ans = AppendTo[ans, {x, y, Sqrt[x^2 + y^2]}]]],
                    {y, x, bnd}],
                {x, bnd}];
            ans]
```

Here we will try **primpythag** with the same initial bound as above.

```
primpythag[50]
```

As you can see, there are considerably fewer primitive triples in this range. We are now ready for our next Research Question.

■ Research Question 4

> Discover as much as you can about primitive Pythagorean triples.

Research Question 4 is more open ended than any previous Research Question in this course. Do as much as you can with it. We will give you two hints:

1. The strongest known conjecture here is connected to the first two parts of the lab! You may want to start by looking at the numbers which come out as hypotenuses.

2. You may want to consider sums and/or differences between the legs and hypotenuse of a Pythagorean triangle.

■ Sums of Many Squares

We have seen that some positive integers can be written as the sum of two squares, but some cannot. Can all positive integers be written as the sum of three squares, or as the sum of four squares? How many squares would it take to express every integer? Maybe there is no fixed value m so that every positive integer is a sum of m squares.

First, we should clarify what we mean by writing a positive integer as the sum of say three squares. We allow using $0^2 = 0$, so we would say that 5 is the sum of three squares because $5 = 1^2 + 2^2 + 0^2$.

The next function will help in our investigations. The programming is more complicated than usual here, so we will focus on the result the function produces. You get to supply a bound; let's call the bound B. The function tries to write each integer from 1 to B as a sum of as few squares as possible. As it goes along, it prints out information saying how many it was not able to write as a sum of two squares, as a sum of three squares, and so on. Finally, it returns a list of B integers saying how many squares were needed for expressing that integer. Here is the definition of the function:

```
sumofmany[B_] := Module[{j, k, vals, numsqrs, notdone},
        vals = Table[0, {B}];
        (* Fill in the squares *)
        Do[vals[[j]] = 1, {j, Floor[Sqrt[B]]}];
        (* Now we start to do the rest *)
    numsqrs = 1;
        notdone = 1; (* This is to get into the loop *)
    While[notdone > 0,
            notdone = 0;
            Do[
                If[vals[[j]] == 0, notdone = notdone + 1],
                {j, B}];
            Print[notdone, " out of ", B,
    " are not the sum of ", numsqrs, " squares\n"];
            numsqrs = numsqrs + 1;
            Do[
                If[vals[[j]] == numsqrs - 1,
                    Do[
                        If[
        vals[[j + k^2]] == 0, vals[[j + k^2]] = numsqrs],
                        {k, Floor[Sqrt[B - j]]}]],
                {j, B, 1, -1}]];
            vals]
```

♀ Mathematica Note: read this only if you really want to know how this function works.

OK brave soul, here we go.

The function makes an array of length **B** called **vals**, where **B** is the bound supplied by the user. First all of the entries are initialized with the value of 0. We want to fill the *j*th entry with the smallest number of squares for expressing the integer *j*. So, we fill the entries where *j* itself is a square with the value of 1.

Now we head for a **While** loop. We will keep looping until there are no zeros left in the array. The first time through the **While** loop, we are marking the entries which are a sum of two squares. The next time we mark the integers which are a sum of three squares, and so on. We keep track of how many times we have been through the loop in the variable **numsqrs**.

Each time through the loop, we count how many entries are still zero, and report this to the user. Suppose we want to mark entries which are the sum of five squares. Then, we search through the list for entries which are the sum of four (but no fewer) squares. For each of these, we add all possible squares (keeping below the bound **B**). If we hit an entry which is zero, then that entry corresponds to an integer which is not the sum of four squares, but is the sum of five squares. So we mark that entry with a 5.

When all of the entries are nonzero, we are done with the **While** loop. All that is left is to print our list **vals**.

Now we try it with a bound of 50:

> **sumofmany[50]**

The output tells us that some of the integers could not be written as a sum of three squares, but all of the integers from 1 to 50 could be written as a sum of four squares. In that final list of numbers, we can see a 3 in the sixth position. This means that 6 is the sum of three squares, but is not the sum of two squares.

Use **sumofmany** to try to answer the following question:

■ Research Question 5

> How many squares are needed to write every integer, or are there integers which need arbitrarily large numbers of squares?
>
> (You should be able to formulate a conjecture here, but the proof of the "right" conjecture is very difficult. In fact, both Fermat and Euler knew the right conjecture, but neither was able to prove it!)

Representation Problems

12.1 Introduction

In this section, we will look at some representation problems where we try to express certain integers as sums of other special types of integers. The problems are of varying difficulty. In some cases you may be able to formulate strong conjectures and provide complete proofs. At other times, finding a good conjecture may be a big challenge for some Research Questions (and proofs may be very hard to come by).

When finding the right conjecture is difficult, remember that it is often useful to consider prime values before working on more general integers.

12.2 Differences of Two Squares

The basic question here is: which integers can be written as the difference of two squares? In other words, we want to know the integer solutions to the equation

$$n = x^2 - y^2$$

Since the equation is already solved for n, a natural way to run experiments is to substitute lots of pairs (x, y) into the right-hand side and see what comes out. The applet below does just that, with some simplifications to improve efficiency. First, since x and y are squared, we will only use nonnegative values for them. Next, we observe that whenever $x = y$, we will get 0, so we omit those combinations. We notice that switching the roles of x and y will negate n. So, it suffices to take $x > y$. The parameter B specifies a bound for the values of x and y, with $0 \leq x \leq B$ and $0 \leq y \leq B$. Finally, the output is sorted. The result will then be that we will only see the positive values of n which are the difference of two squares. Here's what we get for $B = 10$:

Notice that we have not removed duplicates; some numbers appear more than once. This will give you an indication of how many ways certain integers can be written as the sum of two squares. Keep in mind that some numbers may not appear on this list because the bound B is too low. Similarly, increasing the value for B may uncover more ways of writing a given integer as the difference of two squares.

Research Question 1

Which positive integers n can be written as the difference of two squares, and in how many ways can it be done?

12.3 Sums of Two Squares

The idea here is just like the previous section, but for the sum of two squares. In other words, we want to find integer solutions to the equation

$$n = x^2 + y^2.$$

This time we will state the research questions and then give some useful applets. We will only ask the question: how many ways can an integer n be written as the sum of two squares for prime values of n?

Research Question 2

Which primes p can be written as the sum of two squares, and in how many ways can it be done?

Note: This is a case where you will only be able to prove part of your conjecture. The complete proof of an optimal conjecture is beyond the scope of the chapter summary.

Research Question 3

Which positive integers n can be written as the sum of two squares?

Hint: It is probably a good idea to use the standard progression for investigating problems which we have used earlier in the course. Research Question 2 takes care of the case when n is prime. Try products of distinct primes and prime powers before trying to jump to the complete conjecture. Also keep in mind that your answer here needs to be consistent with your conjecture for Research Question 2.

We start with an applet analogous to the one in the previous section. It will compute all values of $x^2 + y^2$ with x and y nonnegative and ranging up to some bound. This time, switching the values of x and y leaves $x^2 + y^2$ unchanged, so we take $x \leq y$. Here is a sample.

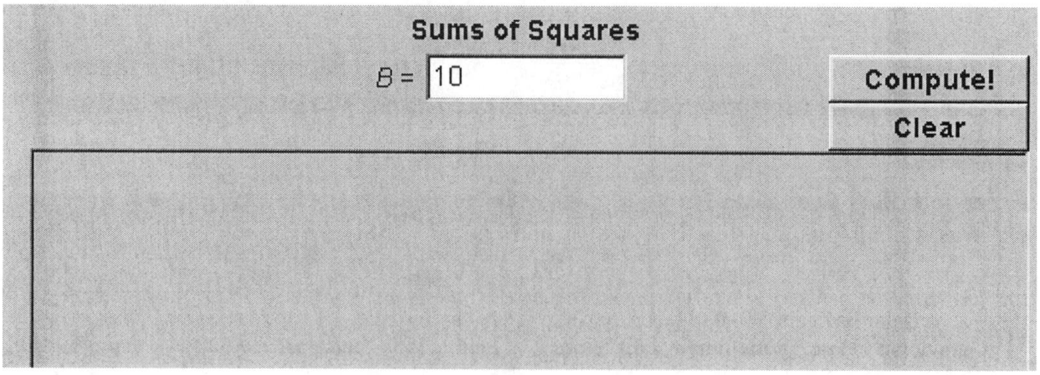

Sums of Squares

$B =$ | 10 | **Compute!**

                                                          Clear

For the problem of sums of squares, if we have a fixed value of n and we want to determine if $n = x^2 + y^2$ for some x and y, there are clearly only finitely many x and y we have to try. In fact, if we take $x \leq y$, then we can be sure that $x \leq (n/2)^{1/2}$. The next applet takes a single value of n as input and uses this observation to find all ways of writing n as a sum of two squares x and y with $x \leq y$. For example, the output from the first applet indicated that 29 could be written as the sum of two squares. Here are the values of x and y with $x \leq y$ that do it:

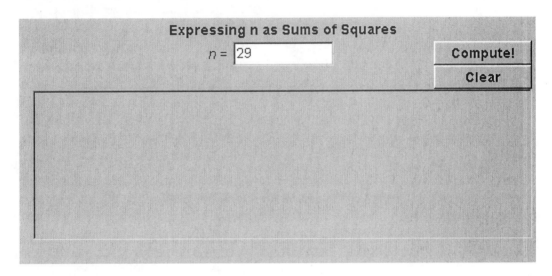

Thus we see that $29 = 2^2 + 5^2$, and that $x = 2$ and $y = 5$ provide the only solution (with $x \leq y$) to the equation $x^2 + y^2 = 29$.

On the other hand, 50 can be written as a sum of two squares in two different ways:

Expressing n as Sums of Squares

$n =$ 50   Compute!

Clear

So, $50 = 1^2 + 7^2 = 5^2 + 5^2$. From the output of the first applet, it appeared that 15 was not the sum of two squares. We can verify this below:

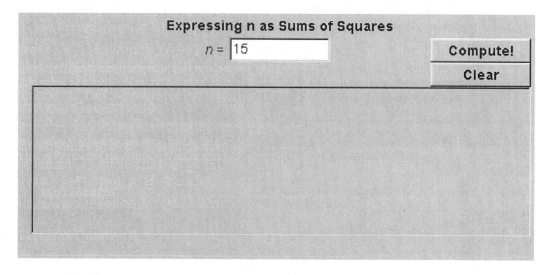

12.4 Pythagorean Triples

Finding Pythagorean triples can be thought of as a special case of determining when an integer is the sum of two squares; a Pythagorean triple consists of three positive integers x, y, and z such that

$$x^2 + y^2 = z^2.$$

You may recognize this equation as being the relation satisfied by the lengths of the sides of a right triangle. Historically, interest in Pythagorean triples started from this geometric connection.

We can search for Pythagorean triples with Java. For the following applet, you provide a bound B and the applet will find all Pythagorean triples with x and y less than or equal to B. Let's try it out.

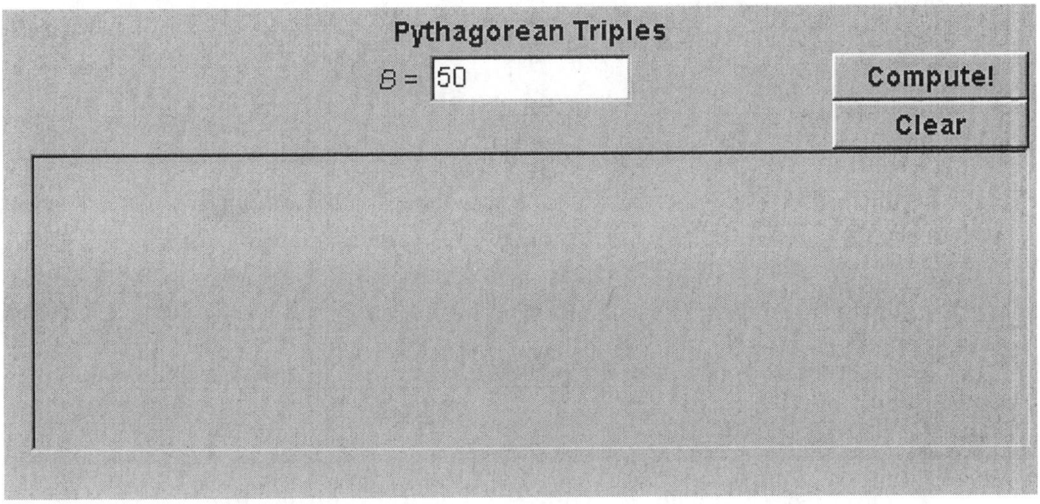

Pythagorean Triples

$B =$ `50` | Compute! |
                                                             | Clear |

As you start to look at these triples, you may notice that some are related to others. For example, we have the first triple, (3, 4, 5), and also (6, 8, 10) which can be gotten from (3, 4, 5) by multiplying each number by 2. We also have (9, 12, 15), which can be gotten by multiplying each term of (3, 4, 5) by 3.

More generally, if (x, y, z) satisfies $x^2 + y^2 = z^2$ and k is an integer, then

$$(kx)^2 + (ky)^2 = k^2(x^2 + y^2) = k^2z^2 = (kz)^2.$$

So, (kx, ky, kz) satisfies the same equation. By reversing this argument, we see that if (x, y, z) is a Pythagorean triple, then we can cancel a factor of $\gcd(x, y, z)$ to get a triple with $\gcd(x, y, z) = 1$. A Pythagorean triple with $\gcd(x, y, z) = 1$ is called a *primitive* Pythagorean triple. Every Pythagorean triple is a multiple of a primitive triple.

The next applet works just like the one above, but it only outputs the primitive triples. Here we will try the new applet with the same initial bound as earlier.

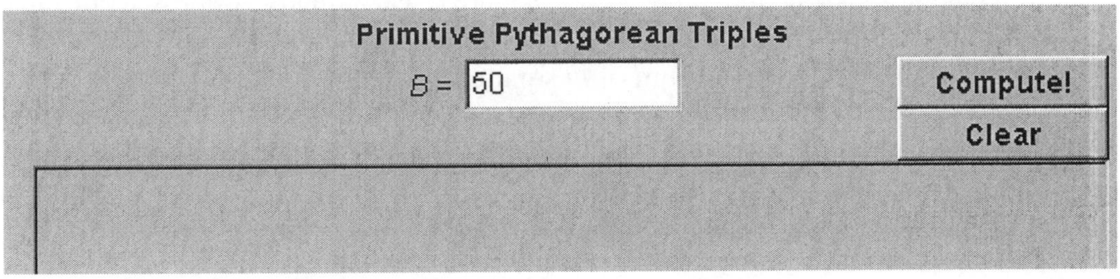

Primitive Pythagorean Triples

$B =$ `50` | Compute! |
                                                             | Clear |

As you can see, there are considerably fewer primitive triples in this range. We are now ready for our next Research Question.

> **Research Question 4**
>
> Discover as much as you can about primitive Pythagorean triples.

Research Question 4 is more open ended than any previous Research Question in this course. Do as much as you can with it. We will give you two hints:

1. The strongest known conjecture here is connected to the earlier parts of the lab! You may want to start by looking at the numbers which come out as hypotenuses.

2. You may want to consider sums and/or differences between the legs and hypotenuse of a Pythagorean triangle.

12.5 Sums of Many Squares

We have seen that some positive integers can be written as the sum of two squares, but some cannot. Can all positive integers be written as the sum of three squares, or as the sum of four squares? How many squares would it take to express every integer? Maybe there is no fixed value m so that every positive integer is a sum of m squares.

First, we should clarify what we mean by writing a positive integer as the sum of say three squares. We allow using 0^2, so we would say that 5 is the sum of three squares because $5 = 1^2 + 2^2 + 0^2$.

The next applet will help in our investigations. You get to supply a bound *B*. The applet tries to write each integer from 1 to *B* as a sum of as few squares as possible. As it goes along, it prints out information saying how many it was not able to write as a sum of two squares, as a sum of three squares, and so on. Finally, it returns a list of *B* integers stating the minimum number of squares needed for expressing that integer. Here's what we get for *B* = 50:

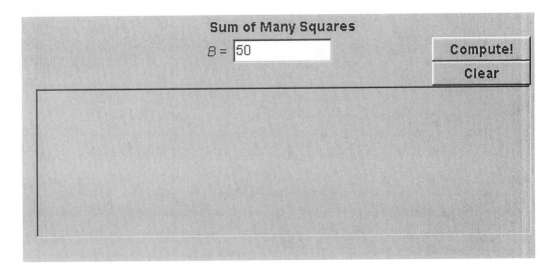

The output tells us that some of the integers could not be written as a sum of three squares, but all of the integers from 1 to 50 could be written as a sum of four squares. In that final list of numbers, we can see a 3 in the sixth position. This means that 6 is the sum of three squares, but is not the sum of two squares.

Use the applet to try to answer the following question:

Research Question 5

How many squares are needed to write every integer, or are there integers which need arbitrarily large numbers of squares?

(You should be able to formulate a conjecture here, but the proof of the "right" conjecture is very difficult. In fact, both Fermat and Euler knew the right conjecture, but neither was able to prove it!)

Homework

1. Write $2665 = 5 \cdot 13 \cdot 41$ as the sum of two squares.

2. Find all Pythagorean triples whose hypotenuse is less than 35.

3. Find all primitive Pythagorean triples whose hypotenuse is less than 50.

4. Find the unique primitive Pythagorean triple with all three sides having length between 2000 and 3000.

5. Prove that if $n \equiv 7 \pmod{8}$, then n cannot be written as the sum of three squares.

6. Prove that if (x, y, z) is a primitive Pythagorean triple, then x, y, and z are pairwise relatively prime. (In general, for three integers a, b, and c, $\gcd(a, b, c) = 1$ does not imply that a, b, and c are pairwise relatively prime.)

7. Suppose p is a prime and z is an integer.

   (a) Prove that if $p^4 \mid z^2$, then $p^2 \mid z$.

   (b) Prove that if $p^3 \mid z^2$, then $p^2 \mid z$.

8. Prove that if m and n can each be written as the sum of two squares and $m \mid n$, then n/m is the sum of two squares.

9. Suppose (x, y, z) is a Pythagorean triple.

   (a) Prove that x or y is a multiple of 3.

   (b) Prove that x or y is a multiple of 4.

   (c) Prove that xyz is a multiple of 5.

10. In each part, either produce an example of the requested triple or prove that no such triple exists.

    (a) A Pythagorean triple with hypotenuse $3^3 5^4 13^2$

    (b) A Pythagorean triple with hypotenuse $3^2 5^4 13^2$

    (c) A primitive Pythagorean triple with hypotenuse $3^2 5^4 13^2$

11. Prove that the number of ways of expressing an integer m as $x^2 + y^2$ where x and y are positive relatively prime integers equals the number of solutions, modulo m, to $a^2 \equiv -1 \pmod{m}$.

12. Prove that for each $j \geq 1$, the equation $x^2 + y^2 + z^2 = 2^j xyz$ has no solutions in positive integers. (Hint: Show that if x, y, and z are a solution to $x^2 + y^2 + z^2 = 2^j xyz$ with $j \geq 1$, then x, y, and z are all even. Use this to produce a solution to a similar equation with a smaller z value.)

13. Prove that in a primitive Pythagorean triple, the sum of the length of the hypotenuse and one leg is either a perfect square or twice a perfect square.

14. Prove that there are infinitely many primitive Pythagorean triples such that the length of the hypotenuse is a perfect square.

15. Find a primitive Pythagorean triple such that the length of the hypotenuse is a fourth power.

16. Suppose (x, y, z) is a primitive Pythagorean triple expressed in terms of r and s as in Research Question 4.

    (a) Prove that the perimeter of the corresponding right triangle is a square if and only if $2r$ and $r + s$ are both perfect squares.

    (b) Find three primitive Pythagorean triples so that the perimeter of the triangle is a perfect square.

17. (a) Prove that if a is an integer and p is a prime such that $p \mid 4a^2 + 1$, then $p \equiv 1 \pmod 4$.

    (b) Prove that there are infinitely many primes of the form $4k + 1$. (Hint: Given any list of primes of the form $4k + 1$, say $p_1, p_2, \ldots, p_n$, what can you say about primes dividing $4(p_1 p_2 \cdots p_n)^2 + 1$?)

13. Continued Fractions

Prelab

Approximation of real numbers by rationals has always been important in mathematics. Indeed, whenever you truncate a decimal representation for a number, you are generating a rational approximation. For example, consider $\pi = 3.14159265\ldots$. Truncating π to three decimal places yields 3.141, which expressed as a fraction is

$$\frac{3141}{1000}.$$

While this type of approximation may be useful in some applications, sometimes it is necessary to have an approximation that has a small denominator. For instance, consider the following problem:

> A clock manufacturer needs two gear wheels with teeth in the ratio $\sqrt[3]{6}:1$.
> It is impractical to have more than 500 teeth on any wheel. Find some
> possibilities for the number of teeth that will approximate the desired
> ratio.[1]

If the wheels have m and n teeth, then we want $m/n \approx \sqrt[3]{6}$. Thus, we need a rational number with relatively small numerator and denominator which gives a good approximation to $\sqrt[3]{6}$.

We now consider a procedure that will allow us to produce good rational approximations that have relatively small denominators. In particular, this method will almost always be superior to simply rounding a decimal expansion. To illustrate the procedure, we will generate a series of approximations of π. The first approximation consists of taking the integer part of π, giving us $\pi \approx 3$. To obtain the second approximation, we start by writing $\pi = 3 + 0.14159265\ldots$, and then expressing the term $0.14159265\ldots$ as 1 divided by its reciprocal:

$$\pi = 3 + 0.14159265\ldots$$
$$= 3 + \frac{1}{7.0625133\ldots}$$
$$\approx 3 + \frac{1}{7} = \frac{22}{7}.$$

---

[1]You do not have to solve this problem now. We will return to this problem in the homework exercises for the chapter.

You probably recognize 22/7 as a famous approximation of π. How good of an approximation is 22/7? See for yourself.

$$\pi = 3.14159265\ldots.$$

$$\frac{22}{7} = 3.14285714\ldots.$$

Not bad at all, considering that the denominator is a single-digit number.

If we iterate the procedure used to find the first two approximations, we get

$$\pi = 3 + \cfrac{1}{7 + 0.0625133\ldots}$$

$$= 3 + \cfrac{1}{7 + \cfrac{1}{15.9965944\ldots}}$$

$$= 3 + \cfrac{1}{7 + \cfrac{1}{15 + 0.9965944\ldots}}$$

$$= 3 + \cfrac{1}{7 + \cfrac{1}{15 + \cfrac{1}{1.0034172\ldots}}}$$

$$= 3 + \cfrac{1}{7 + \cfrac{1}{15 + \cfrac{1}{1 + 0.0034172\ldots}}}$$

$$= 3 + \cfrac{1}{7 + \cfrac{1}{15 + \cfrac{1}{1 + \cfrac{1}{292.6345910\ldots}}}}$$

$$= \ldots.$$

This process generates the *continued fraction expansion*. The numbers 3, 7, 15, 1, 292, ... are called the *quotients*. As a notational convenience, we write $\pi = [3, 7, 15, 1, 292, \ldots]$ to denote the continued fraction expansion. The quotients are indexed starting with 0, so that in general we have

$$x = a_0 + \cfrac{1}{a_1 + \cfrac{1}{a_2 + \cfrac{1}{a_3 + \cdots}}}$$

$$= [a_0, a_1, a_2, a_3, \dots].$$

It turns out that continued fraction expansions converge very quickly. We will see an example of this below, and more examples in the lab. One consequece of this fact is that it is possible for a few terms of a continued fraction expansion to exhaust the precision of a calculator. This means that one might get erroneous results from a hand calculator when computing continued fraction expansions. However, exercises in this Prelab have been choosen so that they can be easily completed using a hand calculator.

1. **Find the first 4 quotients for $\sqrt[3]{2}$.**

If we chop off a continued fraction expansion at a_n we get a *finite continued fraction*. Thus $[a_0, a_1, a_2, a_3, \dots, a_n]$ denotes

$$a_0 + \cfrac{1}{a_1 + \cfrac{1}{a_2 + \cfrac{1}{a_3 + \cfrac{1}{\ddots + \cfrac{1}{a_n}}}}}.$$

It is not hard to see that a finite continued fraction will simplify to a rational number. For example, if we evaluate $x = [3, 7, 15, 1]$ we find

$$x = 3 + \cfrac{1}{7 + \cfrac{1}{15 + \cfrac{1}{1}}} = 3 + \cfrac{1}{7 + \cfrac{1}{16}} = 3 + \cfrac{1}{\frac{113}{16}} = 3 + \frac{16}{113} = \frac{355}{113}. \tag{13.1}$$

Now you try one.

2. **Evaluate $[1, 2, 3, 4]$.**

The example in equation (13.1) was not selected at random. The finite continued fraction $[3, 7, 15, 1]$ came from taking the infinite continued fraction expansion for π

and chopping it off at a_3, giving us the third convergent for the expansion. In general, the *kth convergent* of $[a_0, a_1, \dots]$ is $c_k = [a_0, a_1, \dots, a_k]$.

The table below provides the first few convergents for π.

$$
\begin{aligned}
c_0 &= & [3] &= & 3 & \approx 3.0000000000 \\
c_1 &= & [3, 7] &= & \tfrac{22}{7} & \approx 3.1428571429 \\
c_2 &= & [3, 7, 15] &= & \tfrac{333}{106} & \approx 3.1415094340 \\
c_3 &= & [3, 7, 15, 1] &= & \tfrac{355}{113} & \approx 3.1415929204 \\
c_4 &= & [3, 7, 15, 1, 292] &= & \tfrac{103993}{33102} & \approx 3.1415926530 \\
\hline
& & & & \pi & \approx 3.1415926536
\end{aligned}
$$

At the bottom of the table is π given to 10 decimal places. As you can see, the convergents provide successively better approximations.

3. **Compute the first four convergents $c_0, \dots, c_3$ for $\sqrt[3]{2}$, both as fractions and as decimal approximations.**

Next, let's take a look at what happens if we form the continued fraction expansion of a rational number. The procedure is the same as illustrated earlier. For example, if we expand $88/23$ as a continued fraction, we get

$$
\begin{aligned}
\frac{88}{23} &= 3 + \frac{19}{23} \\
&= 3 + \cfrac{1}{23/19} \\
&= 3 + \cfrac{1}{1 + \frac{4}{19}} \\
&= 3 + \cfrac{1}{1 + \cfrac{1}{19/4}} \\
&= 3 + \cfrac{1}{1 + \cfrac{1}{4 + \frac{3}{4}}} \\
&= 3 + \cfrac{1}{1 + \cfrac{1}{4 + \cfrac{1}{4/3}}} \\
&= 3 + \cfrac{1}{1 + \cfrac{1}{4 + \cfrac{1}{1 + \frac{1}{3}}}}
\end{aligned}
$$

At this point the expansion stops, giving us $88/23 = [3, 1, 4, 1, 3]$. Earlier we noted that finite continued fraction expansions simplify to rational numbers. The preceding example illustrates the converse, that the continued fraction expansion for a rational number is always finite. The proof of this assertion is included as a homework exercise.

4. **Find the complete continued fraction expansion for** $123/32$.

Maple electronic notebook `13-fractions.mws`

Mathematica electronic notebook `13-fractions.nb`

Web electronic notebook Start with the web page `index.html`

Maple Lab: Continued Fractions

■ Approximation with Continued Fractions

■ Introduction

As we saw in the Prelab, continued fractions provide a good means for approximating real numbers by rational numbers. There are a number of ways of looking at continued fractions, and we will have a function for computing each one. Some of these functions come with Maple, and others are easy to produce from what Maple provides. Hit Enter with the cursor in the group below to load the number theory package and our functions. (Do not worry about the warning it gives about there being a new definition of order.)

```
> with(numtheory):

  quotients := (value, num) -> cfrac(value,num,'quotients'):

  fracs := proc(value, num)
   local cf, numout, j;
   cf := quotients(value,num);
   numout := min(num, nops(cf)-1);
   seq(nthconver(cf,j), j=0..numout);
  end:

  fracsf := (value, num) -> evalf(fracs(value,num)):

  errors := proc(value,num)
   local fr,j;
   fr := fracs(value,num);
   seq(evalf(value-fr[j],25),j=1..nops([fr]));
  end:
```

Each function takes two arguments as input: the first is the real number we want to approximate, and the second is the number of terms to use from that real number's continued fraction expansion.

To see the first 12 terms of the continued fraction expansion for π, we use the function `cfrac` as shown:

```
[ > cfrac(Pi,12);
```

When we want to see many terms of a continued fraction expansion, the nicely formatted output from `cfrac` tends to get in the way. You can see just the quotients by using the function `quotients`. Here it is for the same computation as above, 12 terms of π:

```
[ > quotients(Pi,12);
```

Recall that the *convergents* are the fractions we get when we chop off the continued fraction expansion at each step. To see the convergents, we use the function `fracs`. Below are the 0th through 12th convergents for π.

604

```
[ > fracs(Pi,12);
```

The function `fracsf` provides floating point decimal approximations for the output from `fracs`. Below are decimal approximations for, you guessed it, the 0th through 12th convergents for π.

```
[ > fracsf(Pi,12);
```

Due to round-off, a number of the decimal approximations in the above output from `fracsf` are the same. The function `errors` shows how close the convergents come to the original input value by subtracting off the input value from each convergent. The error values are given with 25 decimal places of accuracy.

```
[ > errors(Pi,12);
[ >
```

As you can see, the error values do appear to be headed for 0.

◼ Exercise 1

Compute some continued fraction quotients for the number e from calculus, and report any patterns that you observe. Maple uses `exp(x)` for e^x, so you can get e by typing `exp(1)`.

Note: You need only report the pattern in the quotients, not prove that the pattern continues.

◼ Identifying Rationals from Decimal Approximations

Suppose we had a decimal approximation of an unknown rational number, and we wanted to recover the original rational number that produced the decimal approximation. One approach would be to use the fact that the digits of the decimal approximation of a rational number are ultimately periodic. If we have enough decimal places, we can identify the repeating part and figure out the original rational number from there.

For example, if we had the repeating decimal

$$0.121212...,$$

we could let $x = 0.121212...$, multiply by 100, and then subtract x:

$$100x - x = 12.121212... - 0.121212... = 12.$$

Thus we see that $x = 12/99 = 4/33$.

The above method works fine if we have enough of the decimal expansion to spot the repeating part. But suppose we don't have enough decimal places available to discern any repetition. For example, 0.46017699115044247787 is the first 20 decimal places of a rational number with a moderately small denominator. Can you guess the number?

A good approach is to look at the convergents from the continued fraction expansion of our number, because they will be rational numbers which give good approximations to the value. Let's start with the quotients from the continued fraction expansion:

```
[ > num := 0.46017699115044247787;
    quotients(num,10);
```

The sixth quotient (remember that we start numbering with 0) is huge compared to the others. That means that the fifth quotient of 2 was almost exactly right -- our next best approximation for that quotient was

$$2 + \cfrac{1}{12825445684237526}.$$

It looks like our rational number is [0, 2, 5, 1, 3, 2]. We can compute this convergent directly, but it is also helpful to look at all of the convergents:

```
[ > fracs(num,10);
[ >
```

We can see here the effect of the large quotient. Keep in mind that our value for num is only an approximation to the rational number. On the basis of the above output, it looks like our number is 52/113. We can check our guess by computing a decimal approximation of 52/113. Rather than examining just this one value, let's look at approximations for the previous list of convergents. If our guess of 52/113 is correct, then the sixth entry in the list should be the first one that is correct.

```
[ > printf('num = %.20f', num);
[   evalf(fracs(num,10),20);
[ >
```

In fact, the sixth entry matches except for the last digit, which is a consequence of rounding.

▨ *Exercise 2*

The first 20 decimal places of a rational number are given:

```
[ > num := 0.54260089686098654708;
[ >
```

The denominator of this number is less than 10^{10}. Find the original rational number.

▨ The Calendar

Current astronomical data tells us that it takes the Earth roughly 365 days, 5 hours, 48 minutes, and 42.2 seconds to complete one orbit about the Sun. If we fix the position of the Earth relative to the Sun at noon on January 1, then after exactly 365 days (at noon on January 1 the next year) the Sun has not quite reached its original position. Indeed, since the orbital period is approximately 365 and 1/4 days, then four years later at noon on January 1, the Sun's position has fallen one day behind with respect to a 365-day calendar. With a calendar comprising years solely of 365 days, the synchronization of the Sun with respect to the calendar will fall further and further behind. After 4 (180) = 720 years, the position of the Sun will have fallen 180 days behind, and so midsummer will occur in January (at least in the Northern Hemisphere)!

Julius Caesar was aware of this problem with the calendar, and is responsible for the introduction of *leap years*. Since the Sun falls behind approximately one day every four years, one solution is to add one day to the calendar every four years. The resulting calendar, with 366 days every fourth year, is known as the Julian Calendar.

The Julian Calendar is only approximately correct. A decimal approximation to the orbital period of the Earth, based on the time quoted above, is 365.24219 days. Thus, the Julian Calendar overcompensates by an average of 0.00781 days each year. The calendar will be off by a whole day after 1/0.00781 years. Let's compute that:

```
[ > 1/0.00781;
[ >
```

In the sixteenth century, it was determined that the Sun's position had advanced ten days with respect to the Julian Calendar, and steps were taken to modify the calendar once again. In 1582, Pope Gregory devised a new calendar, which is still in use today. Different countries adopted the new *Gregorian Calendar* at different times. In the United States, this occured in 1752, by which time the error had grown to 11 days. The dates September 3-13, 1752 were skipped entirely as a one-time correction! September 2 was followed immediately by September 14.

To further improve accuracy, the rule for leap years in the Gregorian Calendar was changed slightly from the Julian Calendar:

- every year numerically divisible by 4 would have an extra day (February 29) . . .

- except for century years (1800, 1900, 2000, . . .) which would not have the extra day . . .

- except for years numerically divisible by 400 (2000, 2400, . . .) which would have an extra day after all!

With the Gregorian calendar, an extra day is added 97 out of every 400 years. So, the calendar year averages $365\frac{97}{400}$ days. Since $365\frac{97}{400}$ is equal to 365.2425, this is still not quite right. In practice, this error is managed by using leap seconds which are "skipped" from time to time at midnight on New Year's eve.

We can use continued fractions to look for a simpler scheme for fixing the calendar. We start with the basic continued fraction information for the length of Earth's orbit around the Sun.

Exercise 3

The difference between the orbital period of the Earth and 365 days is computed in the next execution group:

```
[ > orbiterr := 365.24219 - 365;
[ >
```

If we used continued fraction convergents to approximate this number, we could make up new calendars which compensate for the difference between the orbital period of the Earth and 365 days. What would those new calendars be like? Give as many answers as seem reasonable for the precision we have for the orbital period of the Earth. For each answer, describe where you would put the leap days.

Continued Fractions for Square Roots

Let's start by looking at the quotients for the continued fraction expansions of $\sqrt{n}$ for the first few integers that are not squares.

```
[ > quotients(sqrt(2), 20);
[ > quotients(sqrt(3), 20);
[ > quotients(sqrt(5), 20);
[ > quotients(sqrt(6), 20);
```

[>

These sequences have some properties in common. Experiment with other integers, and try to form a conjecture.

■ Research Question 1

Form a conjecture about the continued fraction quotients of $\sqrt{n}$, where n is a positive integer. (You do not have to prove your conjecture, so concentrate on making it as strong as you can.)

■ "Pell's Equation"

Suppose that d and N are fixed integers. The equation

$$x^2 - d\,y^2 = N$$

is known as *Pell's Equation*, and is named after John Pell, a mathematician who searched for integer solutions to equations of this type in the seventeenth century. Ironically, Pell was not the first to work on this problem, nor did he contribute to our knowledge for solving it. Euler, who brought us the ϕ-function, accidentally named the equation after Pell, and the name stuck.

In this section we will concentrate on solutions to Pell's equation for the case where $N = 1$ and $0 < d$. If $y = 0$, then Pell's equation is simply $x^2 = N$, which is not very interesting. Hence we will look for solutions to Pell's equation in *positive* integers x and y. Suppose that we had such a solution. Then we can divide by y^2 to obtain

$$\left(\frac{x}{y}\right)^2 - d = \frac{1}{y^2}.$$

If we factor the left side, we get

$$\left(\frac{x}{y} - \sqrt{d}\right)\left(\frac{x}{y} + \sqrt{d}\right) = \frac{1}{y^2},$$

which implies that

$$\frac{x}{y} - \sqrt{d} = \frac{1}{y^2\left(\dfrac{x}{y} + \sqrt{d}\right)}.$$

Note that the right-hand side of the last expression above will be small, especially if y is large. Hence the left-hand side must also be small, which implies that x/y is close in value to $\sqrt{d}$. Therefore a solution to Pell's equation will give us a good rational approximation to $\sqrt{d}$.

In view of the preceding observations, it is natural to try to reverse the process and search for solutions to Pell's equation by looking among the rational approximations to $\sqrt{d}$ generated from the continued fraction expansion. We will let x be the numerator and y be the denominator of a convergent, and check to see if we have a solution to our equation.

The function `guesssolns` lets you input a value for d, and a bound `bnd` that specifies the number of convergents to be tested. The function takes the kth convergent for the continued fraction expansion of $\sqrt{d}$, which we denote by p_k/q_k, and then computes

$$p_k^2 - d\, q_k^2 .$$

The output is a list of the terms of the form $[[p_k, q_k], p_k^2 - d\, q_k^2]$. Those terms with second entry equal to 1 indicate solutions to Pell's equation.

```
> guesssolns := proc(d, bnd)
    local cf, nums, dens, j;
    cf := cfrac(sqrt(d), bnd);
    nums := seq(nthnumer(cf,j),j=1..bnd);
    dens := seq(nthdenom(cf,j),j=1..bnd);
    seq([[nums[j], dens[j]], nums[j]^2-d*dens[j]^2], j=1..bnd);
  end:
```

Below we search for solutions to $x^2 - 5\, y^2 = 1$:

```
> guesssolns(5, 10);
>
```

The very first term shows us that $(9, 4)$ is a solution to $x^2 - 5\, y^2 = 1$. Let's double check it:

```
> 9^2- 5 * 4^2;
```

There are several other solutions as well: $(161, 72)$, $(2889, 1292)$, $(51841, 23184)$, and $(930249, 416020)$. Let's try another example, say $x^2 - 7\, y^2 = 1$.

```
> guesssolns(7, 10);
>
```

This time we found two solutions, $(8, 3)$ and $(127, 48)$. Now it's your turn.

Exercise 4

Find the smallest solution in positive integers to:

(a) $x^2 - 19\, y^2 = 1$.

(b) $x^2 - 46\, y^2 = 1$.

(c) $x^2 - 61\, y^2 = 1$.

In each of the examples we have considered, it's clear that we could examine more convergents with the hope of finding more solutions. Remarkably, it turns out that if d is not a perfect square, then *every* solution (x, y) of $x^2 - d\, y^2 = 1$ arises as the numerator and denominator of a convergent from the continued fraction expansion of $\sqrt{d}$. (You do not need to prove this assertion.) The question for you to consider is, which convergents should we use? Narrowing down the possibilities is the subject of the next Research Question. You may find the commands guesssolns and quotients helpful in conducting your search for a pattern.

Research Question 2

On the basis of the continued fraction expansion of $\sqrt{d}$, narrow down the possibilities as to which convergents yield solutions to $x^2 - d\, y^2 = 1$.

(Again, focus on formulating a good conjecture. A proof is not reasonably accessible with the knowledge we have developed of continued fractions thus far.)

Mathematica Lab: Continued Fractions

■ Approximation with Continued Fractions

■ Introduction

As we saw in the Prelab, continued fractions provide a good means for approximating real numbers by rational numbers. There are a number of ways of looking at continued fractions, and we will have a function for computing each one.

Each function takes two arguments as input: the first is the real number we want to approximate, and the second is the number of terms to use from that real number's continued fraction expansion.

To see the first 12 terms of the continued fraction expansion for π, we use the function **cfrac** as shown:

```
cfrac[Pi, 12]
```

When we want to see many terms of a continued fraction expansion, the nicely formatted output from **cfrac** tends to get in the way. You can see just the quotients by using the function **quotients**. Here it is for the same computation as above, 12 terms of π:

```
quotients[Pi, 12]
```

Recall that the *convergents* are the fractions we get when we chop off the continued fraction expansion at each step. To see the convergents, we use the function **fracs**. Below are the 0th through 12th convergents for π.

```
fracs[Pi, 12]
```

The function **fracsf** provides floating point decimal approximations for the output from **fracs**. Below are decimal approximations for, you guessed it, the 0th through 12th convergents for π.

```
fracsf[Pi, 12]
```

Due to round-off, a number of the decimal approximations in the above output from **fracsf** are the same. The function **errors** shows how close the convergents come to the original input value by subtracting off the input value from each convergent. The error values are given with 25 decimal places of accuracy.

```
errors[Pi, 12]
```

As you can see, the error values do appear to be headed for 0.

■ Exercise 1

Compute some continued fraction quotients for the number e from calculus, and report any patterns that you observe. (Mathematica uses **E** for e.)

Note: You need only report the pattern in the quotients, not prove that the pattern continues.

611

■ Identifying Rationals from Decimal Approximations

Suppose we had a decimal approximation of an unknown rational number, and we wanted to recover the original rational number that produced the decimal approximation. One approach would be to use the fact that the digits of the decimal approximation of a rational number are ultimately periodic. If we have enough decimal places, we can identify the repeating part and figure out the original rational number from there.

For example, if we had the repeating decimal

$$0.121212...,$$

we could let $x = 0.121212...$, multiply by 100, and then subtract x:

$$100x - x = 12.121212... - 0.121212... = 12.$$

Thus we see that $x = 12/99 = 4/33$.

The above method works fine if we have enough of the decimal expansion to spot the repeating part. But suppose we don't have enough decimal places available to discern any repetition. For example, 0.46017699115044247787 is the first 20 decimal places of a rational number with a moderately small denominator. Can you guess the number?

A good approach is to look at the convergents from the continued fraction expansion of our number, because they will be rational numbers which give good approximations to the value. Let's start with the quotients from the continued fraction expansion:

```
num = 0.46017699115044247787;
quotients[SetAccuracy[num, 50], 10]
```

♥ Mathematica Note:

The function **SetAccuracy** increases the precision used by the computer in storing values. We use it here to decrease round-off error in our computations.

The sixth quotient (remember that we start numbering with 0) is huge compared to the others. That means that the fifth quotient of 2 was almost exactly right —our next best approximation for that quotient was

$$2 + \frac{1}{12825445684237522}.$$

It looks like our rational number is [0, 2, 5, 1, 3, 2]. We can compute this convergent directly, but it is also helpful to look at all of the convergents:

```
fracs[SetAccuracy[num, 50], 10]
```

We can see here the effect of the large quotient. Keep in mind that our value for **num** is only an approximation to the rational number. On the basis of the above output, it looks like our number is 52/113. We can check our guess by computing a decimal approximation of 52/113. Rather than examining just this one value, let's look at approximations for the previous list of convergents. If our guess of 52/113 is correct, then the sixth entry in the list should be the first one that is correct.

```
Print["num = ", num]
N[fracs[SetAccuracy[num, 50], 10], 20]
```

■ **Exercise 2**

The first 20 decimal places of a rational number are given:

num = 0.54260089686098654708;

The denominator of this number is less than 10^{10}. Find the original rational number.

■ The Calendar

Current astronomical data tells us that it takes the Earth roughly 365 days, 5 hours, 48 minutes, and 42.2 seconds to complete one orbit about the Sun. If we fix the position of the Earth relative to the Sun at noon on January 1, then after exactly 365 days (at noon on January 1 the next year) the Sun has not quite reached its original position. Indeed, since the orbital period is approximately $365\frac{1}{4}$ days, then four years later at noon on January 1, the Sun's position has fallen one day behind with respect to a 365–day calendar. With a calendar comprising years solely of 365 days, the synchronization of the Sun with respect to the calendar will fall further and further behind. After $4 \cdot 180 = 720$ years, the position of the Sun will have fallen 180 days behind, and so midsummer will occur in January (at least in the Northern Hemisphere)!

Julius Caesar was aware of this problem with the calendar, and is responsible for the introduction of leap years. Since the Sun falls behind approximately one day every four years, one solution is to add one day to the calendar every four years. The resulting calendar, with 366 days every fourth year, is known as the Julian Calendar.

The Julian Calendar is only approximately correct. A decimal approximation to the orbital period of the Earth, based on the time quoted above, is 365.24219 days. Thus, the Julian Calendar overcompensates by an average of 0.00781 days each year. The calendar will be off by a whole day after 1/0.00781 years. Let's compute that:

1 / 0.00781

In the sixteenth century, it was determined that the Sun's position had advanced ten days with respect to the Julian Calendar, and steps were taken to modify the calendar once again. In 1582, Pope Gregory devised a new calendar, which is still in use today. Different countries adopted the new Gregorian Calendar at different times. In the United States, this occured in 1752, by which time the error had grown to 11 days. The dates September 3–13,1752 were skipped entirely as a one-time correction! September 2 was followed immediately by September 14.

To further improve accuracy, the rule for leap years in the Gregorian Calendar was changed slightly from the Julian Calendar:

• every year numerically divisible by 4 would have an extra day (February 29) . . .

• except for century years (1800, 1900, 2000, . . .) which would not have the extra day . . .

• except for years numerically divisible by 400 (2000, 2400, . . .) which would have an extra day after all!

With the Gregorian calendar, an extra day is added 97 out of every 400 years. So, the calendar year averages $365\frac{97}{400}$ days. Since $365\frac{97}{400}$ is equal to 365.2425, this is still not quite right. In practice, this error is managed by using leap seconds which are "skipped" from time to time at midnight on New Year's eve.

We can use continued fractions to look for a simpler scheme for fixing the calendar. We start with the basic continued fraction information for the length of Earth's orbit around the Sun.

■ **Exercise 3**

The difference between the orbital period of the Earth and 365 days is computed in the cell below:

```
orbiterr = 365.24219 - 365
```

If we used continued fraction convergents to approximate this number, we could make up new calendars which compensate for the difference between the orbital period of the Earth and 365 days. What would those new calendars be like? Give as many answers as seem reasonable for the precision we have for the orbital period of the Earth. For each answer, describe where you would put the leap days

■ Continued Fractions for Square Roots

Let's start by looking at the quotients for the continued fraction expansions of $\sqrt{n}$ for the first few integers that are not squares.

```
quotients[Sqrt[2], 20]

quotients[Sqrt[3], 20]

quotients[Sqrt[5], 20]

quotients[Sqrt[6], 20]
```

■ **Research Question 1**

> Form a conjecture about the continued fraction quotients of $\sqrt{n}$, where n is a positive integer. (You do not have to prove your conjecture, so concentrate on making it as strong as you can.)

■ "Pell's Equation"

Suppose that d and N are fixed integers. The equation

$$x^2 - d\,y^2 = N$$

is known as *Pell's Equation*, and is named after John Pell, a mathematician who searched for integer solutions to equations of this type in the seventeenth century. Ironically, Pell was not the first to work on this problem, nor did he contribute to our knowledge for solving it. Euler, who brought us the ϕ–function, accidentally named the equation after Pell, and the name stuck.

In this section we will concentrate on solutions to Pell's equation for the case where $N = 1$ and $d > 0$. If $y = 0$, then Pell's equation is simply $x^2 = N$, which is not very interesting. Hence we will look for solutions to Pell's equation in positive integers x and y. Suppose that we had such a solution. Then we can divide by y^2 to obtain

$$\left(\frac{x}{y}\right)^2 - d = \frac{1}{y^2}.$$

If we factor the left side, we get

$$\left(\frac{x}{y} - \sqrt{d}\right)\left(\frac{x}{y} + \sqrt{d}\right) = \frac{1}{y^2},$$

which implies that

$$\frac{x}{y} - \sqrt{d} = \frac{1}{y^2 \left(\frac{x}{y} + \sqrt{d} \right)}.$$

Note that the right-hand side of the last expression above will be small, especially if y is large. Hence the left-hand side must also be small, which implies that x/y is close in value to $\sqrt{d}$. Therefore a solution to Pell's equation will give us a good rational approximation to $\sqrt{d}$.

In view of the preceding observations, it is natural to try to reverse the process and search for solutions to Pell's equation by looking among the rational approximations to $\sqrt{d}$ generated from the continued fraction expansion. We will let x be the numerator and y be the denominator of a convergent, and check to see if we have a solution to our equation.

The function **guesssolns** lets you input a value for d, and a bound **bnd** that specifies the number of convergents to be tested. The function takes the kth convergent for the continued fraction expansion of $\sqrt{d}$, which we denote by p_k / q_k, and then computes

$$p_k^2 - d\, q_k^2.$$

The output is a list of the terms of the form $\{\{p_k, q_k\}, p_k^2 - d\, q_k^2\}$. Those terms with second entry equal to 1 indicate solutions to Pell's equation.

```
guesssolns[d_, bnd_] := Module[{fraclist, nums, denoms, guesslist},
        fraclist = fracs[Sqrt[d], bnd];
        nums = Numerator[fraclist];
        denoms = Denominator[fraclist];
    guesslist = Table[{{nums[[i]], denoms[[i]]},
            nums[[i]]^2 - d*denoms[[i]]^2}, {i, 2, bnd + 1}];
        guesslist]
```

Below we search for solutions to $x^2 - 5y^2 = 1$:

```
guesssolns[5, 10]
```

The very first term shows us that $(9, 4)$ is a solution to $x^2 - 5y^2 = 1$. Let's double check it:

```
9^2 - 5 * 4^2
```

There are several other solutions as well: $(161, 72)$, $(2889, 1292)$, $(51841, 23184)$, and $(930249, 416020)$. Let's try another example, say $x^2 - 7y^2 = 1$.

```
guesssolns[7, 10]
```

This time we found two solutions, $(8, 3)$ and $(127, 48)$. Now it's your turn.

■ Exercise 4

Find the smallest solution in positive integers to:

(a) $x^2 - 19y^2 = 1$.

(b) $x^2 - 46y^2 = 1$.

(c) $x^2 - 61y^2 = 1$.

In each of the examples we have considered, it's clear that we could examine more convergents with the hope of finding more solutions. Remarkably, it turns out that if d is not a perfect square, then every solution (x, y) of $x^2 - dy^2 = 1$ arises as the numerator and denominator of a convergent from the continued fraction expansion of $\sqrt{d}$. (You do not need to prove this assertion.) The question for you to consider is, which convergents should we use? Narrowing down the possibilities is the subject of the next Research Question. You may find the commands **guesssolns** and **quotients** helpful in conducting your search for a pattern.

■ Research Question 2

On the basis of the continued fraction expansion of $\sqrt{d}$, narrow down the possibilities as to which convergents yield solutions to $x^2 - d\,y^2 = 1$.

(Again, focus on formulating a good conjecture. A proof is not reasonably accessible with the knowledge we have developed of continued fractions thus far.)

Continued Fractions

13.1   Approximation with Continued Fractions

13.1.1   Introduction

As we saw in the Prelab, continued fractions provide a good means for approximating real numbers by rational numbers. There are a number of ways of looking at continued fractions, and the applet below provides a means for computing each one.

The applet takes two arguments as input: the first is the real number we want to approximate, and the second is the number of terms to use from that real number's continued fraction expansion. For example, to see the first 12 quotients of the continued fraction expansion for π = 3.1415926..., we just click on "Quotients" below:

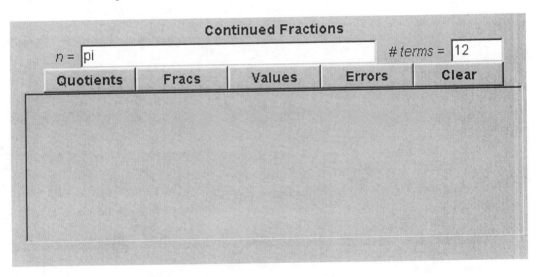

Recall that the *convergents* are the fractions we get when we chop off the continued fraction expansion at each step. To see these fractions, we just click on "Fracs". You can also see decimal approximations for the convergents by clicking on "Values". Give it a try.

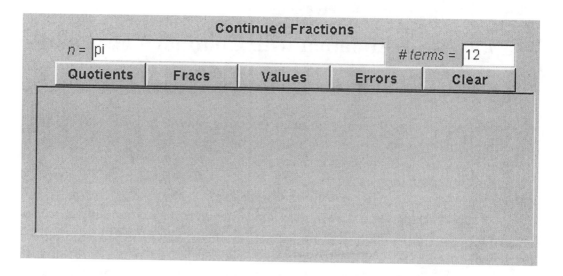

Clicking on "Errors" shows how close the convergents come to the original input value by subtracting off the input value from each convergent.

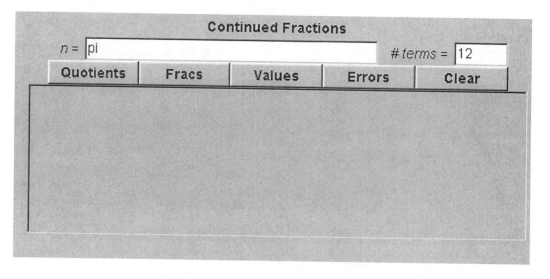

As you can see, the error values do appear to be headed for 0.

Exercise 1

Compute some continued fraction quotients for the number e from calculus, and report any patterns that you observe. (Just type "e" into the applet for e.)

> **Note:** You need only report the pattern in the quotients, not prove that the pattern continues.

13.1.2 Identifying Rationals from Decimal Approximations

Suppose we had a decimal approximation of an unknown rational number, and we wanted to recover the original rational number that produced the decimal approximation. One approach would be to use the fact that the digits of the decimal approximation of a rational number are ultimately periodic. If we have enough decimal places, we can identify the repeating part and figure out the original rational number from there.

For example, if we had the repeating decimal

$$0.121212...,$$

we could let $x = 0.121212...$, multiply by 100, and then subtract x:

$$100x - x = 12.121212... - 0.121212... = 12.$$

Thus we see that $x = 12/99 = 4/33$.

The above method works fine if we have enough of the decimal expansion to spot the repeating part. But suppose we don't have enough decimal places available to discern any repetition. For example, 0.46017699115044247787 is the first 20 decimal places of a rational number with a moderately small denominator. Can you guess the number?

A good approach is to look at the convergents from the continued fraction expansion of our number, because they will be rational numbers which give good approximations to the value. Let's start with the "Quotients" from the continued fraction expansion:

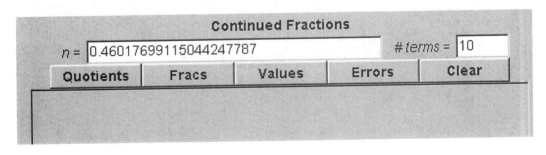

The sixth quotient (remember that we start numbering with 0) is huge compared to the others. That means that the fifth quotient of 2 was almost exactly right -- our next best approximation for that quotient was

$$2 + (1/12825445684237526).$$

It looks like our rational number is [0, 2, 5, 1, 3, 2]. We can compute this convergent directly, but it is also helpful to look at all of the "Fracs":

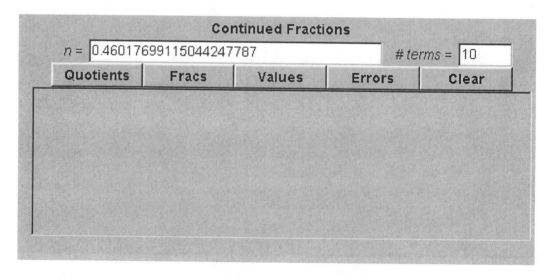

We can see here the effect of the large quotient. Keep in mind that our value is only an approximation to the rational number. On the basis of the above output, it looks like our number is 52/113. We can check our guess by computing a decimal approximation of 52/113. Rather than examining just this one value, let's look at approximations for the previous list of convergents. If our guess of 52/113 is correct, then the sixth entry in the list should be the first one that is correct. Click on "Values" below:

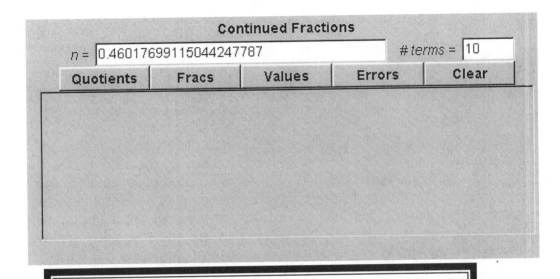

Exercise 2

The first 20 decimal places of a rational number are

0.54260089686098654708.

The denominator of this number is less than 10^{10}. Find the original rational number. The applet below has been primed for your use.

Continued Fractions

$n =$ 0.54260089686098654708   # *terms* = 4

Quotients	Fracs	Values	Errors	Clear

13.2 The Calendar

Current astronomical data tells us that it takes the Earth roughly 365 days, 5 hours, 48 minutes, and 42.2 seconds to complete one orbit about the Sun. If we fix the position of the Earth relative to the Sun at noon on January 1, then after exactly 365 days (at noon on January 1 the next year) the Sun has not quite reached its original position. Indeed, since the orbital period is approximately 365 1/4 days, then four years later at noon on January 1, the Sun's position has fallen one day behind with respect to a 365-day calendar. With a calendar comprising years solely of 365 days, the synchronization of the Sun with respect to the calendar will fall further and further behind. After 4 • 180 = 720 years, the position of the Sun will have fallen 180 days behind, and so midsummer will occur in January (at least in the Northern Hemisphere)!

Julius Caesar was aware of this problem with the calendar, and is responsible for the introduction of leap years. Since the Sun falls behind approximately one day every four years, one solution is to add one day to the calendar every four years. The resulting calendar, with 366 days every fourth year, is known as the Julian Calendar.

The Julian Calendar is only approximately correct. A decimal approximation to the orbital period of the Earth, based on the time quoted above, is 365.24219 days. Thus, the Julian Calendar overcompensates by an average of 0.00781 days each year. The calendar will be off by a whole day after 1/0.00781 years. Let's compute that:

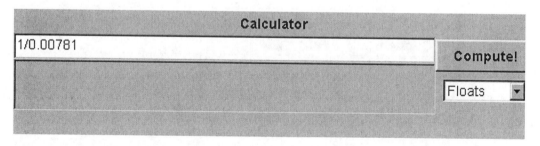

In the sixteenth century, it was determined that the Sun's position had advanced ten days with respect to the Julian Calendar, and steps were taken to modify the calendar once again. In 1582, Pope Gregory devised a new calendar, which is still in use today. Different countries adopted the new Gregorian Calendar at different times. In the United States, this occured in

1752, by which time the error had grown to 11 days. The dates September 3-13, 1752 were skipped entirely as a one-time correction! September 2 was followed immediately by September 14.

To further improve accuracy, the rule for leap years in the Gregorian Calendar was changed slightly from the Julian Calendar:

- every year numerically divisible by 4 would have an extra day (February 29) . . .

- except for century years (1800, 1900, 2000, . . .) which would not have the extra day . . .

- except for years numerically divisible by 400 (2000, 2400, . . .) which would have an extra day after all!

With the Gregorian calendar, an extra day is added 97 out of every 400 years. So, the calendar year averages 365 and 97/400 days. Since 365 + 97/400 = 365.2425, this is still not quite right. In practice, this error is managed by using leap seconds which are "skipped" from time to time at midnight on New Year's eve.

We can use continued fractions to look for a simpler scheme for fixing the calendar. We start with the basic continued fraction information for the length of Earth's orbit around the Sun.

Exercise 3

The difference between the orbital period of the Earth and 365 days is

$$365.24219 - 365 = 0.24219.$$

If we used continued fraction convergents to approximate this number, we could make up new calendars which compensate for the difference between the orbital period of the Earth and 365 days. What would those new calendars be like? Give as many answers as seem reasonable for the precision we have for the orbital period of the Earth. For each answer, describe where you would put the leap days.

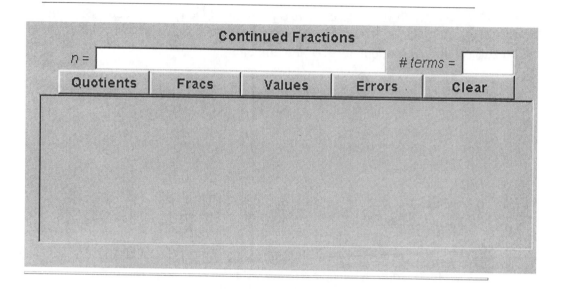

13.3   Continued Fractions for Square Roots

Let's start by looking at the quotients for the continued fraction expansions of $\sqrt{n}$ for the first few integers that are not squares. Click on "Quotients" in each applet below:

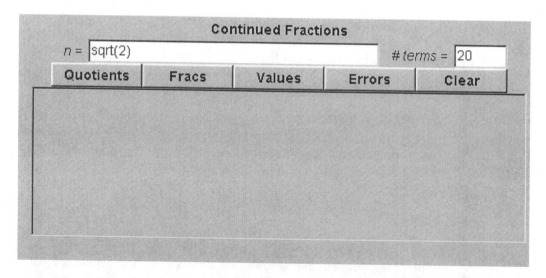

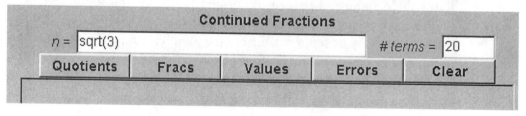

Continued Fractions

$n =$ sqrt(5)   *# terms =* 20

| Quotients | Fracs | Values | Errors | Clear |

Continued Fractions

$n =$ sqrt(6)   *# terms =* 20

| Quotients | Fracs | Values | Errors | Clear |

Research Question 1

Form a conjecture about the continued fraction quotients of $\sqrt{n}$, where n is a positive integer.

(You do not have to prove your conjecture, so concentrate on making it as strong as you can.)

13.4 "Pell's Equation"

Suppose that d and N are fixed integers. The equation

$$x^2 - dy^2 = N$$

is known as *Pell's Equation*, and is named after John Pell, a mathematician who searched for integer solutions to equations of this type in the seventeenth century. Ironically, Pell was not the first to work on this problem, nor did he contribute to our knowledge for solving it. Euler, who brought us the ϕ-function, accidentally named the equation after Pell, and the name stuck.

In this section we will concentrate on solutions to Pell's equation for the case where $N = 1$ and $d > 0$. If $y = 0$, then Pell's equation is simply $x^2 = N$, which is not very interesting. Hence we will look for solutions to Pell's equation in positive integers x and y. Suppose that we had such a solution. Then we can divide by y^2 to obtain

$$(x/y)^2 - d = 1/y^2.$$

If we factor the left side, we get

$$((x/y) - \sqrt{d})((x/y) + \sqrt{d}) = 1/y^2,$$

which implies that

$$(x/y) - \sqrt{d} = 1/(y^2((x/y) + \sqrt{d})).$$

Note that the right-hand side of the last expression above will be small, especially if y is large. Hence the left-hand side must also be small, which implies that x/y is close in value to $\sqrt{d}$. Therefore a solution to Pell's equation will give us a good rational approximation to $\sqrt{d}$.

In view of the preceding observations, it is natural to try to reverse the process and search for solutions to Pell's equation by looking among the rational approximations to $\sqrt{d}$ generated from the continued fraction expansion. We will let x be the numerator and y be the denominator of a convergent, and check to see if we have a solution to our equation.

The applet below lets you input a value for d, and a bound B that specifies the number of convergents to be tested. The function takes the kth convergent for the continued fraction expansion of $\sqrt{d}$, which we denote by p_k/q_k, and then computes

$$p_k^2 - dq_k^2.$$

The output is a list of the terms of the form $((p_k, q_k), p_k^2 - dq_k^2)$. Those terms with second entry equal to 1 indicate solutions to Pell's equation. Here's what we get for $d = 5$ and $B = 10$:

The very first term shows us that (9, 4) is a solution to $x^2 - 5y^2 = 1$. Let's double check it:

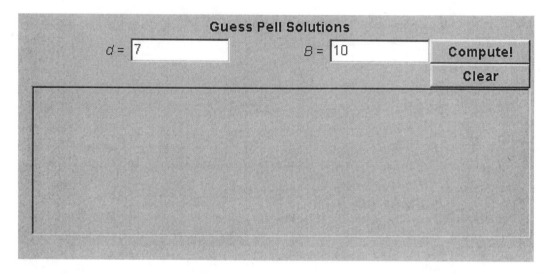

There are several other solutions as well: (161, 72), (2889, 1292), (51841, 23184), and (930249, 416020). Let's try another example, say $x^2 - 7y^2 = 1$.

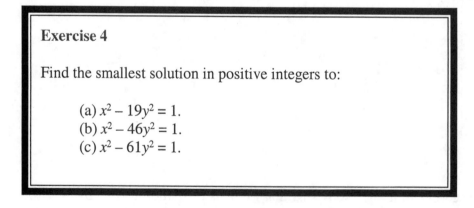

This time we found two solutions, (8, 3) and (127, 48). Now it's your turn.

Exercise 4

Find the smallest solution in positive integers to:

    (a) $x^2 - 19y^2 = 1$.
    (b) $x^2 - 46y^2 = 1$.
    (c) $x^2 - 61y^2 = 1$.

In each of the examples we have considered, it's clear that we could examine more convergents with the hope of finding more solutions.

Remarkably, it turns out that if d is not a perfect square, then every solution (x, y) of $x^2 - dy^2 = 1$ arises as the numerator and denominator of a convergent from the continued fraction expansion of $\sqrt{d}$. (You do not need to prove this assertion.) The question for you to consider is, which convergents should we use? Narrowing down the possibilities is the subject of the next Research Question. You may find the applets provided below helpful in conducting your search for a pattern.

Research Question 2

On the basis of the continued fraction expansion of $\sqrt{d}$, narrow down the possibilities as to which convergents yield solutions to $x^2 - dy^2 = 1$.

(Again, focus on formulating a good conjecture. A proof is not reasonably accessible with the knowledge we have developed of continued fractions thus far.)

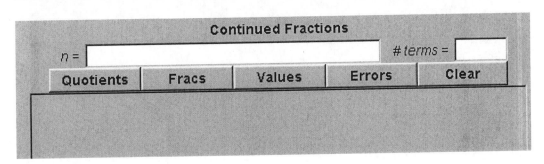

Homework

1. Find the continued fraction expansions of each rational number. Express your answers in the form $[a_0, a_1, \ldots, a_n]$.

   (a) 123/4567   (b) 4567/123

2. Let $x = 13838/373653$.

   (a) Find the continued fraction expansion of x written in the form $[a_0, a_1, \ldots, a_n]$.

   (b) Use formulas (13.7) to find the corresponding sequence of convergents for x.

   (c) Use your answer to part (b) to find integers r and s such that $13838r + 373653s = 1$.

3. Let $x = 11115/5430$.

   (a) Find the continued fraction expansion of x written in the form $[a_0, a_1, \ldots, a_n]$.

   (b) Use the formulas 13.7 to find the corresponding sequence of convergents for x.

   (c) Use your answer to part (b) to write x as a fraction in lowest terms.

4. Find the continued fraction expansions of each rational number. Express your answers in the form $[a_0, a_1, \ldots, a_n]$.

   (a) 5/8   (c) 21/34
   (b) 8/13   (d) 55/89

5. Find the best rational approximation to $\sqrt[3]{3}$ coming from continued fractions whose denominator is less than 100.

6. Find the best rational approximation to $\ln(2)$ coming from continued fractions whose denominator is less than 100.

7. Find a rational number with denominator less than 100 whose decimal expansion begins

   3.4918032786885245901639344262295081967213114754098360655737... .

8. Find a rational number with denominator less than 200 whose decimal expansion begins

$$1.95575221238938053097345132743362831858407079646017699115044247787610619469\ldots.$$

(This is one continuous decimal expansion, but it is broken up because it does not fit on one line.)

9. Let $d > 1$ be an integer which is not a perfect square. Let $\mathbf{Q}(\sqrt{d})$ denote the set $\{a + b\sqrt{d} \mid a, b \in \mathbf{Q}\}$.

   (a) Show that $\mathbf{Q}(\sqrt{d})$ is closed under addition, subtraction, and multiplication (i.e., that the sum, difference, and product of two numbers of the form $a + b\sqrt{d}$ with $a, b \in \mathbf{Q}$ is again of the same form).

   (b) Show that $c + e\sqrt{d} = 0$ if and only $c = 0$ and $e = 0$.

   (c) Show that $\mathbf{Q}(\sqrt{d})$ is closed under division, except for division by zero (i.e., if $a, b, c, e \in \mathbf{Q}$ with $c \neq 0$ or $e \neq 0$, then $\frac{a+b\sqrt{d}}{c+e\sqrt{d}}$ is equal to a number in the right form for $\mathbf{Q}(\sqrt{d})$).

   (d) Show that every element of $\mathbf{Q}(\sqrt{d})$ satisfies a quadratic equation with integer coefficients.

   (e) Show that $a + b\sqrt{d} \in \mathbf{Q}(\sqrt{d})$ is a quadratic irrational if and only if $b \neq 0$.

10. Use exercise 9 to prove that if x has an ultimately periodic continued fraction expansion, then x is a quadratic irrational.

11. Let $x = \sqrt{11}$.

   (a) Find the continued fraction expansion of x. Keep track of the compuations algebraically (i.e., do not rely on only a numerical approximation to $\sqrt{11}$). Go far enough so that you can specify the entire expansion (by knowing the repeating part).

   (b) Find the first four continued fraction convergents to x, as rational numbers.

   (c) Find floating point approximations to the four convergents from part (b).

12. Let $x = \sqrt{14}$.

   (a) Find the continued fraction expansion of x. Keep track of the compuations algebraically (i.e., do not rely on only a numerical approximation to $\sqrt{14}$). Go far enough so that you can specify the entire expansion (by knowing the repeating part).

(b) Find the first four continued fraction convergents to x, as rational numbers.

(c) Find floating point approximations to the four convergents from part (b).

13. Write each of the following numbers in the form $a + b\sqrt{d}$ where $a, b, d \in \mathbf{Z}$.

    (a) $[\overline{1}]$ (b) $[\overline{2}]$

14. Write each of the following numbers in the form $a + b\sqrt{d}$ where $a, b, d \in \mathbf{Z}$.

    (a) $[\overline{1, 2}]$ (b) $[\overline{2, 1}]$

15. Write each of the following numbers in the form $a + b\sqrt{d}$ where $a, b, d \in \mathbf{Z}$.

    (a) $[1, \overline{2}]$ (b) $[2, \overline{1}]$

16. Suppose h_n and k_n are convergents from a simple continued fraction expansion $[a_0, a_1, \dots]$. Prove equation (13.9) by showing that for $n \geq 1$,

$$h_{n-1}k_n - h_n k_{n-1} = (-1)^n.$$

17. Suppose h_n and k_n are convergents from a simple continued fraction expansion $[a_0, a_1, \dots]$.

    (a) Prove that for $n \geq 2$,

$$h_n k_{n-2} - h_{n-2}k_n = a_n(-1)^n.$$

    (b) Prove that for $n \geq 2$,

$$\frac{h_n}{k_n} - \frac{h_{n-2}}{k_{n-2}} = (-1)^n \frac{a_n}{k_n k_{n-2}}.$$

    (c) Use part (b) to prove that

$$\frac{h_1}{k_1}, \frac{h_3}{k_3}, \frac{h_5}{k_5}, \dots$$

    is a decreasing sequence, and

$$\frac{h_0}{k_0}, \frac{h_2}{k_2}, \frac{h_4}{k_4}, \dots$$

    is an increasing sequence.

18. (a) Suppose $a_i \in \mathbf{Z}$ and $a_i > 0$ for $i > 0$, and let $x = [a_0, a_1, a_2, \dots]$. Prove that $\lfloor x \rfloor = a_0$ and $\{x\} = [0, a_1, a_2, \dots]$. (Note: In this problem, x is defined as the value which the continued fraction converges too. Here you are proving that first term in the continued fraction expansion of x is as expected.)

(b) Prove that the infinite continued fraction expansion of an irrational number is unique.

19. Use continued fractions to find a solution to $x^2 - 10y^2 = 1$.

20. Use continued fractions to find a solution to $x^2 - 11y^2 = 1$.

21. Consider the equation $x^2 - 5y^2 = 1$. Using that $(4, 1)$ is one solution to this equation, find three more solutions.

22. Consider the equation $x^2 - 21y^2 = 1$. Using that $(55, 12)$ is one solution to this equation, find two more solutions.

Index

$\equiv$, 146
%, 44

additive order, 177, 182–185, 187, 188, 190, 249, 303, 383, 507, 508
algebraic number, 340
aliquot cycle, 94
alternating series test, 638
amicable pair, 93
axiom, 6

bar code, 244
base b expansion, 245
bijective, 373
Binomial Formula, 516
binomial formula, 517
Brutus, 10

Caesar, Julias, 10
Carmichael number, 417–418, 422, 461, 520, 521
check bit, 193
check digit, 193, 194, 243
Chinese Remainder Theorem, 345–377, 422, 456, 459, 460, 463, 509, 519
    n congruences, 370
    explicit formulas, 371
    in terms of functions, 372–375
    two congruences, 367
cichlids, South American, 96, 143
cipher
    affine, 313–314
    RSA, 238
    shift, 237
congclasses[n], 177

congruence arithmetic
    well defined, 179
congruence class, 146
congruent, 146
conjecture, 1, 7
constructible number, 340
continued fraction
    kth convergent, 602
    efficient method to compute convergents, 636
    expansion, 600
    finite, 601
    palindromic, 631
    periodic, 631, 640
    quotients, 600
    ultimately periodic, 644
corollary, 8
cosmic radiation, 418
counterexample, 8
credit card, 243
CRT, *see* Chinese Remainder Theorem

$d(n)$, *see* divisors, number of
Darrow, Clarence, 10
definition, 6
    advanced use of, 11
    use of, 9
difference of two squares, 587–588
Diophantus, 95
Dirichlet, L., 337
discrete logarithm, 509, 511
divides, 41
divisibility test
    for 10, 237
    for 11, 236

for 2, 237
for 3, 192
for 37, 235
for 5, 237
for 7, 235
for 7, alternative, 237
for 9, 234
length, 240
lengths for small primes, 240
Division Algorithm, 16, 43, 95, 131, 178, 638
for polynomials, 512
divisors, 41
number of, 86, 91, 93, 587, 588
sum of, 88–91, 93, 94, 335
double the cube, 341

Enterprises, Whoopdedoo, 41
Euclid, 131
Euclidean Algorithm, 131, 138, 143, 144, 302, 311, 343, 638–639
reversed, 132, 135, 138, 302, 374, 460
Euler ϕ-function, 183
Euler and even perfect numbers, 334
Euler's Criterion, 553–556, 561
Euler's Theorem, 413, 415, 418, 455, 460
Euler, L., 334, 336, 423

Factor Theorem, 9
Factor Theorem modulo p, 512, 513
Fermat numbers, 336
Fermat prime, 336, 338, 339, 342
Fermat test, 417, 456–458, 460
Fermat's Last Theorem, 343, 594
exponent 4, 595
Fermat's Little Theorem, 415, 417, 423, 516, 553
Fibonacci, *see* Pisano, Leonardo
Fibonacci numbers, 246, 344
findkillers[a,n], 182

Fundamental Theorem of Arithmetic, 43
proof of, 141
Fundamental Theorem of Calculus, 338

Gates, Bill, 238
Gauss, C., 146, 337
GCD Trick, 132, 140, 374–376, 421
geometric constructions, 339–342
Germain prime, 343
Germain, Sophie, 343
greatest common divisor, 14, 42, 137
computation of, 85

Hadamard, J., 337

identity for multiplication, 42
indirect proof, 10
induction, 12
infinite descent, 595
injective, 373
integer, 6, 41
even, 6
ISBN number, 194, 233–234

Julian Calendar, 633

Lagrange's Theorem, 513, 514
Lagrange, Joseph, 307
least common multiple, 93, 368
Legendre symbol, 524, 556
lemma, 8
$li(x)$, 338–339
linear combination, 133
linear diophantine equation, 95, 132–139, 143, 144, 301–303

Mersenne number, 333–339
Mersenne primes, 333–339
(mod n), 146
modcongclasses[n], 177
multiplicative function, 93, 335
multiplicative group, 413

multiplicative inverse, 301, 303, 305, 306, 308, 309, 311, 312, 371, 372, 374, 375, 421, 559, 560, 590
multiplicative order, 383

orbital period of the Earth, 632
ord(a), 383
order, 383
    additive, *see* additive order
    multiplicative, *see* multiplicative order
ounces
    avoirdupois, 379

pairwise relatively prime, 370
parity bit, 193
Pell's equation, 633, 640
perfect number, 93, 333–335, 343
periodic sequences, 189–191
    minimal period, 190
    minimal period, 422
    period of, 189
    purely periodic, 189, 422
    ultimately periodic, 191, 416
$\phi(n)$, 183, 342, 413, 467
ϕ-function, 423–465
Pigeonhole Principle, 239, 416
Pisano, Leonardo, 246
plunge-o-death, 41
Pohlig-Hellman Algorithm, 511
prime, 7, 42
Prime Number Theorem, 337–339
prime-power factorization, 42
primitive root, 467, 552
proposition, 8
Pythagorean triple, 591, 597
    primitive, 591, 592, 595, 597, 598

quadratic congruences
    general, 558
    simple, *see* square roots modulo p

quadratic formula, 558, 559
quadratic irrational, 640, 644
quadratic nonresidues, 523
quadratic reciprocity, 553, 556, 560, 562
quadratic residues, 523
quotient, 43

Rabin-Miller compositeness test, 457, 458, 460
reduction modulo n, 179
regular n-gon, 340
relatively prime, 42, 369
    pairwise, 370
remainder, 43, 177
Remainder Theorem, 512
rep-unit, 244, 343
rock game, 234
RSA cryptosystem, 454

$\sigma(n)$, *see* divisors, sum of
spring water, Artesian, 96, 143
square roots modulo p, 551
square the circle, 341
sum of
    many squares, 593
    two cubes, 565
    two squares, 565, 588–591
surjective, 373

theorem, 7
transcendental number, 340
trisect the angle, 341

Vallée-Poussin, C., 337
value mod n, *see* reduction modulo n

Well-Ordering Principle, 595
Wilson's Theorem, 307, 520
Wilson, John, 307

$\mathbf{Z}_n^*$, 413, 423